Management and Sustainable Development of Coastal Zone Environments

Management and Sustainable Development of Coastal Zone Environments

Edited by

AL. Ramanathan

School of Environmental Sciences
Jawaharlal Nehru University, New Delhi, India

P. Bhattacharya

Department of Land and Water Resource Engineering
Royal Institute of Technology (KTH), Stockholm, Sweden

T. Dittmar

Max Planck Research Group for Marine Geochemistry
University of Oldenburg, ICBM, Germany

M. Bala Krishna Prasad

Earth System Science Interdisciplinary Center
University of Maryland, USA

B.R. Neupane

UNESCO, New Delhi, India

Springer

A C.I.P. Catalogue record for this book is available from the Library of Congress.

ISBN 978-90-481-3067-2 (HB)
ISBN 978-90-481-3068-9 (e-book)

Copublished by Springer,
P.O. Box 17, 3300 AA Dordrecht, The Netherlands
with Capital Publishing Company, New Delhi, India.

Sold and distributed in North, Central and South America by Springer,
233 Spring Street, New York 10013, USA.

In all other countries, except India, sold and distributed by Springer,
Haberstrasse 7, D-69126 Heidelberg, Germany.

In India, sold and distributed by Capital Publishing Company,
7/28, Mahaveer Street, Ansari Road, Daryaganj, New Delhi, 110 002, India.

www.springer.com

Cover photo credit: Photograph of Pichavaram mangroves of Tamilnadu taken by Rajesh Kumar Ranjan.

Printed on acid-free paper

Printed in India.

Foreword

Coastal zones are considered as an interface between sea and land. Life forms adapted to the special environmental conditions have evolved. The natural habitats of the shoreline are very varied but also very small in area, and have been shrinking fast for several decades. Coastal ecosystems tend to have very high biological productivity. The reproduction and nursery grounds of most fish and shellfish species of economic value are in the coastal strip, and a significant proportion of the catch of these species comes from this area, which accounts for almost half of the jobs in the fisheries sector. The quality of coastal waters is a major cause for concern. The two most spectacular phenomena in recent years, oil slicks and algal blooms, are illustrations of the fact that coastal communities frequently suffer the consequences of events or developments occurring inland or offshore and therefore beyond their control. Human settlement of the coastal zones and utilisation of their natural resources since early times has created unique forms of rural and urban landscapes, reflecting cultures centred on trade and largely oriented towards the outside. Unfortunately, urbanisation and uniform agricultural and industrial developments have considerably reduced the biological diversity and cultural distinctness of the landscapes in many parts of the world. A wide range of human activities takes place in the coastal zones (industry, tourism, fishing, aquaculture, etc.), although not necessarily more diverse than in other areas. When these activities develop together on the narrow coastal strip, problems tend to arise, creating conflicts between activities and with the goal of conservation of the nature. Many of the coastal areas might get severely impacted as a result of the general trends of growing population, and the complex socioeconomic conditions. Recent research shows that climate change could involve a rise in sea level of several millimetres per year, and an increase in the frequency and intensity of coastal storms. In addition, growth of the tourism industry in particular will increase human pressure on natural, rural and urban environments around the coastal areas.

Coastal zones are widely managed for a wide variety of utilities in a traditional or more ad-hoc manner. The sector-based management efforts need to be developed to more coordinated and integrated manner for the overall development of the coastal zones. In general, no appropriate framework for integrated coastal zone management exists that could be used as a global model. This volume tries to address these issues with cooperation from various authors from national and international agencies with an aim to create a platform where the most recent researches and experiences and applications of new tools for reconstructing healthy coastal zone are discussed, with particular emphasis on integrated scientific strategies for managing the most densely populated zone in our earth.

The chapters in this book demonstrate the need for the governmental initiatives and commitment for developing programmes for sustainable and balanced coastal zone development. This book presents case studies and examples from various parts of the world and provides a broad overview of various coastal management approaches and tools used to promote coastal zone sustainability in the context of global environmental change and natural hazards. These case studies provide an insight into present day issues, challenges and opportunities; and highlight the key features, principles and new approaches that need to be considered in future efforts for efficient coastal management. It is necessary to address the scientific strategies which underpins the implementation of the successful implementation of the plans for coastal zone development regardless whether the problems at hand are dominated by development issues, safety issues or ecological issues. Further, integrated coastal zone management is a dynamic process driven by societal needs and demands due to increasing population and climate change effects.

Effective integrated coastal zone management (ICZM) means the establishment and support of better ecological and socioeconomic governance that would assist us to think and act beyond short term and small scale outputs to long-term sustainable coastal management strategies. The contents of this book will also help in developing future management strategies for safe drinking water supplies in the rural coastal regions. The overall aim of an integrated coastal zone development and management programme must therefore emphasize an integrated approach, particularly in the developing countries involving groundwater quality monitoring, hydrology, and other environmental indicators. Several critical groundwater pollution problems particularly arise due to the impact of natural hazards like tsunami, excessive groundwater abstraction in the coastal zones which lead to groundwater salinity. A holistic approach of inter-disciplinary and multi-disciplinary character must be developed involving participation of geologists, coastal engineers, geochemists, medical practitioners, social scientists and other stakeholders and policy makers to address the increasingly complex issues of sustainable coastal zone management.

I hope that this book will serve the people to acquire the scientific information and experiences required for ensuring coastal groundwater quality and quantity aspects to protect coastal zone and its aquifers from over-exploitation and consequent salinity problems and stimulate future work for sustainable development and management of the coastal zones with rich biodiversity and other natural resources.

I would like to congratulate the great efforts of the editorial team of this volume which will serve as a scientific information base for future planning of the management of coastal zones around the world and energize synergy among academicians, researchers, stakeholders, NGOs, policy makers and the corporate sector for documentation and dissemination of knowledge in the coastal groundwater management.

Gunnar Jacks
Professor Emeritus
Royal Institute of Technology (KTH)
Stockholm, Sweden

Preface

The ecosystem of coastal regions has rich biodiversity due to the presence of the estuaries, mangroves and coral reefs. The sea level changes, global warming, changes in coastal region, geomorphology, mining activities and natural hazards like floods, tsunami and earthquake add to the magnitude of the coastal problems. Keeping all these factors in mind, the formulation of this book has been conceived. India has a long coast-line of about 7560 km with east and west coasts. The monsoon-dependent part of the country in most of the years depends on the groundwater supply to fulfill the water requirements. The situation becomes more critical during the monsoon failure periods. However, the rising demand for fresh groundwater, mostly for drinking, irrigation and industrial use leads to problems in coastal area and force the over-exploitation of these resources. But the ground waters are increasingly becoming saline or polluted due to these anthropogenic stresses. There is a need for sustainable exploitation to yield fresh water if the flow mechanisms are well established to protect, conserve and restore the coastal aquifer. So the groundwaters of coastal aquifer are to be described, evaluated and explained primarily by application of principles of hydrogeochemistry to understand the migration of solutes using field data, isotopes and the numerical models. The salinity is not only due to seawater intrusion but also due to soil salinization, palaeo salinity, secondary salt precipitation pollution, etc. Human pressure on the coast zone from urbanization, industrialization, aquaculture and agricultural activities is responsible for this situation.

A coastal zone is the interface between the land and water. These zones are important because a majority of the world's population inhabit such zones. Coastal zones are continually changing because of the dynamic interaction between the oceans and the land. Further, coastal zones are highly vulnerable to the natural and anthropogenic perturbations. In order to protect the coastal zones for sustainable earth system management, Coastal Zone Management (CZM) practices have been evolved in all coastal countries. With the adoption of Chapter 17 of the Agenda 21 of the United Nations Conference on Environment and Development at Rio de Janeiro in 1992, integrated coastal zone management practices acquired a new facet. Various developmental activities prevailing in the coastal zone started affecting the biogeochemical, ecological, economic and sociological status. In recent years, the sustainability of coastal region communities began to be considered as manageable with an integrated approach to coastal zone environment.

In spite of rapid development in coastal science over the past decades, most of the major coastal issues of the Agenda 21 have remained unaddressed. Control of coastal erosion, protecting from sea level increase and control of coastal pollution continue to be major issues which need greater scrutiny. Nutrient pollution causes serious problems in the coastal environment by reducing dissolved oxygen, a critical parameter for the biological productivity, and thereby increases the occurrence of harmful algal blooms (HAB), hypoxic/anoxic zones, etc. For over a century, coastal waters have been utilized as a convenient dumping ground for waste materials, which caused not only serious environmental degradation but also direct threat to human health. In order to manage the coastal ecosystems, sustainable management practices are to be designed which would be derived from the regional earth system models, ecological and biogeochemical models.

This book represents the inter- and multi-disciplinary view of authors for a meaningful and practical guidance for the better management of coastal zones which comprises contributions from distinguished scientists from both academia and institutions from all over the world and also a number of eminent Indian scientists who have been actively working in this area. The seventeen chapters in the volume are grouped under four sections covering almost all aspects of the issues related to sustainable coastal zone management in national and global perspectives including health and technological concerns such as: Evaluation, status prediction, modelling and developments of coastal zones: Management issues; Coastal zone water resources (quantity and quality): Challenges for sustainability; Biodiversity of coastal zones and its sustainability; and Threats to coastal aquatic ecosystems: Developmental and sustainability issues.

The volume presents case studies and examples from various parts of the world and provides a broad overview of various coastal management approaches and tools used to promote coastal zone sustainability in the context of global environmental change and natural hazards. These case studies provide an insight into present day issues, challenges and opportunities; and highlight the key features, principles and new approaches that need to be considered in future efforts for efficient coastal management. It is necessary to address the scientific strategies which underpin the successful implementation of the plans for coastal zone development regardless of the problems at hand which are dominated by development issues, safety issues or ecological issues. Further, integrated coastal zone management is a dynamic process driven by societal needs and demands due to increasing population and climate change effects. Finally each chapter in this book demonstrates the need for the governmental initiatives and commitment for developing programme for sustainable and balanced coastal zone development.

We would like to thank all the contributors to this volume and take this opportunity to acknowledge all our colleagues for their time-consuming efforts to review the manuscripts of the chapters for this volume. Their efforts with high quality review of the manuscripts contributed significantly to keep the high scientific contents of the book. AL. Ramanathan would like to thank his collaborators and research scholars for supporting his research activities over two decades which helped him in bringing out this volume successfully on coastal ecosystems. Prosun Bhattacharya would like to thank Sida-SAREC and Sida-Department of Development Partnerships and MISTRA, Sweden for their support to the KTH on arsenic research for sustainability of safe drinking water supplies.

The main goal of the book is to focus attention of all affected parties worldwide on deteriorating pristine nature of this environment and increasing man-marine conflict problems and to present some challenges for safe coastal groundwater production in order to invoke appropriate actions in efficient innovative directions. We hope that this book will be useful for coastal scientists, environmental scientists, engineers, managers and administrators in both academia and industries and for government and regulatory bodies dealing with coastal issues by providing an opportunity to acquire relevant scientific information and experiences in "Management and Sustainable Development of Coastal Zone Environments".

Lastly, the editors wish to thank the publishers for bringing out this volume successfully.

AL. Ramanathan
P. Bhattacharya
T. Dittmar
M. Bala Krishna Prasad
B.R. Neupane

About the Editors

Dr. AL. Ramanathan, PhD in sedimentary geochemistry from Centre of Advanced Studies, Panjab University, worked as faculty in Geology Department, Annamalai University, for one decade. He is currently Professor in the School of Environmental Sciences, Jawaharlal Nehru University, New Delhi, India. He has specialized in coastal biogeochemistry and hydro geochemistry aspects and worked extensively on the mangroves, estuaries and coastal ground waters of India for the past two decades. He is engaged in coastal research with different parts of the world, such as nutrient dynamics, paleo environment, source identification etc., with a number of universities and research organisations in India, Australia, Russia, USA, etc. He has guided several PhDs in the above subject and published more than sixty papers in referred reputed journals like *Estuaries; Wetland Ecology and Management; Estuarine, Coastal and Shelf Science; Journal of Coastal Research; Bull. of Marine Sciences; Indian Journal of Marine Sciences; Marine Pollution Bulletin; Hydrogeology Journal; Water Soil and Air Pollution; Hydrological Process; Geofluids; Applied Geochemistry; Journal of Geochemical Exploration* etc. He has also published five books and several chapters in many books. He has got three times IFS Sweden (Project) Awards to work on mangrove biogeochemistry besides working in national and international projects from Indo-Australian, Indo-Russian, DST, MoEF, MoWR, etc. He was a Post-doctoral fellow under STA Japan programme, UGC-Russian Academy of Science Program, CSIR, INSA, DST, etc. He is a member of editorial board in *Indian Journal of Marine Sciences* and served as referee for many national and international journals as well.

Prof. P. Bhattacharya holds a PhD in Sedimentary Geochemistry from University of Delhi, India and is an expert on groundwater chemistry. He is currently Professor of Groundwater Chemistry and leader of the KTH-International Groundwater Arsenic Research Group, at the Department of Land and Water Resource Engineering at KTH, Stockholm. He is engaged with research on groundwater quality in different parts of the world, such as arsenic and fluoride contamination in collaboration with a number of universities and research organisations in India, Bangladesh, China, Australia, Argentina, Ghana,

Costa Rica, Bolivia and USA. He has authored/coauthored over 170 international publications in peer-reviewed journals and conference proceedings and has organised several international workshops on natural arsenic in groundwaters. He has edited five books on various aspects of natural arsenic in groundwater and groundwater for sustainable development until 2009 and two more books are currently in the pipeline to be published. He is an editor of the book "Arsenic in Soil and Groundwater Environment" (Elsevier, 2007) and of a number of special issues of the peer-reviewed journals e.g., *Science of the Total Environment* entitled "Arsenic in the Environment: Biology and Chemistry" (Elsevier, 2007), *Journal of Contaminant Hydrology* with the title "Geogenic Arsenic in Hydrologic Systems: Controls and Challenges" (Elsevier, 2008); *Environment International* entitled "Arsenic Geochemistry, Transport Mechanism in the Soil-Plant System, Human and Animal Health Issues" (Elsevier, 2009); *Environmental Geochemistry and Health* entitled "Arsenic in the Environment—Risks and Management Strategies" (Springer, 2009). He has also edited a number of special issues on arsenic. He is also Associate Editor of the two reputed journals, *Journal of Hydrology and Environment International* published by Elsevier. He has been at the advisory board of the EU Marie Curie Training network, Aqua Train (since July 2008) and also in the panel of national and international committees of research councils.

Dr. T. Dittmar is working in Max Planck Research Group for Marine Geochemistry, University of Oldenburg, Institute for Chemistry and Biology of the Marine Environment (ICBM), Germany. He did his Masters in Geoecology from University of Bayreuth, Germany and PhD in Marine Chemistry, University of Bremen, Germany. He has occupied various academic positions such as Research Associate in Alfred Wegener Institute for Polar and Marine Research, Bremerhaven, Germany; Assistant Professor, Florida State University, Tallahassee, USA; Leader of the Max Planck Research Group for Marine Geochemistry, University of Oldenburg and Max Planck Institute for Marine Microbiology, Germany. He received prestigious academic awards such as: Wilhelmshaven-Award for Marine Research, Fellow of the German Research Society, University of Washington, USA. He has got more than forty peer-reviewed scientific papers (including four book chapters).

Dr. M. Bala Krishna Prasad is working in Earth System Science Interdisciplinary Center, University of Maryland, USA. He has done his PhD in Biogeochemistry, School of Environmental sciences, Jawaharlal Nehru University, India. He is at present a Research Associate in University of Maryland, USA. He has worked earlier as Lecturer in Environmental Science, University of Hyderabad, India and Post doctoral fellow in School of Environmental Science, Jawaharlal Nehru University, India. He has received Young Scientist Award, Ministry of Science & Technology, India; Linnaeus–Palme Research Fellowship, Department of Earth Sciences, Uppsala University, Sweden; Berkner Fellowship of AGU, USA and Junior Research Fellowships as well. He has more than dozen papers in reputed international journals like Wetland Ecology and Management, Estuarine, Coastal and Shelf Science, Journal of Coastal Research, Bull. of Marine Sciences, Indian Journal of Marine Sciences, etc.

Dr. Bhanu R. Neupane is the Regional Programme Specialist for UNESCO's Cluster Office in New Delhi, where he has been working as the Programme Specialist for Water Resources and Hydrology for South Asian Region. In this capacity, he is responsible for the implementation of the International Hydrological Programme (IHP) and the preparation and design of regional projects and activities on hydrology and water resources. Prior to his current posting in 2005, Dr. Neupane worked in Nepal, consultant in the US and Europe and served in several Southeast Asian countries. He joined UNESCO in the year 2000 as a Programme Specialist and served with the UN system-wide World Water Assessment Programme (WWAP) until his transfer to New Delhi. Dr. Neupane earned an M.S. from Thailand and a Post Graduate degree in water quality modeling from the USA. He earned his PhD degree in Water Resources Management from Canada and also a Doctorate degree in Business Administration from the International School of Management, France. Dr. Neupane has over 100 publications in different forms including seven edited books. He is referee to 11 international journals.

Contents

Section 2 Coastal Zone Water Resources (Quantity and Quality): Challenges for Sustainability

Section 3 Biodiversity of Coastal Zones and Its Sustainability

Section 4 Threats to Coastal Aquatic Ecosystems: Developmental and Sustainability Issues

SECTION 1

Evaluation, Status Prediction, Modelling and Developments of Coastal Zones: Management Issues

1

Observational Needs for Sustainable Coastal Prediction and Management

Raghu Murtugudde
Earth System Science Interdisciplinary Center
Department of Atmospheric and Oceanic Science
University of Maryland, College Park, MD 20740, United States

1. INTRODUCTION

The urgency of actions needed for avoiding the tipping points in the functioning of the Earth System is now becoming more and more obvious (Lovelock, 2009). The main objective here is to highlight the observational needs for regional Earth System predictions and projections, where such predictions and projections are assumed *a priori* as the main decision-making tools for sustainable management of the Earth System. This is especially so in the context of coastal zones since ever increasing migrations to the coastal zone are having unique and unprecedented impacts on the Earth System. Prediction in this context carries a closer correspondence to reality than projection with an intrinsically larger uncertainty in the latter. I will focus on regional Earth System prediction keeping in mind that coastal zones are but a specific application of this concept. I will thus use the sustainable coastal management and adaptive management of a regional Earth System interchangeably.

A large number of coastal observing systems exist (see examples for the U.S. at http://www.csc.noaa.gov/cots/ and http://www.usnfra.org/) and despite the growing realization that monitoring more and more physical, biological, and chemical parameters is crucial, a more holistic approach is needed to consider the natural-human systems as continuously interacting components of the Earth System that are defining the future evolution of the system itself and the time for reliable decision-making tools for sustainable management and for avoiding catastrophic regimes or tipping points in the future is upon the global human (Schellnhuber, 1998; Rial et al., 2004). The main objective of this chapter is to provide the Earth System perspective for sustainable management of the coastal zones with a focus on the observational needs for delivering usable and skillful regional Earth System predictions and projections for decision-making. I am clearly starting with an *a priori* assumption that regional Earth System predictions and projections have the best potential for offering a decision-making tool under uncertainty (Murtugudde, 2009).

1.1 Defining the Earth System

Schellnhuber (1998; 1999) has led the efforts to provide the overarching definition for the Earth System as being comprised of the ecosphere and the anthroposphere. The ecosphere here is the geosphere-biosphere complex and includes the more well-known components such as the ocean, atmosphere, cryosphere, etc. along with the biosphere, whereas the anthroposphere puts man at the helm from where he appears to be watching the consequences of his actions weighing the consequences and corrective actions (Murtugudde, 2009). The comprehensive piping diagram of the Earth System, the so-called Bretherton diagram (Schellnhuber, 1999), illustrates the enormous range of spatio-temporal scales of interactive components we must deal with but this should also serve as the roadmap for integrated Earth System observations. This is clearly recognized in coordinating the global efforts under the Global Earth Observing System of Systems (GEOSS) which I will return to later. The critical new argument I would like to make is that future observational systems must be built on the interactive nature of the ecosphere and the anthroposphere and also must recognize that sustainable management will require that we do not forget the distinction between the global governance issues and the 'place-based' specificity of various regions, especially the coastal domains (Mitchell and Romero Lankao, 2003). It is thus the interdisciplinary and process observations in addition to observing characteristic parameters or variables of the system (e.g., temperature, humidity, carbon, productivity, etc.) and the global and regional scales that I would like to emphasize here.

1.2 What is Sustainability?

Intuitively, sustainability simply implies the ability of one generation to use the resources without jeopardizing the ability of the future generations to access the same resources (Clark et al., 2003), while a mathematical form would state that the local rate of change for all the resources by all organisms be zero (Falkowski and Tchernov, 2003). The literature is quite vast on sustainability in various contexts and the need for institutional structures required to address the related research and practical issues including the intrinsically multi-disciplinary nature of the very concept (Gallopin, 2003). Sustainability is essentially adaptive management with participatory decision making and learning-by-doing in the context of regional or coastal management (Gallopin, 2003; Kinzig et al., 2003; Murtugudde, 2009).

Global governance issues under climate change are a major focus of the IPCC process but the basic issue of sustainable management of the Earth System offers two distinct scales of human control of the Earth System, viz., regional and global. Observational needs for sustainable management then must consider human actions and responses also at these scales, especially in the evolving coastal regions under multiple-stressors such as population growth, land use change, and eutrophication (Committee on Earth-Atmosphere Interactions, 2007). Schellnhuber (1998) offers a few paradigms for such a control, at least to avoid the catastrophic domains where human existence is possible but subsistence will be miserable. He does offer a prioritized list to achieve sustainable development, as optimization (achieving the best Earth System performance), stabilization (achieve a desirable Earth System state), and pessimization (simply manage to avoid the worst Earth System states). The caveat ofcourse is that putting these paradigms into operation via the Earth System Models is highly non-trivial but the task of avoiding catastrophes cannot be abandoned (Kinzig et al., 2003; Committee on Earth-Atmosphere Interactions, 2007; Lovelock, 2009). As monumental a task as it is to provide useful and usable Earth System predictions with validation, uncertainties, and skill assessment, not attempting to build viable decision-making tools would be criminal negligence of the global human. Hence, it is imperative for us to consider the observational systems that will make this feasible and avoid overwhelming nature (Steffen et al., 2007).

1.3 Modelling the Earth System

The concept of Earth System modelling and prediction clearly evolved from the pioneering efforts in weather and climate prediction (Richardson, 1922; Phillips 1960; Nimias, 1968). Climate forecast has taken a complex trajectory compared to weather prediction since climate has many modes of variability such as the monsoons and the El Niño-Southern Oscillation (ENSO), with their own spatio-temporal scales and predictabilities (Kelly, 1979; Charney and Shukla, 1981; Cane et al., 1986). The envelope of climate prediction continues to be pushed with new advances in decadal time-scale predictions (Keenlyside et al., 2008). The natural evolution of climate modelling towards Earth System models was motivated by some of the most fascinating Earth System feedbacks, such as the potential role of bio-physical feedbacks on droughts over Sahara (Charney et al., 1975), and more recently, feedbacks from marine biogeochemistry and ecosystems (Ballabrera-Poy et al., 2007a; Huntingford et al., 2008). Another major new direction of development of relevance to Earth System prediction was the early dynamic downscaling to regional scales (Dickinson et al., 1989; Giorgi and Bates, 1989). Even as the spatial resolutions of the Earth System models improve with each IPCC assessment, they are expected to remain at ~10 km scales for many years if not decades. It is evident that adaptive management of resources, especially coastal management issues, demand Earth System information at the order of 1 km or less and the only way to reach these goals is via dynamical and statistical downscaling. Dynamical downscaling through regional climate modelling has now been applied to various Earth System issues such as human health, agriculture, and water resources (Graham et al., 2001; Mearns et al., 2003; Giorgi and Diffenbaugh, 2008). The Earth System is indeed a system of systems and the regional specificity of the ecosphere and the anthroposphere must be seen as an integrated global Earth System with nested regional Earth Systems with their own idiosyncrasies. Coastal zones offer a unique set of human-nature interactions where the attraction of their services in terms of natural beauty, terrestrial and marine ecosystems, and so on, are the very reason for the stampede of new migrations and the uniqueness of the interactions between man and nature. In terms of observational challenges also, the coastal zones are unique for both remote and in situ techniques but also offer opportunities for driving technological innovations (Christian et al., 2006).

1.4 Earth System Prediction

While there is no unique approach to an Earth System modelling framework, the International Geosphere Biosphere Project (IGBP), DIVERSITAS, the World Climate Research Program (WCRP), and the International Human Dimensions Program (IHDP) have created a new Earth System Science Partnership focussed on energy and carbon cycles, food systems, water resources and human health as the most critical issues for human well-being (http://www.essp.org). Along these lines, the WCRP launched a new strategic framework for Coordinated Observation and Prediction of the Earth System (COPES), which lists the following as one of its aims: to facilitate analysis and prediction of Earth System variability and change for use in an increasing range of practical applications of direct relevance, benefit and value to society (http://wcrp.ipsl.jussieu.fr/). Any skillful Earth System prediction must be reliably useful for decision-making, keeping in mind the holistic principles of sustainable management of the future trajectories of the Earth System evolution (Schellnhuber, 1998). The enormity of the task is daunting considering the complexity of the interactions and feedbacks between humans and natural systems with the coupling dependent on space, time, and organizational structures (Liu et al., 2007). Observing these feedbacks themselves is very crucial for sustainable coastal management.

1.5 Observing the Earth System for Sustainable Management

The most well-known mode of climate variability, viz., the El Niño-Southern Oscillation (ENSO), with its global reach offers an excellent analogy for Earth System interactions and a set of predictable targets with applications from agriculture to fisheries to human health (McPhaden et al., 2006). As the gold-standard for successful climate prediction, ENSO also offers one of the best examples of the role of observations in improving process understanding and translating it into successful predictions. The Tropical Ocean Global Atmosphere-Tropical Atmosphere Ocean (TOGA-TAO) array of moored buoys in the tropical Pacific combined with a number of satellites offered a clear demonstration of how well-coordinated and integrative observing systems do lead to routine, operational and usable climate and Earth System predictions (McPhaden et al., 1998). Sustained observational arrays are since established in the tropical Atlantic and the Indian Oceans (Bourles et al., 2008; McPhaden et al., 2009).

The question of uncertainties in Earth System predictions at both short and long time-scales are crucial with the former requiring quantitative measures of skill in addition, whereas projections of future trajectories of the Earth System at longer time-scales will need to offer a more solid understanding of the known unknowns or irreducible uncertainties (Schellnhuber, 1998; Cox and Nakicenovic, 2003; Biermann 2007; Dessai et al., 2009). The need for sustained observations for continuously validating and assessing uncertainties in our Earth System models will need global and regional scale Earth System monitoring such as the Global Earth Observing System of Systems (GEOSS), being co-coordinated by the Group on Earth Observations (GEO; http://www.earthobservations.org/index.html). The stated vision for GEOSS is to realize a future wherein decisions and actions for the benefit of human kind are informed by coordinated, comprehensive and sustained earth observations and information. The GEO plan defines nine societal benefit areas of disasters, health, energy, climate, water, weather, ecosystems, agriculture and biodiversity which is nearly comprehensive enough for the monitoring and nowcast-forecast vision of Earth System prediction models.

Hard decisions on Earth System management and policy will be made by experts in 'soft' sciences with some of the softest information coming from the 'hard' sciences such as climate physics (Mitchell and Romero Lankao, 2003). Reliability of the climate and Earth System information can be enhanced and more quantifiable success can be achieved at regional scales and shorter lead-times (days to seasons), in high resolution regional Earth System models with the boundary conditions provided by the global Earth System models. The advantages of local and regional understanding of natural-human system interactions or the "place-based" Earth System predictions and decision-making are evident in a number of success stories (Mitchell and Romero Lankao, 2003).

The observations for regional Earth System prediction must begin to consider the monitoring of the natural system as it is constantly being kicked around by the human system. The Earth System does span the range from microbes to man and while one should be skeptical of models, it is imperative to remember that the situation is clearly not as rosy as modellers tend to believe but neither it is as hopeless as social scientists assume.

2. OBSERVATIONS FOR SUSTAINABLE COASTAL MANAGEMENT

Instead of offering a shopping list of observations and data management-distribution strategies needed for sustainable coastal management, I will focus on an example of a practical application, viz., regional Earth System prediction for human health, which is inseparable from the environment, water, and agriculture (Bell et al., 1993). This is an ideal example for considering the coastal zones since no other regions experience closer interactions between these three components of the Earth System. Even though the environmental connection to human health has been known since the time of Hippocrates (Franco and Williams, 2000), remarkably few mechanistic models have been developed to exploit weather and climate predictions for human health applications.

The traditional approach or the old paradigm of climate prediction for human health tends to find correlations between climatic variables and disease incidences, outbreaks, or indicators that are precursors to an outbreak (Kelly-Hope and Thompson, 2008). However, climate change is expected to alter not only the environmental conditions but also population growth and movement which will clearly affect the transmission dynamics of any disease we can think of. The impacts of global change are clearly manifest in global indicators such as temperature and sea level rise but the impacts on humans are often associated with local changes in weather, ecology, hydrology, etc. Any observational system that purports to be a part of the prediction system for human health must capture the linkages from climate change to human health with the intermediate steps of microhabitat selection by the relevant microbes, transmission dynamics, socioeconomics, and adapt to the advances in and needs for research and also to human feedbacks and modulating influences. A succinct way to illustrate the potential range of observations for this one particular application can be illustrated by a schematic shown in Fig. 1. Note that the author has deliberately mixed the observational platforms with the drivers (climate change, regional weather changes), and with processes (transmission), impacts (human health), responses (adaptation) and feedbacks (modulating influences). The motivation is to highlight again the integral nature of natural-human system and the need to avoid disciplinary boundaries in developing our observing systems.

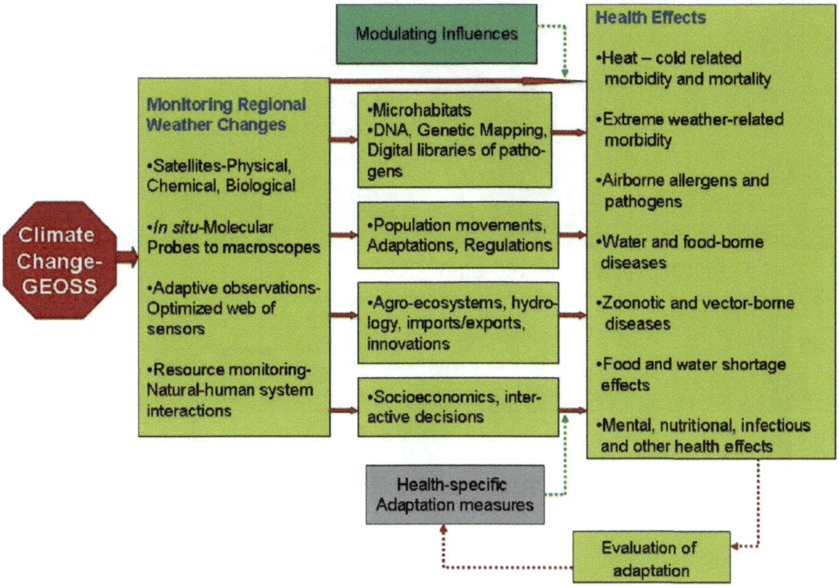

Figure 1: Schematic of linkages from climate change driver to human health to illustrate the need for interdisciplinary observations. A comprehensive and integrated approach to monitor the Earth System at regional and local scales will have to capture the natural-human system behaviour and their interactions to be able to provide effective decision-making tools based on regional Earth System predictions and projections for human health. Similar pathways exist for other Earth System components and resource managements. Note that observations are not only necessary for physical, chemical, and biological parameters but also for processes and natural-human system interactions.

2.1 Observing the Natural-human System

The connections in Fig. 1 are self-evident if one recognizes that the ultimate reliability and success of a prediction system will depend on filling the gaps in mechanistic linkages from changes in climate to human health (McMitchael et al., 2003). Climatic variables such as temperature, precipitation, humidity, and the frequency of their occurrences via changes in extreme events will affect human health through associated changes in ecological responses and transmission dynamics with a whole host of socioeconomic and demographic factors exerting many complex modulating influences (Stewart et al., 2008). The role of the microbial contamination pathways can be exemplified by considering the example of human infections by toxic algal blooms in the marine or lacustrine environment. The algae or the microbes in these water-bodies exploit a microhabitat for their own competitive edge and not to genetically render themselves toxic or virulent to humans since infected persons do not necessarily return to the water-body to provide feedback to the microbes

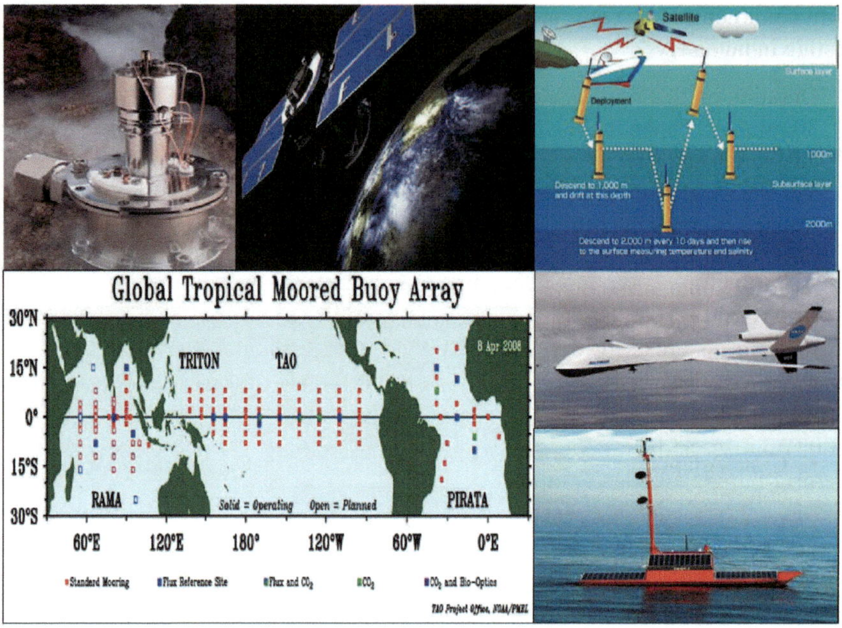

Figure 2: Observational systems for regional Earth System prediction and monitoring will obviously rely on the traditional platforms such as satellites, moored and drifting buoys, unmanned aerial vehicles, flux towers and so on. New innovations will most likely involve miniaturized detectors that will not only monitor environmental variables in the open and indoor spaces but also record genetic level information to facilitate understanding, modelling, and prediction of Earth System interactions from microbes to man. Top left corner shows a 4-inch tall mass spectrometer as an example of miniaturization (http://aemc.jpl.nasa.gov/activities/mms.cfm).

(Stewart et al., 2008). Thus, using a climatic habitat index has severe limitations in forecasting the incidences or toxicity of such harmful algal blooms or pathogen levels without also considering the genetic, chemical, and biological factors, the microbial contamination pathways, human behaviour and exposure. Figure 1 is modified from Murtugudde (2009) to demonstrate that climate change indicators at global and regional scales are the targets of GEOSS and many of the local meteorological and related networks monitor the regional weather manifestations of climate change.

The concept being emphasized here is the need to think beyond the traditional platforms for observational instruments such as satellites, in situ moored, drifting, and robotic buoys, flux towers, volunteer ships, air-borne manned and unmanned vehicles, and so on. These platforms will obviously continue to play foundational roles in our observational needs and will evolve to be more efficient, more accurate, faster, cheaper, and better. The data assimilative approaches to observational system simulation experiments

(OSSEs), which have thus far been mainly disciplinary (Ballabrera-Pay et al., 2007b), have served well for optimizing observational systems and reducing or building in redundancies. The artificial dichotomy of disciplinary boundaries has been a handicap in sustainable coastal management (Lubchenco, 1998) and must begin to be shed even in the context of existing interdisciplinary observations (Dickey, 2003) to consider Earth System OSSEs. Defining the indicators to monitor the impact of natural-human interactions will thus be integral aspects of the observational system designs and the OSSEs will be ongoing exercises to attain adaptability in the observing system (Christian and Mazzilli, 2007). The latter is especially important to observe the feedback of the human system to the evolving natural system (Hibbard, 2007). The coastal observational needs in terms of the need to integrate the marine observations with the terrestrial and freshwater observations is already recognized (Christian, 2003) but we still have a long way to go in making it a monitoring of the natural-human system.

2.2 Observing the Coast from Microbes to Man

It is now known that microbes modify the ocean environment (Rohwer and Thurber, 2009) and their influence cascades into ecosystem levels. Such a concept is not as obvious on land but it is known that the abundance and diversity of microbes in soil are just as large if not larger than the aquatic environments (Srinivasiah et al., 2008) and certain symbiont cyanobacteria in the terrestrial and marine biosphere produce neurotoxins (Cox et al., 2005). In the meantime, the science and technology of sequencing genetic make-up of living organisms and detecting them continue to grow by leaps and bounds (DeLong, 2009), including barcoding of floral and faunal DNA (Jakupciak and Colwell, 2009). This is an opportunity to drive technological innovation to not only using acoustic and other techniques to monitor the food webs and biomass but also include monitoring of DNA and RNA on observing platforms such as Argo or have miniaturized probes that go from genetics and genomics to ecology to human health and all other aspects of Earth System prediction (Janzen et al., 2009; Bowler et al., 2009).

The levels of most of the harmful algae and pathogens are related to human activity such as agriculture, waste water treatments, and land use change (Diaz and Rosenberg, 2008; Patz et al., 2004). Combined with the fact that coasts continue to get denser in human occupation and sea level continues to rise, the ocean observing systems cannot be designed in isolation anymore. More importantly, these disparate observations have to be integrated into Earth System models, especially in the high resolution regional Earth System models. What shape would such a regional Earth System prediction system take? As stated earlier, there is no unique framework and regional specificity is crucial. One example of such a system is underway at the University of Maryland and it is briefly described here, purely to provide the context for this entire discussion.

2.3 A Prototype Regional Earth System Prediction for an Estuary

A nascent but quite a comprehensive effort on regional Earth System prediction is underway within the Earth System Science Interdisciplinary Center (http://essic.umd.edu) of the University of Maryland with dynamic downscaling of the seasonal to interannual climate forecasts and IPCC projections for the Chesapeake watershed with a regional atmosphere, watershed, and a regional ocean model. Routine forecasts of the Chesapeake airshed, watershed, and the estuary include seasonal predictions and decadal projections of such linked products as pathogens, harmful algal blooms, sea nettles, water and air quality, fisheries, dissolved oxygen, inundation and storm-surge, and so on. A prototype decision-making tool has been developed where the user can change the land use types (urban, wetlands, different crops, forests, livable habitat and smart growth concepts) and choose the time period of interest from the past, present, or the future to compute the nutrient loading in the Chesapeake Bay, dissolved oxygen levels, harmful algal blooms, sea nettles, fisheries habitat suitability, etc. The tool is being made fully three-dimensional with the Google Earth and Google Ocean concepts to provide an integrated assessment and education tool for terrestrial and marine ecosystems and other resources. A unique, new approach is being attempted where specific users are being directly given the Earth System forecasts and the flow of information in their decision-making process is being monitored to obtain quantitative feedbacks. A larger context for this prototype effort is provided by a programme called Climate Information: Responding to User Needs (CIRUN) which organizes workshops of users varying from agricultural to insurance sectors to national security (http://climateneeds.umd.edu), as a pioneering effort to drive a demand-pull for specific Earth System information instead of the old paradigm of supply-push where a vast number of model products are placed on the loading-dock hoping for users to pick them up, simply because modellers think they are useful (Murtugudde, 2009). The early returns on the use of our sea nettle forecasts by the recreational boaters are quite encouraging but we eagerly await the feedback from the watermen, river keepers, forest conservators, etc.

While most of the observational input into this prediction system has been for empirical modeling, validation, skill assessment and some optimization of parameters, efforts are underway to carry out physical and biogeochemical OSSEs with the ensemble transform Kalman filter techniques (Takemasa and Aranami, 2006). Natural-human system interactions are being considered in the health sphere by gathering environmental and pollution data along with mortality data to develop health indicators and more importantly, computational social science approaches to understand transmission dynamics of infectious diseases and environmental contributions (Ferguson, 2007; Lazer et al., 2009). The motivation is to drive a health observation system to generate personalized, predictive, and pre-emptive Earth System information for human health. While the system is being built for the Chesapeake watershed, it is a prototype that can be implanted for any part of the world.

2.4 Model-data Synthesis for Sustainable Management

It is the day-to-day management of the regional Earth System that will be crucial for global sustainability since global aggregation, which is meaningful for certain indicators such as greenhouse gas concentrations and global temperature increase, most environmental stressors are local such as land cover change and the impacts also tend to be local such as human health (Patz et al., 2005). The ongoing saga of the swine flu clearly illustrates the natural-human system interactions in its full glory and the opportunities out of this potential disaster from the regional Earth System prediction is that there have been giant strides in improving process understanding of the natural-human system behaviour (Vespignani, 2009; Cho, 2009). We thus have a much better knowledge of what observations we need to capture these interactions.

The onus is on us to drive technological innovations for creations of global digital libraries of air and water quality including pathogens and their genetic information and also instrumentation so that decision-makers on the ground carrying detectors such as hand-held bacterial counters or optimally distributed web of sensors that monitor environmental factors and bacterial levels can instantly validate the Earth System forecasts against the digital libraries (Stewart et al., 2008; DeLong, 2009; Jakupciak and Colwell, 2009). As stated above, novel advances in computational social science can capture transmission dynamics by using human movement and behaviour which should drive macro-scale human ecological observations as a part of the Earth System monitoring. The prediction models must be effective decision-making tools for specific mitigation and adaptation measures and response training such that the evaluation of the impacts of policy and management decisions in modulating climate change, regional weather changes, resource distributions and allocations, population growth and movements and the associated cascades to human health must be a continuous feedback to the Earth System observation and prediction.

Much gets said about the need to effectively communicate the uncertainties in predictions (Allen et al., 2000; Stainforth et al., 2007), but it is very crucial to ensure that data quality across disciplines is clearly understood and communicated in the sustainable management and regional Earth System prediction context (Costanza, 2007). The 'degree of goodness' as defined by Costanza (2007) will clearly not be uniform across disciplines in going from microbes to man and it is absolutely critical to establish the impacts of these differences on the information being extracted from these data and the synthesis of these data into models.

3. CONCLUDING THOUGHTS

The conceptual framework being offered is simply an extension of existing ideas. It is assumed that high resolution regional Earth System models offer the best hope for effective decision-making tools to adaptively manage the Earth System under climate change pressures and these are also the best tools for sustainable coastal management. This presents a monumental challenge but an unprecedented opportunity to develop integrated Earth System observation strategies and drive technological innovation. The author further advocates that the global Earth System observational needs are being effectively coordinated under GEOSS but regional Earth System prediction will require additional regional specificities. Figure 2 depicts it as a drive towards miniaturization of instruments needed to capture the details at the microbial level which have always been important but now will need to be resolved to understand the consequences of climate change on microbial dynamics and their feedback to the natural-human system and its interactions. In addition to the traditional observational platforms, observations in more and more details with smaller and smaller instruments will play a major role and they will need to observe not just the physical, chemical, and biological parameters and processes but also human ecology and the natural-human system interactions. They will have to fully exploit these disparate data from every component of the Earth System, to reduce uncertainties and improve skills of our decision-making tools based on regional Earth System predictions.

ACKNOWLEDGEMENTS

The Chesapeake Bay Forecast System is funded by NOAA and involves partners from across campus and also from HPL, NOAA, DNR, EPA, and NASA. Part of my time was also funded by the NASA Indian Ocean grant and the NOAA mesoscale grant. The author gratefully acknowledges the input from M. Bala Krishna Prasad which was crucial for the completion of the draft.

REFERENCES

Allen, M.R., Stott, P.A., Mitchell, J.F.B., Schnur, R. and Delworth, T.L. (2000). Quantifying the uncertainty in forecasts of anthropogenic climate change. *Nature*, **407**, 617-620.

Ballabrera-Poy, J., Hackert, E., Murtugudde, R. and Busalacchi, A.J. (2007b). An Observation System Simulation Experiment for an optimal moored instrument array in the Indian Ocean. *Journal of Climate*, **20**, 3284-3299.

Ballabrera-Poy, J., Murtugudde, R., Zhang, R-H. and Busalacchi, A.J. (2007a). Coupled ocean-atmosphere response to seasonal modulation of ocean color: Impact on interannual climate simulations in the tropical Pacific. *Journal of Climate*, **20**, 353-374.

Bell, D.E., Clark, W.C. and Ruttan, V.W. (1993). Global Research Systems for Sustainable Development: Agriculture, Health, and Environment. *In:* Ruttan, V.W. (ed.) Agriculture, Health and Environment. University of Minnesota Press, MN, pp. 358-379.

Biermann, F. (2007). Earth System governance as a crosscutting theme of global change research. *Global Environmental Change*, **17**, 326-337.

Bourles, B., Lumpkin, R., McPhaden, M.J., Hernandez, F., Nobre, P., Campos, E., Yu, L., Planton, S., Busalacchi, A.J., Moura, A., Servain, J. and Trotte, J. (2008). The PIRATA program : History, accomplishments, and future directions. *Bulletin of American Meteorological Society*, **89**, 1111-1123.

Bowler, C., Karl, D.M. and Colwell, R.R. (2009). Microbial oceanography in a sea of opportunity. *Nature*, **459**, 207-212.

Cane, M., Zebiak, S.E. and Dolan, S.C. (1986). Experimental forecasts of El Nino. *Nature*, **321**, 827-832.

Charney, J.G. and Shukla, J. (1981). Predictability of monsoons. *In:* Lighthill, S.J., Pearce, R.P. (eds.) Monsoon dynamics. Cambridge University Press, pp. 99-109.

Charney, J.G., Stone, P.H. and Quirk, W.J. (1975). Drought in the Sahara: A Biogeophysical Feedback Mechanism. *Science*, **187**, 434-435.

Cho, A. (2009). Ourselves and our interactions: The ultimate physics problem? *Nature*, **325**, 406-408.

Christian, R.R. (2003). Coastal initiative of the Global Terrestrial Observing System. *Ocean and Coastal Management*, **46**, 313-321.

Christian, R.R., DiGiacomo, P.M., Malone, T.C. and Talaue-McManus, L. (2006). Opportunities and Challenges of Establishing Coastal Observing Systems. *Estuaries and Coasts*, **29**, 871-875.

Christian, R.R. and Mazzilli, S. (2007). Defining the coast and sentinel ecosystems for coastal observations of global change. *Hydrobiologia*, **577**, 55-70.

Clark, W.C., Crutzen, P.J. and Schellnhuber, H.J. (2003) .Science for global sustainability: Toward a new paradigm. *In:* Schellnhuber, H.J., Crutzen, P.J., Clark, W.C., Claussen, M., Held, H. (eds.) Earth system analysis for sustainability. The MIT Press, Cambridge, MA, pp. 1-28.

Committee on Earth-Atmosphere Interactions (2007). Understanding and Responding to Multiple Environmental Stresses. National Academies Press, Washington DC, p. 139.

Costanza, R. (2007). Assessing and communicating data quality: Toward a system of data quality grading. *In:* Costanza, R., Graumlich, L.J., Steffen, W. (eds.) Sustainability or Collapse? An integrated history and future of people on Earth. The MIT Press, Cambridge, MA, pp. 39-48.

Cox, P.A., Banack, S.A., Murch, S.J., Rasmussen, U., Tien, G., Bidigare, R.R., Metcalf, J.S., Morrison, L.F., Codd, G.A. and Bergman, B. (2005). Diverse taxa of cyanobacteria produce â-N-methylamino-L-alanine, a neurotoxic amino acid. *Proceedings of National Academy of Science USA*, **102**, 5074-5078.

Cox, P.M. and Nakicenovic, N. (2003). Assessing and simulating the altered functioning of the Earth System in the anthropocene. *In:* Schellnhuber, H.J., Crutzen, P.J., Clark, W.C., Claussen, M., Held, H. (eds.) Earth system analysis for sustainability. The MIT Press, Cambridge, MA, pp. 293-311.

DeLong, E.F. (2009). The microbial ocean from genomes to biomes. *Nature*, **459**, 207-212.

Dessai, S., Hulme, M., Lempert, R. and Pielke Sr, R. (2009). Do we need better predictions to adapt to a changing climate? *EOS*, **90**, 111-112.
Diaz, R.J. and Rosenberg, R. (2008). Spreading dead zones and consequences for marine ecosystems. *Science*, **321**, 926-929.
Dickey, T.D. (2003). Emerging ocean observations for interdisciplinary data assimilation systems. *Journal of Marine Systems*, **40**, 5-48.
Dickinson, R.E., Errico, R.M., Giorgi, F. and Bates, G.T. (1989). A regional climate model for the western U.S. *Climate Change*, **15**, 383-422.
Falkowski, P. and Tchernov, D. (2003). Human footprints in the ecological landscape. *In:* Schellnhuber, H.J., Crutzen, P.J., Clark, W.C., Claussen, M., Held, H. (eds.) Earth system analysis for sustainability. The MIT Press, Cambridge, MA, pp 211-226.
Ferguson, N. (2007). Capturing human behavior. *Nature*, **446**, 733.
Franco, D.A. and Williams, C.E. (2000). Airs, Waters, Places and Other Hippocratic Writings: Inferences for Control of Food-borne and Water-borne Disease. *Journal of Environmental Health*, **62**, 2000.
Gallopin, G.C. (2003). What kind of system of science (and technology) is needed to support the quest for sustainable development? *In:* Schellnhuber, H.J., Crutzen, P.J., Clark, W.C., Claussen, M., Held, H. (eds.) Earth system analysis for sustainability. The MIT Press, Cambridge, MA, pp 367-386.
Giorgi, F. and Bates, G.T. (1989). The climatological skill of a regional model over complex terrain. *Monthly Weather Review*, **117**, 2325-2347.
Giorgi, F. and Diffenbaugh, N.S. (2008). Developing regional climate change scenarios for use in assessment of human health and disease impacts. *Climate Research*, **36**, 141-151.
Graham, L.P., Rummukainen, M., Gardelin, M. and Bergström, S. (2001). Modelling Climate Change Impacts on Water Resources in the Swedish Regional Climate Modelling Programme. *In:* Brunet, M., López, D. (eds.) Detecting and Modelling Regional Climate Change and Associated Impacts. Springer-Verlag, Berlin/Heidelberg/New York, 567-580.
Hibbard, K.A., Rapporteur (2007). Decadal-scale interactions of humans and the environment. *In:* Costanza, R., Graumlich, L.J., Steffen, W. (eds.) Sustainability or Collapse? An integrated history and future of people on Earth. The MIT Press, Cambridge, MA, pp. 341-375.
Huntingford, C., Fisher, R.A., Mercado, L., Booth, B.B., Sitch, S., Harris, P.P., Cox, P.M., Jones, C.D., Betts, R.A., Malhi, Y., Harris, G.R., Collins, M. and Moorcroft, P. (2008). Towards quantifying uncertainty in predictions of Amazon "dieback". *Philosophical Transaction of Royal Society*, **B 363(1498):** 1857-1864.
Jakupciak, J.P. and Colwell, R.R. (2009). Biological agent detection technologies. *Molecular Ecology Resources*, **9** (Suppl. 1), 51-57.
Janzen, D.H. et al. (2009). Integration of DNA barcoding into an ongoing inventory of complex tropical biodiversity. *Molecular Ecology Resources*, **9** (Suppl. 1), 1-26.
Keenlyside, N.S., Latif, M., Jungclaus, J., Kornblueh, L. and Roeckner, E. (2008). Advancing decadal-scale climate prediction in the North Atlantic sector. *Nature* **453**, 84-88.
Kelly, P.M. (1979). Towards the prediction of climate. *Endeavour*, **3(4)**, 176-182.
Kelly-Hope, L. and Thompson, M.C. (2008). Climate and infectious diseases. *In:* Thompson, M.C., Garcia-Herrera, R., Beniston, M. (eds.) Seasonal Forecasts, Climatic Change, and Human Health. Springer Science+Business Media, 31-70.

Kinzig, A.P., Clark, W.C., Edenhofer, O., Gallopín, G.C., Lucht, W., Mitchell, R.B., Romero Lankao, P., Sreekesh, S., Tickell, C. and Young, O.R. (2003). Group Report: Sustainability. *In:* Schellnhuber, H.J., Crutzen, P.J., Clark, W.C., Claussen, M., Held, H. (eds.) Earth system analysis for sustainability. The MIT Press, Cambridge, MA, pp. 409-434.

Lazer, D., Pentland, A., Adamic, L., Aral, A., Barabasi, A-L., Brewer, D., Christakis, N., Contractor, N., Fowler, J., Gutmann, M., Jebara, T., King, G., Macy, M., Roy, D. and Van Alstyne, M. (2009). Computational Social Science. *Science*, **323**, 721-723.

Liu, J., Dietz, T., Carpenter, S.R., Alberti, M., Folke, C., Moran, E., Pell, A.N., Deadman, P., Kratz, T., Lubchenco, J., Ostrom, E., Ouyang, Z., Provencher, W., Redman, C.L., Schneider, S.H. and Taylor, W.W. (2007). Complexity of coupled human and natural systems. *Science*, **317**, 1513-1516.

Lovelock, J. (2009). The vanishing face of Gaia: A final warning. Basic Books, Philadelphia, PA, 263 p.

Lubchenco, J. (1998). Entering the century of the environment: A new social contract for science. *Science*, **279**, 491-497.

McMichael, A.J., Campbell-Lendrum D.H., Corvalan C.F., Ebi, K.L., Githelo, A., Scheraga, J.D. and Woodward, A. (2003). Climate change and human health: Risks and Responses. World Health Organization, Geneva, 2003. 332 p.

McPhaden, M. J., Busalacchi, A. J., Cheney, R., Donguy, J.-R., Gage, K. S., Halpern, D., Ji, M., Julian, P., Meyers, G., Mitchum, G. T., Niiler, P. P., Picaut, J., Reynolds, R. W., Smith, N. and Takeuchi, K. (1998). The Tropical Ocean-Global Atmosphere (TOGA) observing system: A decade of progress. *Journal of Geophysical Research*, **103**, 14169-14240.

McPhaden, M.J., Zebiak, S.E. and Glantz, M.H. (2006). ENSO as an integrating concept in Earth Science. *Science*, **314,** 1740-1745.

McPhaden, M.J., Meyers, G., Ando, K., Masumoto, Y., Murty, V.S.N., Ravichandran, M., Syamsudin, F., Vialard, J., Yu, L. and Yu, W. (2009). RAMA: The Research Moored Array for African-Asian-Australian Monsoon Analysis and Prediction. *Bulletin of American Meteorological Society*, **90,** 459-480.

Mearns, L.O., Carbone, G., Doherty, R.M., Tsvetsinskaya, E., McCarl, B.A., Adams, R.M. and McDaniel, L. (2003). The uncertainty due to spatial scale of climate scenarios in integrated assessments: An example from U.S. agriculture. *Integrated Assessment*, **4(4)**, 225-235.

Mitchell, R.B. and Romero Lankao, P. (2003). Institutions, science, and technology in the transition to sustainability. *In:* Schellnhuber, H.J., Crutzen, P.J., Clark, W.C., Claussen, M., Held, H. (eds.) Earth system analysis for sustainability. The MIT Press, Cambridge, MA, pp. 387-408.

Murtugudde, R. (2009). Regional Earth System Prediction: A decision-making tool for Sustainability? *Current Opinion in Environmental Sustainability*, in press.

Namias, J. (1968). Long Range Weather Forecasting—History, Current Status, and Outlook. *Bulletin of American Meteorological Society*, **49**, 438-470.

Patz, J.A., Campbell-Lendrum, D., Holloway, T. and Foley, J.A. (2005). Impact of regional climate change on human health. *Nature*, **438**, 310-316.

Patz, J.A., Daszak, P., Tabor, G.M., Aquirre, A.A., Pearl, M., Epstein, J., Wolfe, N.D., Kilpatrick, A.M., Foufopoulos, J., Molyneux, D., Bradley, D.J., Members of the working group on Land use change and disease emergence (2004). Unhealthy landscapes: Policy recommendations on land use change and infectious disease emergence. *Environmental Health Perspectives*, **112**, 1092-1098.

Phillips, N.A. (1960). Numerical Weather Prediction. *In:* Advances in Computers. New York Academic Press.

Rial, J.A., Pielke Sr, R.A., Beniston, M., Claussen, M., Canadell, J., Cox, P., Held, H., Noblet-Ducoudré, N., Prinn, R., Reynolds, J.F. and Salas, J.D. (2004). Nonlinearities, feedbacks, and critical thresholds within the Earth's climate system. *Climate Change*, **65**, 11-38.

Richardson, L.F. (1922). Weather Prediction by Numerical Process. Cambridge University Press, Cambridge.

Rohwer, F. and Thurber, R.V. (2009). Viruses manipulate the marine environment. *Nature*, **459**, 207-212.

Schellnhuber, H.J. (1998). Discourse: Earth System Analysis - The Scope of the Challenge. *In:* Schellnhuber, H.-J., Wenzel, V. (eds.) Earth System Analysis - Integrating science for sustainability. Springer, Heidelberg, pp. 5-195.

Schellnhuber, H.J. (1999). 'Earth system' analysis and the second Copernican revolution. *Nature*, **402**, C19-C23.

Srinivasiah, S., Bhavsar, J., Thapar, K., Liles, M., Schoenfeld, T. and Wommack, K.E. (2008). Phages across the biosphere: contrasts of viruses across soil and aquatic environments. *Research in Microbiology*, **159,** 349-357.

Stainforth, D.A., Allen, M.R., Tredger, E.R. and Smith, L.A. (2007). Confidence, uncertainty, and decision-support relevance in climate predictions. *Philosophical Transactions of Royal Society*, **A365**, 2145-2161.

Steffen, W., Crutzen, P.J. and McNeill, J.R. (2007). The Anthropocene: Are humans now overwhelming the great forces of Nature? *Ambio*, **36**, 614-621.

Stewart, J.R., Gast, R.J., Fujioka, R.S., Solo-Gabriele, H.M., Meschke, J.S., Amaral-Zettler, L.A., del Castillo, E., Polz, M.F., Collier, T.K., Strom, M.S., Sinigalliano, C.S., Moeller, P.D.R. and Holland, A.F. (2008). The coastal environment and human health: microbial indicators, pathogens, sentinels, and reservoirs. *Environmental Health*, **7** (Suppl 2):S3, p. 14, doi: 10.1186/1476-069X-7-S2-S3.

Takemasa, M. and Aranami, K. (2006). Applying a Four-dimensional Local Ensemble Transform Kalman Filter (4D-LETKF) to the JMA Nonhydrostatic Model (NHM). *SOLA*, **2**, 128-131.

Vespignani, A. (2009). Predicting the behavior of techno-social systems. *Nature*, **325**, 425-428.

2

Ecological Modelling as a Tool for Coastal Ecosystem Management

Wen Long
University of Maryland, Center for Environmental Science
Cambridge, Maryland, 21613, USA

1. INTRODUCTION

Ecosystem in general refers to a complex natural community made of living resources and their relationship with the natural environment. Ecosystem management, as a much debated concept over the past few decades, was well defined in Grumbine (1994) as: "It is the managerial actions that humans take on nature with the goals of maintaining ecological integrity above all with other specific goals including maintaining viable populations, ecosystem representation, ecological processes, protecting evolutionary potential of species and accommodating human exploitation of the nature". Levin et al. (2009) summarized that ecosystem-based-management (EBM) has evolved from a vague principle to a central paradigm underlying living resource policies. Ecosystem-based-management takes into account interactions of ecosystem components and management sectors, thus defining management strategies for entire system instead of individual components of the system. Levin et al. (2009) also pointed out that early ecosystem research and management were dominated by reductionist investigations, and to date, we have come to a consensus that synthesis and integration approaches are necessary to conserve healthy ecosystems and restore impaired ecosystems.

Ecological modelling, as a useful tool for implementing management actions, evaluating risks and assessing ecosystem status, has been under rapid development in recent decades. In this article, the most recent advances in ecological modelling in relation to management and its cycle of planning, implementation, evaluation and adaptation are reviewed. The subject deserves an entire book but this article deals only with most recent advancement to outline the frontal areas of this rapidly advancing field. It also concentrates on coastal, estuarine systems.

The subject matter of this article is organized as the following. In Section 2, the concepts of ecosystem based management and ecological modelling are introduced. In Section 3, the statistically based models are briefly introduced with a few examples. Section 4 gives in-depth description of ecological models from the perspectives of hydrodynamics, lower trophic levels, higher trophic levels and sediment diagenesis. In Section 5, several key issues of ecological modelling are discussed with details and focus on recent advancements. The article concludes outlining future directions, followed by an extensive references list for readers with further interests in this subject.

2. COASTAL ECOSYSTEM MANAGEMENT AND MODELLING

Coastal ecosystem covers wetlands, estuaries, lower river reaches and coastal seas. Coastal ecosystem is of great importance for it is one of the most exploited areas for human activities and also one of the most affected areas by anthropogenic changes such as pollution, excessive nutrient loading and municipal developments. Confronting problems in coastal ecosystems are diverse, making management difficult to plan and implement. Some of the pressing issues are: (a) climate change, (b) eutrophication and associated hypoxia and harmful algal blooms (HABs) etc. Many research activities have been carried out to tackle these problems. With the anticipation of increased variability of weather events under the global warming reality, we face serious challenges in predicting seasonal and long-term phytoplankton production and compositional responses to nutrient input reductions. Sherman and Duda (1999) specifically proposed a globally allied and modular strategy for assessment and management of the declining global coastal ecosystem.

Leslie and McLeod (2007) reviewed the challenges of implementing marine ecosystem-based management. The real challenges come in moving beyond concepts towards implementation in specific settings. From ecological science point of view, a large part of difficulties in management roots from our limited understanding of the complex nature. Many existing problems remain to be understood and many new challenges show up with the evolution of the environment and our society. Modelling of coastal systems has been a very rapidly developing field with information technology and its benefit of being cost effective in analyzing risks, evaluating ecohealth status, examining hypothesis.

Coastal ecosystem modelling has gained tremendous amount of achivements over the last decade. The progress is due to the importance of the subject enforced by increasing public attention and consensus of the need for systematic management and much of the breakthrough benefited from rapid development of hydrodynamic modelling of coastal currents, tides, river plumes, surface and internal waves and turbulence mixing, with sediment transport processes taking a slower pace but coming along. The relatively more complete understanding of aquatic chemistry and its modelling practices have also been on the leading facet of the coastal ecosystem modelling. For example, Park et al. (2008) reviewed the most comprehensive chemical model of aquatic systems to date, AQUATOX, which consists of 450 mathematical equations of chemicals and combines aquatic ecosystem, chemical fate and ecotoxicological constructs for a truly integrative fate and effects model. AQUATOX is a general, public domain, mechanistic ecological risk assessment model to be used to evaluate effects from various environmental stressors including nutrients, organic wastes, sediments, toxic organic chemicals, flow and temperature in aquatic ecosystems including streams, rivers, ponds, lakes, reservoirs and estuaries and can be linked to watershed models like HSPF (Hydrological Simulation Program Fortran) and SWAT (Soil & Water Assessment Tool).

The use of ecological modelling for management is world wide, not only in more economically developed countries. There are numerous examples and an important point is that lessons learned by managers and scientists in more industrialized or environmental stressed areas can be shared to those in less developed areas. Gattie et al. (2007) contrasted the traditional reductionists' view of simplifying problems to be mathematically and technically easy for understanding of interested questions under sought, by proposing systems and engineering ecology, as an integrated science comprised of principles from environmental theory, ascendancy theory, exergy theory, energy theory, ecological network analysis and ecological modelling. The author is skeptical to the idea and holds reservations on whether such all-involving systems can be comprehensible or analyzable using our traditional tools in solving problems in physics, chemistry and biology etc. fields. But he also shares the vision of an emerging field called ecological engineering, emphasizing practical means in implementing managerial actions on ecosystems.

Technically, ecosystem models can be cast into two broad categories: statistical models and dynamical models and they are introduced in the next two sections.

3. STATISTICAL ECOSYSTEM MODELLING

Statistical models often emphasize on the use of observed data to predict forthcoming changes in relative short time under the assumption that the causal variables/forcing are included in the model. Statistical models often provide quicker implementation and easier application to managers than fully dynamic models. Statistics of observed data are ways to represent the characteristics of the system or system change, and they are usually connected to the structure of the system and functions of the system components. Statistical models are methods to relate external changes such as nutrient loading change to the statistical measures of observed data. Traditional statistical methods include simple correlation, various forms of regression etc. Modern statistical methods include neural networks, clustering algorithms etc. Statistical models are usually highly site specific and cannot be readily applied to other sites. In practice, Baysian network methods, regression methods, artificial neural network (ANN) methods are commonly adopted.

Greene et al. (2009) developed a new suite of multiple regression models for hindcasting and forecasting midsummer hypoxia in the Gulf of Mexico, with model inputs derived from load estimating methods. Variability in mid-summer hypoxic area was described by models that incorporated May discharge, May nitrate, and February total phosphorus (TP) concentrations or their spring (discharge and nitrate) and winter (TP) averages. The regression models predicted the observed hypoxic area within ±30%, yet model residuals (model-data difference) showed an increasing trend with time. Model forecasts predicted that a dual 45% reduction in nitrate and TP concentration would likely reduce hypoxic area to approximately 5000 square kilometre. Monte Carlo simulations predicted that five years after an instantaneous 50% nitrate reduction or dual 45% nitrate and TP reduction it would be possible to resolve a significant reduction in hypoxic area. The multiple regression models and statistical approaches applied provide improved capabilities for evaluating dual nutrient management strategies to address Northern Gulf of Mexico (NGOM) hypoxia.

Lee et al. (2003) used an artificial neural network (ANN), as a data driven modelling approach, to predict the algal bloom dynamics of the coastal waters of Hong Kong. They used the back-propagation learning algorithm for training the ANN. An ANN is commonly divided into three or more layers: an input layer, a hidden layer(s), and an output layer. The input layer contains the input nodes (neurons), i.e. the input variables for the network. The output layer contains the desired output of the system and the hidden layer usually contains a series of nodes associated with transfer functions. Each layer of the ANN is linked by weights that have to be determined through a learning algorithm. For details please see Lee et al. (2003) and Haykin (1994). The study suggested that the algal concentration in the eutrophic sub-tropical coastal water is mainly dependent on the antecedent algal concentrations in the previous 1~2 weeks. The study shows that an ANN model with a small number of input variables is able to capture trends of algal dynamics, but data with a minimum sampling interval of one week is necessary.

Bandelj et al. (2009) also used a back-propagation neural network (BPNN, Zupan and Gasteiger, 1999) model in conjunction with fuzzy *k*-means clustering algorithm (Legendre and Gallagher, 2001) to explore the existence of complex and non-linear relationships between environmental factors and benthic communities in the Lagoon of Venice, and then used the model to derive functional response of these communities to changing environmental conditions. Through the above mentioned ANN applications in recent years, it appears that ANN can be a powerful predictive alternative to traditional modelling techniques.

4. DYNAMIC ECOSYSTEM MODELLING

Compared to statistic models, dynamic ecosystem models are constructed based on fundamental principles in physics, chemistry and to a lesser degree biology, with insights obtained through identified processes that control the structure of the ecosystem at question and the functions and connectivity of its components. Dynamic ecosystem models are derived from the understanding of processes within the system and can also help to improve understanding. Most dynamic models are also process-based, meaning that they quantify biomass as a function of computed biomass and nutrients and computed processes that may include growth, respiration, predation, and transport. Parameters in process-based models are usually assigned so that computations match one or more observed quantities (Cerco and Noel, 2004). Dynamic models are much less site specific and can usually apply to ecosystems with similar composition and functionality. Dynamic models are limited by (a) number of significant processes considered; (b) mathematical difficulties in formulating certain processes, especially for nonlinear interactions; (c) constraints in obtaining required input fields and (d) uncertainties associated with many model parameters. Owning the overarching complexity of ecosystem and the goal of integrated and systematic approach for the purpose of EBM, it takes significant or sometimes unrealistic amount of time and effort to develop a well formulated, coded, calibrated and validated dynamic ecosystem model.

A fully developed dynamic model would involve some form of coupling among physical system (usually hydrodynamics), lower trophic biological community, higher trophic biological communities, sediment and wetlands. The recent advances of these modelling aspects are given below in more detail.

4.1 Hydrodynamic Modelling

With respect to coastal ocean and estuarine systems, many comprehensive hydrodynamic models now exist and are under tremendous amount of improvement, among which public domain models POM (Princeton Ocean Model) from Princeton University, ROMS (Regional Ocean Modelling System, Wilkin et al., 2005, Shchepetkin et al., 2005) from the Rutgers University, FVCOM (Finite Volume Coastal Ocean circulation Model, Chen et al. 2006) from University of Massachusetts-Dartmouth and Woods Hole Oceanographic Institution and ELCIRC/SELFE (Zhang et al. 2004, 2008) represent the most widely accepted models.

4.1.1 Hydrostatic Models

ROMS (Regional Ocean Modelling System, www.myroms.org) and POM (Princeton Ocean Model, www.aos.princeton.edu/WWWPUBLIC/htdocs.pom/) are the two most widely used, and freely available, hydrostatic, 3-dimensional, primitive variable based, prognostic (time integrating) physical ocean models used by scientists and engineers in the world. POM was initially developed by Blumberg and Mellor (1987) with numerous improvements and a fully documented manual by Mellor (2004). Recent developments of POM include wetting and drying by Oey et al. (2005), parallelization by New York State University Stony Brook group (sbPOM) and data assimilation. ROMS inherited much formulations and numerics from POM with significant improvements in programming structure and has recently drawn a larger community with dedicated software management, preprocessing and post-processing, and peer experience share. Numerically, both models are based on finite difference methods on staggered orthogonal grid system in horizontal dimensions. A significant feature of ROMS is the s coordinate transformation in the vertical direction, with which the model horizontal coordinates follow the topography by stretching the vertical layers to fixed sigma levels. A long-standing difficulty in terrain-following s coordinate ocean modelling is the numerical instabilities associated with pressure gradient calculation, and this is largely solved in Shchepetkin and McWilliams (2003) based on the reconstruction of the density field and the physical z coordinate as continuous functions of transformed coordinates with subsequent analytical integration to compute the pressure-gradient force. Multilateral coupling of ROMS with other modelling systems such as coastal wind wave model, sea ice model, meteorological models and sediment transport models are also in rapid development. An NPZD (nutrient-phytoplankton-zooplankton-detritus) based pelagic system ecological model (Fennel et al., 2006) is also nested in ROMS. A few adjoint models for data assimilation studies are also under rapid development. Several turbulence schemes for modelling turbulence are included in ROMS, they are tested and discussed in Warner et al. (2005). Parallelization with MPI (message passing interface) and OpenMP are both implemented in ROMS, so that modern parallel computers with distributed memory or shared memory architectures can be utilized to reduce computational time.

FVCOM (Finite Volume Coastal Ocean circulation Model (Chen et al., 2006) is entirely similar to ROMS and POM in formulation, except that numerically it is based on finite volume methods on triangular grid system (unstructured grid), enabling the modelling ability of following complex coastal lines and zooming in for fine scale simulations of interested areas. FVCOM is coupled with biogeochemical cycling model (Ji et al., 2002), wave model (Qi et al., 2009), sea ice model and Kalman filtering ensemble model. Most recent advances include the extension to non-hydrostatic modelling capabilities in Lai et al. (2009). Parallelization is based on MPI and 3D wetting and drying is also available. ELCIRC and SELFE are a suite of models developed by Zhang et al. (2004, 2008) at the Center for Coastal Margin Observation & Prediction, Oregon Health and Science University, originally for specific modelling challenges for the Columbia River, and later tested against standard benchmarks and become widely used for estuaries and continental shelves around the world. The models are based on unstructured grid with an Eulerian-Lagrangian algorithm to solve hydrostatic equations. SELFE has recently been extended to have non-hydrostatic capability as well. ELCIRC is volume conservative with low order numerical scheme, and SELFE employs a high-order scheme with volume conservation not rigorously guaranteed. Limited efforts have been made to extend SELFE with sediment transport models and ecological applications.

Other widely used 2D/3D hydrostatic models based on similar methods include CH3D, EFDC, ECOMSED, ADCIRC etc. with ECOMSED and ADCIRC freely available publicly.

4.1.2 Non-hydrostatic Models

For cases in which vertical motion scale is comparable to horizontal motion scale, the hydrostatic pressure assumption is invalid and non-hydrostatic models are required. It is important to include vertical acceleration and solve pressure field dynamically to capture short waves for fine scale inundation, transport and mixing, surface and internal wave breaking etc. phenomena. The solution in general involves solving for pressure field from a so-called Poisson equation, which could be numerically demanding due to a large, sparse matrix linear equation system to be encountered.

Lai et al. (2009) extended the FVCOM model into FVCOM-NH which becomes a non-hydrostatic model. The model is tested against idealized cases of solitary wave separation on a sill, Kelvin-Helmholtz instability of a lock exchange problem with excellent agreement to analytical solutions and direct numerical simulations (DNS). The SUTANS model (http://suntans.stanford.edu/,f) (Fringer et al., 2006), developed at Stanford University, is a non-hydrostatic, unstructured-grid, parallel, coastal ocean simulation tool based formulations outlined in Casulli (2002). The free-surface and vertical diffusion are discretized with the theta-method, which eliminates the Courant condition associated with fast free-surface waves and the friction term associated with small vertical grid spacing at the free-surface and bottom boundaries. The grid employs *z*-levels in the vertical and triangular cells in the horizontal plane. Advection of momentum is accomplished with the second-order accurate unstructured-grid scheme of Perot (2000), and scalar advection is accomplished semi-implicitly using the method of Gross (2002), in which continuity of volume and mass are guaranteed when wetting and drying is employed. The wetting and drying capabilities of SUNTANS enable its use for coastal as well as estuarine domains. The theta-method for the free-surface yields a two-dimensional Poisson equation, and the non-hydrostatic pressure is governed by a three-dimensional Poisson equation, which are both solved with the preconditioned conjugate gradient algorithm with diagonal preconditioning. SUNTANS is written in the C programming language, and MPI is employed for use in a distributed memory parallel computer system.

4.1.3 Challenges of Current Hydrodynamic/Physical Models

Challenges remain in fundamental turbulence modelling and also in coupling hydrodynamic models to less rigorously formulated systems such as physical hydrology, groundwater flow, sediment transport and ice models. There are also challenges in coupling ocean models to meteorological models for they are often formulated differently and developed separately, and both systems demand a lot of computational power. Expertise and experience in different fields, which take significant amount of effort to obtain, are also required to facilitate seamless coupling of these comprehensive modelling systems. Nevertheless, integrated regional earth system models are now appearing and will soon change the landscapes of ecosystem modelling greatly. Systematic and integrational approaches are coming to the reality.

4.2 Lower Trophic Level Modelling

Among many mechanistic, deterministic, process-based ecological models, a large portion of them can be classified as lower trophic level models, in which the main concerns are nutrient cycling, biogeochemical pathways, primary production and lower trophic level animals such as copepods. Most of these lower trophic level models treat living organisms and plants as functional groups to represent their ecological processes and biological life cycles mainly concerning their growth, mortality as well as behaviours. Some of these models simulate individuals of a particular species and their interaction with the environment (individual-based modelling (IBM), also called agent-based modelling). The so-called NPZD models represent the family of models that use nutrient, phytoplankton, zooplankton and detritus groups to simulate the biogeochemical interactions of the basic components of coastal ecosystems. The most prominent characteristics of these models is that the biological and chemical components can be mathematically represented using "concentration", i.e. amount of material per unit volume, and the concentrations are subject to transport and mixing of circulations predicted from hydrodynamic models or prescribed a priori.

Mitra et al. (2007) pointed out that most NPZD models (e.g. Fasham et al., 1990) had been developed based on description of phytoplankton yet the zooplankton grazing functions are important determinants for plankton system dynamics and should be given more biologically sound descriptions in models. More specifically, they found that models without consideration of the declined grazing loss on phytoplankton due to low quality prey can still reproduce observed phytoplankton biomass, by compensating the error with inadequately formulated non-grazing loss rates. The correction of considering prey quality influence on grazing rates and assimilation efficiencies of zooplankton and using adequate non-grazing loss rates will route more materials through detritus pool, which can have important implications to trophic dynamics. Cerco and Noel (2004) reported success in reproducing primary production in the

Chesapeake Bay (USA), finding good agreement with measured light-saturated carbon fixation, net phytoplankton primary production, and gross phytoplankton primary production, while maintaining realistic calculations of algal carbon, chlorophyll, limiting nutrient, and light attenuation. The modelling system is called Chesapeake Bay Environmental Model Package (CBEMP) which comprises a three-dimensional hydrodynamic model (Johnson et al., 1993), a eutrophication model (Cerco and Cole, 1993) and a sediment diagenesis (aging of sediments) model (Di Toro, 2001).

In Chen and Ji et al. (2002), a coupled physical and biological model was developed for Lake Michigan, with physical circulation and thermal stratification predicted by the Princeton Ocean Model (POM) and the biological model an eight-component, phosphorus-limited, lower trophic level food web model, which included phosphate and silicate for nutrients, diatoms and non-diatoms for dominant phytoplankton species, copepods and protozoa for dominant zooplankton species, bacteria and detritus. The biological model was later modified and embedded within FVCOM (Chen et al., 2006). In a companion paper, Ji et al. (2002) documented fully three dimensional coupled physical-biological modelling of Southern Lake Michigan with reasonable agreement on spatial distributions of nutrients and phytoplankton.

As an extremely deleterious consequence of cultural eutrophication, harmful algal blooms (HABs, also known as red tide) have caused increasing attention for their catastrophic damage to coastal aquaculture. Modelling HABs, especially toxic HABs, is an important and challenging task and so far only limited success has been achieved. Franks (1997) gave an excellent review of early models of HABs by starting from simple mathematical growth and grazing control on phytoplankton that give rise to conditional exponential growth to efforts that couple simple physics (advection and mixing) to more laborious approaches that integrate model and data together. However, it is seen that early HABs modelling attempts were heuristic in nature and lack of insights from true biological processes such as swimming behaviour, feeding behaviour and species specific life cycles. How a species become growing exponentially and how a species turn from non-toxic to toxic to other living organisms are different issues and the details remain poorly understood. In the last decade, techniques to differentiate species and strains of tiny harmful algae have been developed and they aid to the understanding of HABs and promote individual-based models (IBM), within which complex life cycles of the interested species are formulated and modelled. Recently, Hellweger et al. (2008) developed a model to describe the life cycle (including a resting stage called akinete) of cyanobacteria (*Anabaena*) in the shallow Bugach Reservoir in Siberia, Russia. Model analysis show that most of the long-term nutrient uptake for *Anabaena* occurs in the sediment bed, which suggests that the sediment bed is not just a convenient overwintering location but may also be the primary source of nutrients. An in silico tracing experiment showed that most water column cells (90%) originated from cells located in the sediment bed during the preceding winter, indicating that the formation of resting stages is of critical importance to the survival of the population on an annual basis.

Sohma et al. (2008) developed an excellent ecosystem model – Ecological Connectivity Hypoxia Model (ECOHYM)—and applied it to Tokyo Bay, a hypoxic estuary in Japan. The model couples hydrodynamics, pelagic ecosystem and benthic ecosystem together. Model biogeochemical state variables are selected based on lower trophic levels: phytoplankton, zooplankton, detritus, dissolved organic matter, NH_4–N, NO_3–N, PO_4–P, benthic algae, suspension feeders, deposit feeders (benthic faunas), dissolved oxygen (DO) and oxygen demand units (ODU, representing the stoichiometric expression of Mn^{2+}, Fe^{2+} and S^{2-} oxygen demand). In the pelagic system, detritus was divided into fast-labile, slow-labile and refractory parts. Bacterial mineralization processes were divided into oxic, suboxic and anoxic fractions. One hundred years of simulation with seasonal forcing for the Tokyo Bay was carried out to obtain periodic steady state solutions. The model succeeded in reproducing the vertical DO profiles during the hypoxic season. Although no fundamentally new ecological processes are introduced and we'd like to see more of sensitivity analysis to identify key processes in the integrated system, the organization of the modelling structure was very elucidating and the paper was well presented; hence learning the details of the model development in Sohma et al. (2008) is hereby encouraged.

4.3 Higher Trophic Level Food Web Modelling

The lower trophic level models described in the previous subsection are more focussed on nutrient cycles, within which growth and mortality of organisms are pathways of fundamental nutrient elements such as C, N, P and O. The interactions among higher trophic levels and with other components of the aquatic environment call for different approaches and the understanding of the pertinent processes is of great importance to managing coastal ecosystems and in assessing economic values of these systems. Modelling of these higher trophic levels is here described in a general term called "food web modelling".

The Ecopath, Ecosim and Ecospace series of models developed at the Fisheries Center, University of British Columbia, Canada, with increasing complexity, have been so far the most widely used and freely available (www.ecopath.org) food web modelling tool for analysis of exploited aquatic ecosystems and multiple-species management as opposed to traditional single species management. The Ecopath approach was initiated by Polovina in the early 1980s, and has been under continuous development since 1990, with Ecosim emerging in 1995 (Walters et al., 1997, 2000), and Ecospace in 1998 (Walters et al., 1999), leading to an integrated software package, 'Ecopath with Ecosim', or 'EwE'. Ecopath was originated from steady state assumption. It is a model for mass balance and energy balance of various trophic groups based on predator-prey actions for a stock assessment of a water body. Ecosim

(Walters et al., 1997, 2000), as opposed to Ecopath, alleviates the steady state assumption by considering the time integration of the biomass of a group. Ecosim also has a component that simulates the bioaccumulation of tracers/contaminants (e.g. mercury) through the food web. The concentration of a particular tracer in group is calculated by accounting the gain and loss through various ways. Ecospace (Walters et al., 1999), as a significant step forward, has extended Ecosim to have spatial distribution for more realistic representations to waters with apparent heterogeneity. The approach is to dynamically allocate fish biomass across spatially gridded map with same equations solved as in Ecosim for each grid cell with the addition of (a) simple equilibrium hydrodynamic models to give velocity field to conserve water mass in the grid cells and advect biomass of user defined groups; and (b) habitat suitability/preferences for each group in a cell to parameterize immigration and emigration of a group in a grid cell to its neighbour grid cells.

Although EwE is a comprehensive tool for multiple species modelling and ecosystem-based management, the application of it sometimes can be difficult due to scarcity of available data to constrain the model and limitation due to financial and man-power issues. For this reason, and also for the purpose of understanding the details of processes involved, studies of a targeted species have also been traditionally carried out and continue to be useful as a complement to large models such as EwE. Recent examples of dynamic modelling of single species include oyster modelling in Hofmann et al. (1992) and Wang et al. (2008), codfish modelling in Cardinale and Svedang (2004), bioenergetics-based model for Pacific herring in Megrey et al. (2007).

4.4 Sediment Flux Modelling

As the ecosystem comprises physical system (water, suspended sediments, light etc.), biological system (living organisms etc.) and chemical system (inorganic nutrients, organic nutrients and many other solutes), it is in general not possible to simulate living organisms without modelling the chemicals that are important to organisms. The sedimentary bottom of coastal oceans and estuaries are one of the most important players in an ecosystem, for it serves as a reservoir for nutrients that are important for both the pelagic system and the benthic system. From eutrophication point of view, remineralization processes (part of diagenesis, i.e. sediment-aging process) in the sediments are fuelled by the organic matter settling and the resulting inorganic nutrients that return to the water column are as important as inorganic nutrients introduced from river loading, wetland releasing, sewage treatment effluents and atmospheric deposition. The chemical processes of organic matter remineralization within the sediments are relatively well understood (Burdige, 1991; Dhakar and Burdige, 1996; Wang and Van Cappellen, 1996; Van Cappellen and Wang, 1996; Morse and Eldridge, 2007), yet quantifying the reactions remains to be challenging. Challenges come from difficulties in measuring returning fluxes, effects of bioturbation on pore water mixing rates and from determining the sequences of redox reactions based on available oxidants (electron acceptors). A central theme is the oxygen regulation on remineralization processes.

Di Toro (2001) made significant and sound simplifications to the processes by partitioning the sediments into a thin surface aerobic layer and a thicker and deeper anaerobic layer and assuming concentrations within the two layers can be represented by their average values within each layer. Mass balance equations based on chemical reactions of pore water species and sediment particle binding species are made and solved simultaneously forced by overlying water concentrations and settling of organic matter. Both steady state solutions and time-dependent (dynamic) solutions are available. Returning pore water fluxes (dissolved nutrients and oxygen consumption fluxes etc.) are calculated using diffusive mass transfer coefficients between the aerobic layer and overlying water. The approaches by Di Toro (2001) have been adopted in many comprehensive ecological modelling systems including WASP by US EPA, EFDC by Tetratech, ECOMSED by HydroQual, CE-QUAL-W2 (Cole and Buchak, 1995), and most recently AQUATOX. Many other modellers have developed diagenesis models that are based on differential equations of redox reactions and pore water vertical diffusion within the sediments. Wang and Van Cappelen (1996) provided steady state and vertically dependent solutions which fully incorporates the reaction couplings among the elements of C, O, N, S, Fe, and Mn. Hantush (2007) developed a model framework for simulating nitrogen and carbon cycling at the sediment-water interface, and predicting oxygen consumption by oxidation reactions inside the sediments. In Hantush (2007), two-layer structure of sediment (aerobic/oxygenated and anaerobic/anoxic) as in Di Toro (2001) was again adopted, except that vertical profiles of pore water concentrations of NH_4, O_2, NO_3 and CH_4 gas are solved analytically. The following assumptions are made to simplify the formulations: (1) thin aerobic layer; (2) negligible mineralization of organic nitrogen in the aerobic layer; (3) negligible advection; (4) first-order reaction rates; (5) completely reduced nitrate in the anaerobic zone; (6) rapid conversion of nitrite to nitrate in the process of nitrification; (7) all dissolved organic carbon is in the form of methane (CH_4); (8) negligible immobilization of mineral nitrogen; and (9) negligible sulfate and other redox species in the sediments. Although some assumptions such as completely reduced nitrate in anaerobic zone and negligible sulfate and other redox species are less justified and they limit the applicability of the model, the research made important step forward by providing analytical solutions that are both time and space dependent.

Soetaert et al. (2000) reviewed the approaches that have been adopted for coupling benthic and pelagic biogeochemical models. They classified the approaches into five levels, separated by the amount of details given to the sediment processes. The most complex approach (level 4) fully couples water column processes to a vertically resolved biogeochemical sediment model. Level 3 approach is achieved by using a vertically integrated dynamic sediment model. Level 2 next is a reflective type of boundary level, where particulate material arriving at the sediment surface is instantaneously transformed into dissolved components. Level 1 models impose bottom-water concentration of dissolved substances or the sediment-water exchange flux. Level 0 models plainly ignore sediment fluxes. By testing all five levels of modelling approaches to a continental shelf ecosystem, they pointed out that level 1 approaches, in which sediment fluxes or concentrations are imposed, are especially badly designed because they fail to assure conservation of mass. It was suggested that the best choice was level 3 approach in which the evolution of sedimentary particulate matter is part of the solution and the bottom fluxes of dissolved constituents are parameterized based on mass budget considerations. Level 3 formulations represent the best balance between computational demand and attained accuracy at the time of the review. In fact, Level 3 modelling approach was adopted in Di Toro (2000), WASP (US EPA), EFDC (TectraTech Inc.), ECOMSED (Hydroqual Inc.), RCA (Hydroqual Inc.), CE-QUAL-W2 (Cole and Buchak, 1995), ECOHYM and AQUATOX etc. models. Wang and Van Cappellen (1996) and Dhakar and Burdige (1996) fall into the category of level 4, yet they only provided steady state solution at that time. Morse and Eldridge (2007) extended the Van Cappellen and Wang (1996) model to give non-steady state solutions, making it fully in the category of level 4. We might think of Hantush (2007) as being somewhat in between level 4 and level 3 for it's vertically resolving but it does not require solving vertical diffusion equations numerically using discrete schemes because analytical solutions are given. Examples of level 2 include Seitzinger (1996) and Fennel et al. (2006) with the advantage of being simple and straightforward, yet with the disadvantages viz. (a) sediment model fluxes should not be an instantaneous response to received settling organic matter and (b) the approach only works in oxgenic conditions and cannot be applied to cases with hypoxia as pointed out by Fennel et al. (2009). With today's computational power, especially with the boom of parallel computing techniques, it can be concluded that level 4 models will serve the best need and not place a severe constraint in terms of computational cost.

5. SPECIAL ISSUES

5.1 Marshes, Wetlands and Mangroves

Many integrated hydrodynamics, water column ecology and benthic models consider point sources such as river loadings as the main source of nutrients. However, for estuaries and coastal waters, often there are marshes and mangroves etc. in wetland systems along the lateral boundaries characterized as being inter-tidal, shallow, brackish and rich of organic and inorganic matter. These wetland systems can store a large amount of nutrients and organics and buffer chemical changes in the ocean. The alternating tidal cycles can induce seepage (ground water) flow between the ocean and the wetland system, and set up mixing/exchange of dissolved organic and inorganic matter between the wetland and the ocean. Coastal storms can bring large amount of episodic rainfall and form surface water running off the lateral boundaries and wash nutrients into the ocean. These processes provide what's called non-point sources to coastal oceans and estuaries. Despite the prominent position of non-point nutrient sources, associated with wetlands under the driving mechanisms of tides, groundwater flow and coastal storms, little effort has been devoted to dynamical modelling of this aspect of coastal ecosystems due to the fact that wetlands are complex and less mathematically amenable. Process-based box models exist, yet many of their focus is on wetland rather than its impact to ocean ecology and many are highly empirical in nature and site specific. For example, Krest et al. (2000) provided radium measurements indicating marsh nutrient export supplied by groundwater discharge is the major nutrient source of the extremely productive ecosystem of North Inlet, South Carolina, USA. Similar results were detected by Moore (1996) and further evidence was documented in Moore et al. (2002).

Retention of inorganic and organic nutrients as well as primary production and remineralization in coastal wetlands are constantly acting; they can interplay with episodic storms to release nutrients and fuel episodic algal blooms. We may hypothesize that re-aeration/mixing/de-stratification that replenish bottom water DO by coastal storms can be rapidly restored to stratified regime after the storm. Yet the nutrients released by coastal wetlands after the stratification is re-established remain in the coastal ocean and can fuel phytoplankton growth leading to hypoxia.

5.2 Data Assimilation

Data assimilation is a set of special methods to dynamically insert observational data into dynamic numerical models so that model predictions will be "clamped" towards observations for continuous nowcasting and forecasting. The "clamp" can involve both changing the initialization fields of next prediction using available observational data and previous prediction and changing important model parameter values to ensure distance between model and data is confined to an optimal level. Application of data assimilation to ocean modelling was pioneered in Ghil (1989), Dombrowsky and De Mey (1992) and Robinson (1996). It is an important step for making sustained nowcasting and forecasting feasible and available in real time application fields to the research and operational communities. Robinson (1996) also clearly defined a set of terms that are commonly used today: nowcast, forecast and hindcast. Nowcast is a present time estimate from melded data and dynamical model. Forecast is future prediction, which is initialized by nowcasts. Hindcasts are best melded estimates utilizing historical data. Robinson and Lermusiaux (2001) reviewed the methods that are available for data assimilation. They classified the methods into three categories: estimation theory, control theory and direct minimization methods.

The estimation theory methods include nudging, Kalman filtering (Kalman, 1960), optimal interpolation (OI), successive corrections, error subspace statistical estimation (ESSE), ensemble/Monte-Carlo methods etc. The control theory methods include adjoint method (also known as variational method), generalized inverse method etc. The direct minimization methods include decent method (DM), simulated annealing method (SAM), genetic algorithm (GM) etc. The optimal interpolation, Kalman filtering and adjoint methods are the most widely used and are briefly introduced in this section. Earlier review and clarification of notations of Kalman filtering, optimal interpolation and adjoint methods are given in Ide et al. (1997).

5.3 Submerged Aquatic Vegetation Modelling

Submerged (sometimes also called "submersed") aquatic vegetations (SAVs) are important biomass for estuarine, coastal and lake ecosystems. They oxygenate and stabilize sediments and control nutrient fluxes to water column. Modelling of SAV as an aid to ecosystem management has long been sought.

Giusti and Marsili-Libelli (2005) developed a coupled model with a 2D hydrodynamic model for water movements and an ecological model describing the interactions between nutrients and SAV. The model outputs daily variations in nutrient concentrations and vegetation biomass and they are used to infer the necessary managerial actions in favour of beneficial macrophytes (representing SAV species) over deleterious macroalgae for the Orbetello Lagoon (west coast of Italy). The ecological model consists of a nitrogen-detritus based nutrient cycling submodel, a macroalgae sub-model and a macrophyte (*Ruppia*, representing *R. maritima* and *R. chirrosa*) submodel. Model results show that with the point-source discharge around that time (2004-2005), the macroalgae and macrophytes tend to complete a cycle in a year with two peaks of macroalgae in spring and autumn and a single peak of macrophyte (*Ruppia*) in late spring. It seems to be a useful tool to aid action-planning such as selective harvesting policies, by finding the best time and locations to remove macroalgae, to avoid harmful blooms of algae.

Cerco and Moore (2001) coupled the Wetzed and Neckles (1986) and Madden and Kemp (1996) SAV models to an eutrophication model of the Chesapeake Bay. The model successfully computed the spatial distribution and abundance of SAV for the period 1985-1994. The spatial distribution is primarily determined by computed light attenuation, which gave the conclusion that restoration of SAV requires solids control, alone or combined with nutrient controls. As a consequence of their modelling effort, the US EPA Chesapeake Bay Program set a specific sediment reduction goal for the Chesapeake Bay, calling for a 29% reduction in solids loading from watershed, compared to 1985 level, to increase water clarity and restore SAV biomass. In Cerco and Moore (2001), three state variables are used to describe the SAV biomass: shoots, roots and epiphytes, where shoots are SAV biomass above ground, roots are biomass below ground and epiphytes are attached biomass growth. The shoots and epiphytes have nutrient exchange with water column nutrient pool and the roots exchange nutrients with sediment diagenetic model. Light is essential for SAV growth and it is modelled with a series of sequential attenuations by colour, inorganic and organic solids in the water and self-shading of shoots and epiphytes. The model was applied to the Chesapeake Bay for the period of 1985-1994 with excellent performance in reproducing the spatial distribution of SAV and reasonable performance in reproducing SAV biomass for the three species considered, i.e. *Zostera*, *Ruppia* and a general freshwater species representing freshwater communities, and epiphytes. Sensitivity analysis showed that for the Chesapeake Bay, reduction of solids loading is more effective than reduction of nutrient loading in promoting SAV gain. Advice to management was that solids loading reduction and nutrient loading reduction should be combined to restore SAV in the region. There are a few limitations in the above SAV model: (a) influence of SAV on hydrodynamics was not included, (b) population processes, including the propagation of roots and extension of eelgrass meadows, aging, reproduction, were not considered; and (c) predation is highly uncertain. Some of the aspects could be borrowed from vegetation dynamics modelling, crop modelling etc. fields for further improvements.

5.4 Model Skill Assessment

All ecological models require some sort of skill assessment to be applicable for practical problems of EBM. Fitzpatrick (2009) pointed out that there is a need to assess the degree of confidence one has in the projections of mechanistic models of nutrients, primary production and dissolved oxygen in assisting natural resource and water quality managers to evaluate effectiveness of nutrient allocation management in reducing eutrophication. Fitzpatrick reviewed existing model skill measures and suggested that no single metric is suitable for assessing model skill. Qualitative skill assessments such as plotting time-series or spatial plots of model versus data are traditionally used extensively. New skill assessment tools include Taylor diagram (Taylor, 2001), target diagram (Jollioff et al., 2009). Determining skill by regressing model versus observation is shown to be able to differentiate errors due to bias of model relative to data and errors due to bad correlation between model and data. It is also emphasized that modellers should conduct sensitivity analysis on key model parameters and inputs. Metrics used for quantifying model skill include (a) coefficient of determination, used to represent the percentage of variability that can be explained by the model; (b) coefficient of efficiency (Nash and Sutcliffe, 1970); (c) root mean square error (RMSE); (d) model bias relative to observation, (e) the Wilmott skill score (Wilmott, 1981); (f) the Murphy and Epstein (1989) skill; (g) the Taylor (2001) skill score and (h) Jolliff skills score (Jolliff et al., 2009).

6. CONCLUSION AND FUTURE DIRECTIONS

In this article, the basic ideas of ecosystem-based management, the use of ecological models as tools for ecosystem management and common approaches that have been developed are introduced. Focus has been given to coastal ocean problems. As a rapidly developing field, many challenges remain unsolved or emerge. In general, models based on mechanistic foundations improve confidence in model predictions, while statistical methods provide empirical support for parameter selection and allow for estimates of predictive uncertainty (Borsuk et al., 2001). Some new problems and directions that are of great concern and require more complete investigation can be summarized as:

(a) *Reversibility of ecosystem:* Recent studies show that some heavily eutrophicated systems have come across some regime shifts and manifest hysteresis behaviour (i.e. the system will not return to its previous states with external environmental stessors removed). For example, Conley et al. (2009) reported that bottom oxygen of a hypoxic system will not restore with nutrient input reduced to the level before hypoxia occurs once the ecosystem reaches a threshold when bottom-living organisms are killed and the system switches to once dominated by anaerobic processes. Such a behaviour is of great concern because once the system is stressed beyond the "threshold", system will not be reversible and restoration will be not possible. How to find the threshold of a system and how to assess its impact are not well understood.

(b) *Mechanistic modelling of harmful algal blooms:* To date, the exact mechanisms of harmful algal blooms are still not well understood although we may roughly attribute the occurrence to excessive nutrient. Hence most HAB models are based on empirical habitat methods. With new researches targeted at studying the physiology, life cycles (Hense et al., 2006) and toxicology etc. of harmful algal species, we expect that the next generation of HAB models will be process based with the life cycle, growth and mortality, generation of toxins modelled. Hood et al. (2007) developed a heuristic population dynamics model of *Pfiesteria* based on diet and growth characteristics of different types of *Pfiesteria*.

(c) *Ecological forecasting:* Ecological forecasting, following the long estabilished meteorological forecasting and recent advances in ocean forecasting, is becoming a field that will improve our capability to prevent, manage and mitigate hazardous events in ecosystems. Operational meteorology, operational oceanography and operational ecology will join together to serve the needs of resource management.

(d) *Climate change and its impacts on ecosystem:* The implications of climate change on ecosystem is complex and of global scale. We are still a long way from being able to quantify the impacts with desirable precision. Research in this direction is pressing and will only make ways forward with small and firm steps.

As a final remark, the debate between reductionist approaches versus system approaches should be settled as the following. Reductionist approaches may conflict system goal for their objectives in simplifying system to emphasis on particular process or components of a system. Yet, reductionist approaches are good for concentrated study on particular parts, for improving understanding on particular processes. Ecosystem based management, by its nature, requires system approach, whereas a system approach may be too complex to handle or may not be readily available. Hence the two approaches should be complementing each other instead of being exclusive.

ACKNOWLEDGEMENTS

The author sincerely thanks the editors Dr. M.B.K. Prasad and Dr. Ramanathan for the invitation to write this review to coastal ecological modelling and for the very pleasant and productive collaboration. Special thanks are also given to reviewers of this article for their proof reading and constructive suggestions.

REFERENCES

Bandelj, Vinko, Daniele Curiel, Sovan Lek, Andrea Rismondo and Cosimo Solidoro (2009). Modelling spatial distribution of hard bottom benthic communities and their functional response to environmental parameters. *Ecological Modeling*, doi:10.1016/j.ecolmodel.2009.04.024.

Blumberg, A.F. and Mellor, G.L. (1987). A description of a three dimensional coastal ocean circulation model. *In:* Heaps, N. (ed.), Three-Dimensional Coastal Ocean Models. American Geophysical Union, Washington, DC, pp. 1-16.

Borsuk, M.E., Stow, C.A., Higdon, D. and Reckhow, K.H. (2001). A Bayesian hierarchical model to predict benthic oxygen demand from organic matter loading in estuaries and coastal zones. *Ecological Modelling*, **143**, 165-181.

Burdige, D.J. (1991). The kinetics of organic matter mineralization in anoxic marine sediments. *Journal of Marine Research*, **49**, 727-761.

Cardinale, Massimiliano and Henrik Svedäng (2004). Modelling recruitment and abundance of Atlantic cod, *Gadus morhua*, in the eastern Skagerrak–Kattegat (North Sea): Evidence of severe depletion due to a prolonged period of high fishing pressure. *Fisheries Research*, **69**, 263-282.

Casulli, V. and Zanolli, P. (2002). Semi-implicit numerical modeling of nonhydrostatic free-surface flows for environmental problems. *Mathematical and Computer Modelling*, **36**, 1131-1149.

Cerco, Carl F. and Moore, K. (2001). System-wide submerged aquatic vegetation model for Chesapeake Bay. *Estuaries*, **24(4)**, 522-534.

Cerco, Carl F. and Noel, Mark R. (2004). Process-based primary production modeling in Chesapeake Bay. *Mar. Ecol. Prog. Ser.*, **282**, 45-58.

Cerco, C. and Cole, T. (1993). Three Dimensional Eutrophication Model of Chesapeake Bay. *Journal of Environmental Engineering*, **119**, 1006-1025.

Chen, Changsheng, Ji, R. et al. (2002). A model study of the coupled biological and physical dynamics in Lake Michigan. *Ecological Modelling*, **152**, 145-168.

Chen, Changsheng, Beardsley, Robert C. and Geoffrey Cowles (2006). An unstructured grid, finite-volume coastal ocean model (FVCOM) system. *Oceanography*, **19(1)**, 78-89.

Cole, T.M. and Buchak, E. (1995). CE-QUAL-W2: A two-dimensional, laterally averaged, hydrodynamic and water quality model, Version 2.0. Technical Report. US Army Corps of Engineers Waterways Experiments Station, Vicksburgh, MS.

Conley, D.J., Carstensen, J., Vaquer-Sunyer, R. and Duarte, C.M. (2009). Ecosystem thresholds with hypoxia. *Hydrobiologia*, **629**, 21-29.

Dhakar, S.P. and Burdige, D.J. (1996). A coupled, non-linear, steady state model for early diagenetic processes in pelagic sediments. *American Journal of Science*, **296(3)**, 296-330.

Di Toro, D.M. (2001). Sediment Flux Modeling (Environmental Science and Technology), Wiley-Interscience.

Dombrowsky, E. and DeMey, P. (1992). Continuous assimilation in an open domain of the Northeast Atlantic. Part I: Methodology and application to AthenA-88. *J. Geophys. Res.*, **97**, 9719-9731.

Fasham, M.J.R., Ducklow, H.W. and Mckelve, S.M. (1990). A nitrogen-based model of plankton dynamics in the oceanic mixed layer. *J. Mar. Res.*, **48**, 591-639.

Fennel, K., Wilkin, J., Levin, J., Moisan, J., O'Reilly, J. and Haidvogel, D. (2006). Nitrogen cycling in the Middle Atlantic Bight: Results from a three-dimensional model and implications for the North Atlantic nitrogen budget. 20, GB32007, doi:10.1029/2005GB002456.

Fennel, Katja, Damian Brady Dominic Di Toro, Fulweiler, Robinson W., Gardner, Wayne S., Giblin, Anne, McCarthy, Mark J., Rao, Alexandra, Sybil Seitzinger, Marie Thouvenot-Korppoo and Craig Tobias (2009). Modeling denitrification in aquatic sediments. *Biogeochemistry*, **93**, 159-178, DOI 10.1007/s10533-008-9270-z.

Fitzpatrick, J.J. (2009). Assessing skill of estuarine and coastal eutrophication models for water quality managers. *Journal of Marine Systems*, **76**, 195-211.

Franks, P.J.S. (1997). Models of harmful algal blooms. *Limnology and Oceanography*, **42(5, part 2)**, I273- I282.

Fringer, O.B., Gerritsen, M. and Street, R.L. (2006). An unstructured-grid, finite-volume, non-hydrostatic, parallel coastal-ocean simulator. *Ocean Modelling*, **14(3-4)**, 139-278.

Gattie, David K., Kellam, Nadia N. and Turk, H. Jeff (2007). Informing ecological engineering through ecological network analysis, ecological modelling, and concepts of systems and engineering ecology. *Ecological modelling*, **208**, 25-40.

Ghil, M. (1989). Meteorological data assimilation for oceanographers. I: Description and theoretical framework. *Dyn. Atmos. Oceans*, **13(3-4)**, 171-218.

Giusti, Elisabetta and Stefano Marsili-Libelli (2005). Modelling the interactions between nutrients and the submersed vegetation in the Orbetello Lagoon. *Ecological Modelling*, **184**, 141-161.

Grumbine, R.E.(1994). What Is Ecosystem Management? *Conservation Biology*, **8(1)**, 27-38.

Greene, R.M., Lehrter, J.C. and Hagy III, J.D. (2009). Multiple regression models for hindcasting and forecasting midsummer hypoxia in the Gulf of Mexico. *Ecological Applications*, **19(5)**, 1161-1175.

Gross, E.S., Bonaventura, L. and Rosatti, G. (2002) Consistency with continuity in conservative advection schemes for free-surface models. *Int. J. Numer. Meth. Fluids*, **38**, 307-327.

Haidvogel, D.B. and Beckmann, A. (1999). Numerical Ocean Circulation Modeling, Imperial College Press, London, UK.

Hantush, Mohamed M. (2007). Modeling nitrogen-carbon cycling and oxygen consumption in bottom sediments. *Advances in Water Resources*, **30**, 59-79.

Haykin, S. (1994). Neural Networks. Prentice-Hall, Englewood Cliffs, NJ.

Hellweger, F.L., Kravchuk, E.S., Novotny, V. and Gladyshev, M.I. (2008). Agent-based modeling of the complex life cycle of a cyanobacterium (Anabaena) in a shallow reservoir. *Limnol. Oceanogr.*, **53(4)**, 1227-1241.

Hense, Inga and Aike Beckmann (2006). Towards a model of cyanobacteria life cycle—effects of growing and resting stages on bloom formation of N_2-fixing species. *Ecological Modelling*, **195**, 205-218.

Hofmann, E.E., Powell, E.N., Klinck, J.M. and Wilson, E.A. (1992). Modeling oyster populations. III: Critical feeding periods, growth and reproduction. *Journal of Shellfish Research*, **11(2)**, 399-416.

Hood, R.R., Zhang, X., Glibert, P.M., Roman, M.R. and Stoecker, D.K. (2006). Modeling the influence of nutrients, turbulence and grazing on Pfiesteria population dynamics. *Harmful Algae*, **5**, 459-470.

Ide, K., Bennett, A., Courtier, P., Ghil, M. and Lorenc, A. (1997). Unified notation for data assimilation: Operational, sequential and variational. *J. Metorolog. Soc. of Japan*, **75**, 181-189.

Ji, R., Chen, C., Schwab, D.J., Beletsky, B., Fahnenstiel, G.L., Johengen, T.H., Vanderploeg, H., Eadie, B., Bundy, M., Gardner, W. and Cotner, J. (2002). Influences of suspended sediments on the ecosystem in Lake Michigan: A 3-D coupled bio-physical modeling experiment. *Ecol. Model.*, **152(2-3)**, 169-190.

Jolliff, Jason K., Kindle, John C., Igor Shulman, Bradley Penta, Marjorie, A.M., Friedrichs, Robert Helber and Arnone, Robert A. (2009). Summary diagrams for coupled hydrodynamic-ecosystem model skill assessment. *Journal of Marine Systems*, **76**, 64-82.

Kalman, R.E. (1960). A new approach to linear filter and prediction problems. *Journal of Basic Engineering*, **82**, 35-45.

Krest, J.M., Moore, W.S. and Gardner, L.R. (2000). Marsh nutrient export supplied by groundwater discharge: Evidence from radium measurements. *Global Biogeochemical Cycles*, **14(1)**, 167-176.

Krivtsov, V. (2004). Investigations of indirect relationships in ecology and environmental sciences: A review and the implications for comparative theoretical ecosystem analysis. *Ecological Modelling*, **174**, 37-54.

Lai, Zhigang, Changsheng Chen, Cowles, Geoffrey W. and Beardsley, Robert C. (2009). A non-hydrostatic version of FVCOM, Part I: Validation Experiments, submitted.

Lee, J.H.W., Yan Huang, Mike Dickman and Jayawardena, A.W. (2003). Neural network modelling of coastal algal blooms. *Ecological Modelling*, **159**, 179-201.

Legendre, P. and Gallagher, E.D. (2001). Ecologically meaningful transformations for ordination of species data. *Oecologia*, **129**, 271-280.

Lermusiaux, P.F.J. (2001). Evolving the subspace of the three-dimensional multiscale ocean variablity: Massachusetts Bay. *Journal of Marine Systems*, **29**, 385-422.

Leslie, H.M. and McLeod, Karen L. (2007). Confronting the challenges of implementing marine ecosystem-based management, *Front. Ecol. Environ.*, **5(10)**, 540-548, doi:10.1890/060093.

Levin, Phillip S., Fogarty, Michael J., Murawski, Steven A. and Fluharty, David (2009). Integrated Ecosystem Assessments: Developing the Scientific Basis for Ecosystem-Based Management of the Ocean. *PLoS Biology*, **7(1)**, e1000014, 0023-0028.

Madden, C.J. and Kemp, W.M. (1996). Ecosystem model of an estuarine submersed plant community: Calibration and simulation of eutrophication responses. *Estuaries*, **19(2B)**, 457-474.

Megrey, B.A., Rose, K.A., Klumb, R.A., Hay, D.E., Werner, F.E., Eslinger, D.L. and Smith, S.L. (2007). A bioenergetics-based population dynamics model of Pacific herring (*Clupea harengus pallasi*) coupled to a lower trophic level nutrient–phytoplankton–zooplankton model: Description, calibration, and sensitivity analysis. *Ecological Modeling*, **202**, 144-164.

Mellor, G.L. (2004). Users guide for a three-dimensional, primitive equation, numerical ocean model. Program in Atmospheric and Oceanic Sciences. Princeton University, Princeton, NJ 08544-0710, June.

Mitra, Aditee, Flynn, Kevin J. and Fasham, Michael J.R. (2007). Accounting for grazing dynamics in nitrogen-phytoplankton-zooplankton (NPZ) models. *Limnol. Oceanogr.*, **52(2)**, 649-661.

Moore, W.S. (1996). Large ground water inputs to coastal waters revealed by ^{226}Ra enrichment. *Nature*, **380**, 612-614.

Moore, W.S., Krest, J., Taylor, G., Roggenstein, E., Joye, S. and Lee, R. (2002). Thermal evidence of water exchange through a coastal aquifer: Implications for nutrient fluxes. *Geophysical Research Letters*, **29(14)**, 1704, 10.1029/2002GL014923.

Morse, J.W. and Eldridge, P.M. (2007). A non-steady state diagenetic model for changes in sediment biogeochemistry in response to seasonally hypoxic/anoxic conditions in the "dead zone" of the Louisiana shelf. *Marine Chemistry*, **106**, 239-255.

Murphy, A.H. and Epstein, E.S. (1989). Skill scores and correlation coefficients in model verification. *Monthly Weather Review*, **117**, 572-581.

Nash, J.E. and Sutcliffe, J.V. (1970). River flow forecasting through conceptual models. Part 1: A discussion of principles. *J. Hydrol.*, **10(3)**, 282-290.

Oey, L.-Y. (2005). A wetting and drying scheme for POM. *Ocean Modelling*, **9**, 133-150.

Park, Richard A., Clough, Jonathan S., Wellman, Marjorie Coombs (2008). AQUATOX: Modeling environmental fate and ecological effects in aquatic ecosystems. *Ecological Modelling*, **213**, 1-15.

Perot, B. (2000). Conservation properties of unstructured staggered mesh schemes. *J. of Computational Physics*, **159**, 58-89.

Qi, Jianhua, Changsheng Chen, Beardsley, Robert C., Will Perrie, Cowles, Geoffrey W. and Lai, Zhigang (2009). An unstructured-grid finite-volume surface wave model (FVCOM-SWAVE): Implementation, validations and applications. *Ocean Modelling*, **28**, 153-166.

RCA: User's guide for RCA (Release 3.0), Hydroqual, Inc., 2004.

Robinson, A.R. (1996). Physical processes, field estimation and an approach to interdisciplinary ocean modeling. *Earth-Science Reviews*, **40**, 3-54.

Seitzinger, S.P. and Giblin, A. (1996). Estimating denitrification in North Atlantic continental shelf sediments. *Biogeochemistry*, **35**, 235-260.

Shchepetkin, A.F. and McWilliams, J.C. (2003). A method for computing horizontal pressure-gradient force in an oceanic model with a nonaligned vertical coordinate. *Journal of Geophysical Research*, **108(C3)**, doi:10.1029/2001JC001047.

Shchepetkin, A.F. and McWilliams, J.C. (2005). The Regional Ocean Modeling System: A split-explicit, free-surface, topography following coordinates ocean model. *Ocean Modelling*, **9**, 347-404.

Sherman, K. and Duda, A.M. (1999). An ecosystem approach to global assessment and management of coastal waters. *Mar Ecol Prog Ser*, **190**, 271-287.

Soetaert, Karline, Middelburg, Jack J., Herman, Peter M.J. and Kerst Buis (2000). On the coupling of benthic and pelagic biogeochemical models. *Earth-Science Reviews*, **51**, 173-201.

Sohma, Akio, Yasuyuki Sekiguchi, Tomohiro Kuwae and Yoshiyuki Nakamura (2008). A benthic-pelagic coupled ecosystem model to estimate the hypoxic estuary including tidal flat—Model description and validation of seasonal/daily dynamics. *Ecological Modelling*, **215**, 10-39.

Taylor, K.E. (2001). Summarizing multiple aspects of model performance in a single diagram. *J. Geophys. Res*., **106(D7)**, 7183-7192.

Van Cappellen, P. and Wang, Y. (1996). Cycling of iron and manganese in surface sediments: A general theory for the coupled transport and reaction of carbon, oxygen, nitrogen, sulfur, iron and manganese. *American Journal of Science*. **296**, 197-243.

Walters, C., Christensen, V. and Pauly, D. (1997). Structuring dynamic models of exploited ecosystems from trophic mass-balance assessments. *Rev. Fish Biol. Fish*., **7**, 139-172.

Walters, C., Pauly, D. and Christensen, V. (1999). Ecospace: Prediction of mesoscale spatial patterns in trophic relationships of exploited ecosystems, with emphasis on the impacts of marine protected areas. *Ecosystems*, **2**, 539-554.

Walters, C., Pauly, D., Christensen, V. and Kitchell, J.F. (2000). Representing density dependent consequences of life history strategies in aquatic ecosystems: EcoSim II. *Ecosystems*, **3**, 70-83.

Wang, Y. and Cappellen, P.V. (1996). A multicomponent reactive transport model of early diagenesis: Application to redox cycling in coastal marine sediments. *Geochimica et Cosmochimica Acta*, **60(16)**, 2993-3014.

Wang, Hongqing, Huang, Wenrui, Harwell, Mark A., Edmiston, Lee, Johnson, Elijah, Ping, Hsieh, Milla, Katherine, Christensen, John, Stewart, Jessica and Liu, Xiaohai (2008). Modeling oyster growth rate by coupling oyster population and hydrodynamic models for Apalachicola Bay, Florida, USA. *Ecological Modelling*, **211**, 77-89.

Warner, J.C., Sherwood, C.R., Arango, H.A. and Signell, R.P. (2005). Performance of four turbulence closure models implemented using a generic length scale method. *Ocean Model*., **8(1-2)**, 81-113.

Wilkin, J.L., Arango, H.G., Haidvogel, D.B., Lichtenwalner, C.S., Durski, S.M. and Hedstrom, K.S. (2005). A regional Ocean Modeling System for the Long-term Ecosystem Observatory. *J. Geophys. Res*., **110**, C06S91, doi:10.1029/2003JC002218.

Wilmott, C.J. (1981). On the validation of models. *Physical Geography*, **2(18)**, 4-194.

Zhang, Y.-L., Baptista, A.M. and Myers, E.P. (2004). A cross-scale model for 3D baroclinic circulation in estuary-plume-shelf systems. I: Formulation and skill assessment. *Cont. Shelf Res.*, **24**, 2187-2214.

Zhang, Y.-L. and Baptista, A.M. (2008). SELFE: A semi-implicit Eulerian-Lagrangian finite-element model for cross-scale ocean circulation. *Ocean Modelling*, **21(3-4)**, 71-96.

Zupan, J. and Gasteiger, J. (1999). Neural Networks in Chemistry and Drug Design. Wiley-VCH, Weinheim, Germany, 380 pp.

3

Current Status of Coastal Zone Management Practices in India

S. Swarna Latha and M. Bala Krishna Prasad[1]

Washington College of Law, American University, 4801 Massachusetts Avenue NW, Washington, DC 20016, USA

[1]Earth System Sciences Interdisciplinary Center, University of Maryland College Park, MD 20740-3823, USA

1. INTRODUCTION

India has a coastline of about 7500 km of which the mainland accounts for 5400 kms, Lakshadweep coast extends to 132 km and Andaman & Nicobar Islands to about 1900 km. Nearly 250 million people live within a distance of 50 km from the coast. The exclusive economic zone (EEZ) of the country has an area of 2.02 million sq km comprising 0.86 million sq km on the west coast, 0.56 million sq km on the east coast and 0.6 million sq km around the Andaman and Nicobar islands. The east coast supports activities such as agriculture and aquaculture while a number of industries are supported on the west coast. Tourism has emerged as a major economic activity in coastal states such as Goa, Kerala and Orissa.

The Indian coastal environment includes various fragile ecosystems of estuaries, mangroves, sea grasses, coral reefs, etc. supporting humans by delivering various goods and services. However, now these ecosystems are in critical condition because of increased human disturbances which make them highly vulnerable. In order to protect and conserve these coastal ecosystems, the Government of India has enacted several laws and ratified several international principles. Hence in this article an attempt has been made to evaluate the current status of Indian coastal environment in the era of the global environmental change and legal mechanisms to protect coastal zone for the future sustainable development.

2. STATUS OF INDIAN COASTAL ENVIRONMENT

The Indian coastal zone is endowed with a very wide range of coastal ecosystems like mangroves, coral reefs, sea grasses, salt marshes, sand dunes, estuaries, lagoons, etc., which are characterized by distinct biotic and abiotic processes. The distribution of mangroves and coral reef ecosystems in India is shown in Fig. 1. Mangrove cover in India has been estimated at approximately 6700 km^2 confined mainly along the east and west coasts and Andaman and Nicobar Islands which is approximately 9% of the South and Southeast Asian mangrove distribution (Fig. 2) and 5% of the world's mangrove distribution (FSI, 2001). According to the latest evaluation by Rao et al. (1999), 67% of the mangroves and associated plant species are endangered, while 97% of the plant species are threatened. Indiscriminate cutting, reclamation for agriculture and urbanization, fuel and overgrazing by domestic cattle have severely degraded mangroves in India. The threat to mangroves in recent years comes mainly

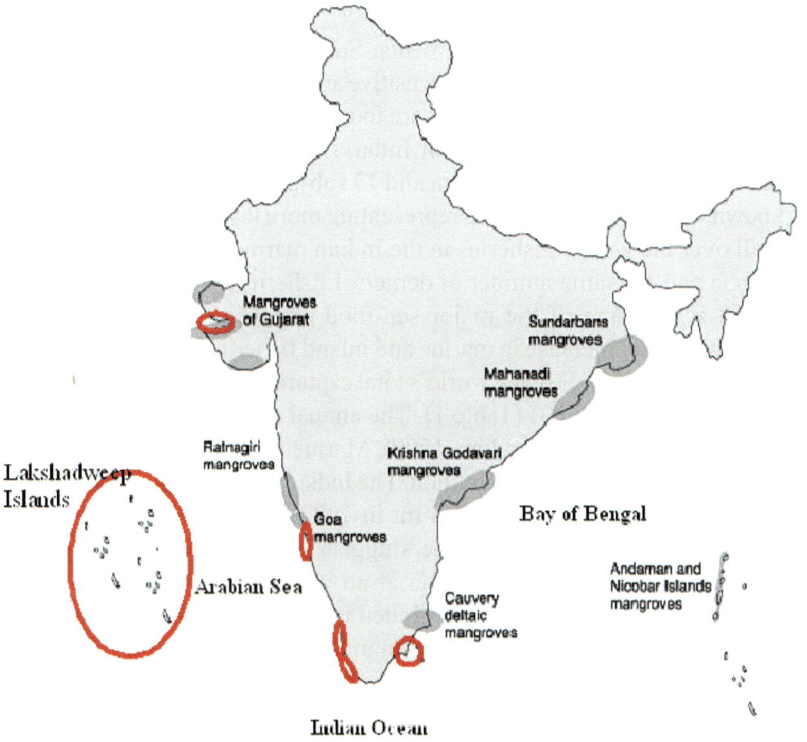

Figure 1: Map of India showing the coastal zone and distribution of mangrove (grey filled circles) and coral reef (red open circles).

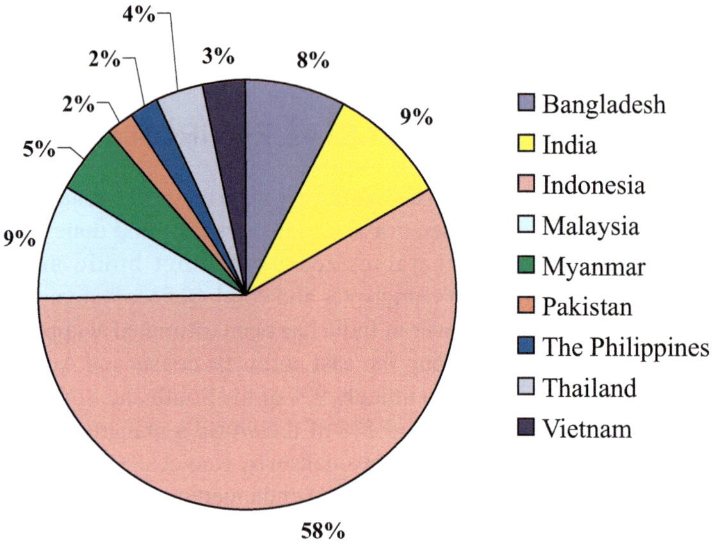

Figure 2: Percentage distribution of mangroves in the South and Southeast Asian region (Total area: 73,683 km^2; Source: Burke et al., 2001).

from aquaculture and urban settlements. Sand dunes which support diverse flora are categorized as ecologically sensitive areas under the Coastal Regulation Zone notification of 1991. Coral reefs are found in the Palk Bay, Gulf of Mannar, Gulf of Kutch, central west-coast of India, Lakshadweep and Andaman and Nicobar islands. A total of 50 genera and 13 sub-genera of reef-building corals are known to occur in Indian reefs representing more than half of those recorded from all over the world. Fisheries in the Indian marine environment comprise 15 pelagic and the same number of demersal fisheries.

India is also one of the major sea-food exporting country. There is a significant decadal increase in marine and inland fisheries in India and it shares 7.75% in Asia and 4.44% in the world's total capture and aquaculture production in 1999-2001 (FAO, 2003) (Table 1). The annual export of fisheries is 0.4 mt worth Rs 47,000 million (Pandian, 1999). Marine fishery exports in 2000 were 421,075 mt valued at Rs 63,965 million. The Indian marine production increased from 0.534 mt in 1950-51 to 2.576 mt in 1992-93. However, the growth of Indian marine fisheries has become sluggish in recent years (Acharya and Thakur, 1999) and reached a plateau at around 2.8 mt by 1995-96 (MOA, 1996). While the inland sector contributed increasingly (6.2% annually since 1980-81) to the growth of fish production in India (5.21% annually since 1980-81), the growth in marine food production decreased to 2.5% during 1990-99 from 3.73% during 1980-90 (Krishnan et al., 2000). The potential harvestable yield of marine fish stock in the Indian EEZ is estimated to be 3.9 mt (Devaraj and Vivekanandan, 1999; Somvanshi, 1999). About one million people in 3651 villages of India situated along the coast are employed in marine capture

Table 1: Capture and aquaculture production totals for marine and inland fisheries

	Total capture and aquaculture production (×1000 mt)			*Marine aquaculture production (×1000 mt)*			*Inland aquaculture production (×1000 mt)*			*Marine capture production (×1000 mt)*			*Inland capture production (×1000 mt)*		
	1979-1981	*1989-1991*	*1999-2001*	*1979-1981*	*1989-1991*	*1999-2001*	*1979-1981*	*1989-1991*	*1999-2001*	*1979-1981*	*1989-1991*	*1999-2001*	*1979-1981*	*1989-1991*	*1999-2001*
World	72,900	99,449	129,429	2,136	5,020	14,240	2,632	8,025	21,372	63,018	80,101	85,153	5,114	6,303	8,665
Asia*	29,517	44,556	74,156	1,404	3,700	11,481	1,973	6,774	19,663	23,602	30,786	37,521	2,539	3,296	5,491
India	2,412	3,830	5,752	4	34	94	363	1,048	1,999	1,493	2,258	2,784	552	490	875

*(except Middle East)
(*Source:* FAO, 2003)

fisheries. Indian fishery also supports several ancillary activities such as boat building, processing plants etc. All these features make this an important sector from the economic and social viewpoint.

Presently, Indian coastline is facing increasing human pressures e.g., over-exploitation of marine resources, dumping of industrial and toxic wastes, oil spills and leaks which have resulted in substantial damage to its ecosystems (Aggarwal and Lal, 2000). Substantial amounts of various pollutants are released directly into the coastal environment (Table 2), which are highly toxic and have serious detrimental affects on the marine biota. These toxic pollutants magnify in the marine biota which subsequently migrate along the food chain and finally reaches human resulting in serious health problems. Reduction of freshwater input is an acute problem in coastal regions of India due to over-extraction of groundwater which has resulted in lowering of groundwater table and salt water intrusion. The lowering of groundwater table has also led to subsidence of coastal land and is highly prone to sea level rise.

Table 2: Fluxes of pollutants into the Indian coastal environment

S. No.	*Pollutant*	*Input annual flux*
1	Sediments	1,600 million tonnes
2	Industrial effluents	50×10^6 m^3
3	Sewage—largely untreated	0.41×10^9 m^3
4	Garbage and other solid waste	34×10^6 tonnes
5	Fertilizer residue	5×10^6 tonnes
6	Synthetic detergents residue	1,30,000 tonnes
7	Pesticide residue	65,000 tonnes
8	Petroleum hydrocarbons (Tar ball residue)	3,500 tonnes
9	Mining and dredging waste	0.2×10^6 tonnes

Source: http://www.teriin.org/teri-wr/coastin/papers/paper2.htm

3. IMPACT OF SEA LEVEL RISE ON THE INDIAN COASTAL ENVIRONMENT

The impact of global warming-induced sea level rise due to thermal expansion of near-surface ocean water has a great significance to India due to its extensive low-lying densely populated coastal zone. Sea level rise is likely to result in loss of land due to submergence of coastal areas, inland extension of saline intrusion and groundwater contamination may have wide economic, cultural and ecological repercussions. Observations suggest that the sea level has risen at a rate of 1 to 2 mm per year along the Indian coastline since 1950s (Church et al., 2001). Computer model simulation studies forced by an ensemble of four Atmospheric Ocean General Circulation Model (AOGCM) outputs indicate that the ocean region adjoining the Indian subcontinent is likely to warm up at its surface by about 1.5°C to 2°C by the middle of this century and by about

2.5°C to 3.5°C by the end of the century (Table 3). The corresponding thermal expansion of Indian coastal waters related sea level rise is expected to be between 15 and 38 cm by the middle of century and 46 to 59 cm by the end of the century (Aggarwal and Lal, 2000) and this simulated sea level rise along the Indian coast line is comparable with the projected global mean sea level rise of 50 cm by the end of this century and may have considerable impacts on the Indian coastal environment. This causes the total area of 5763 km^2 (≈ 0.41%) along the Indian coastal states to be inundated and almost 7.1 million (≈ 4.6%) of coastal population to be directly affected (TERI, 1996; Table 4). The most vulnerable areas are the Rann of Kutch zone in Gujarat, Mumbai and South Kerala. Deltaic regions of Ganges (West Bengal), Kaveri (Tamilnadu),

Table 3: Climate change projections for India based on an ensemble of four AOGCM outputs (Numbers in bracket are for the GHG+aerosol forcing experiments and rest are for GHG forcing experiments)

Year	*Temperature change (°C)*			*Precipitation change (%)*			*SLR (cm)*
	Annual	*Winter*	*Monsoon*	*Annual*	*Winter*	*Monsoon*	
2020	1.36±0.19	1.61±0.16	1.13±0.43	2.9±3.7	2.7±3.7	2.9±3.7	4 to 8
	(1.06±0.14)	(1.19±0.44)	(0.97±0.27)	(1.05±3.7)	(-10.1±10)	(1.05±3.7)	
2050	2.69±0.41	3.25±0.36	2.19±0.88	6.7±8.9	-2.9±26.3	6.7±8.9	15 to 38
	(1.92±0.2)	(2.08±0.85)	(1.81±0.57)	(-2.36±7.1)	(-14.8±18.9)	(-2.36±7.1)	
2080	3.84±0.76	4.52±0.49	3.19±1.42	11±12.3	5.3±34.4	11±12.3	46 to 59
	(2.98±0.42)	(3.25±0.53)	(2.67±1.49)	(-0.13±15.2)	(-11.2±21.2)	(-0.13±15.2)	

Adopted from Aggarwal and Lal, 2000.

Table 4: Potential effects of 1 m sea level rise on Indian coastal states and population

State/Union territory	*Coastal area (million hectares)*			*Population (millions)*		
	Total	*Likely to be inundated*	*Percentage*	*Total*	*Likely to be inundated*	*Percentage*
Andhra Pradesh	27.504	0.055	0.19	66.36	0.617	0.93
Goa	0.37	0.016	4.34	1.17	0.085	7.25
Gujarat	19.602	0.181	0.92	41.17	0.441	1.07
Karnataka	19.179	0.029	0.15	44.81	0.25	0.56
Kerala	3.886	0.012	0.3	29.08	0.454	1.56
Maharastra	30.771	0.041	0.13	78.75	1.376	1.75
Orissa	15.575	0.048	0.31	31.51	0.555	1.76
Tamilnadu	13.006	0.067	0.52	55.64	1.621	2.91
West Bengal	8.875	0.122	1.38	67.98	1.6	2.35
Andaman & Nicobar Islands	0.825	0.006	0.72	0	0	0
Total	139.594	0.571	0.41	416.74	7.1	1.68

Adopted from Aggarwal and Lal, 2000.

Table 5: Percentage of land likely to be affected in case of 1 m sea level rise along the Indian coastal states

State	*Cultivated land*	*Cultivable land*[1]	*Forest land*	*Land not available for agriculture*[2]
Gujarat	0.03	0.08	-	0.89
Maharastra	0.39	0.21	0.09	0.31
Goa	0.65	0.03	-	0.31
Karnataka	0.51	0.13	0.13	0.23
Tamilnadu	0.39	0.39	-	0.21
West Bengal	0.74	0.04	-	0.22

[1] Cultivable land is the land that can be brought under cultivation
[2] Land not available for agriculture is the land under human settlements, commerce, etc.
Adopted from Aggrwal and Lal, 2000.

Krishna and Godavari (Andhra Pradesh), Mahanadi (Orissa) and also islands of Lakshadweep and Andaman & Nicobar would be totally lost. In addition, estimates indicate that the potential impacts of 1 m sea level rise on the land uses affected by the coastal states in terms of the share of total land affected (Table 5) and the cultivated land would be affected by both inundation and intrusion owing to sea level rise.

India has been identified as one amongst 27 countries which are most vulnerable to the impacts of global warming related accelerated sea level rise (UNEP, 1989). The high degree of vulnerability of Indian coasts can be mainly attributed to extensive low-lying coastal area, high population density, frequent occurrence of cyclones and storms, high rate of coastal environmental degradation on account of pollution and non-sustainable development. Most of the people residing in coastal zones are directly dependent on natural resource bases of coastal ecosystems. Any global warming-induced climatic change such as increase in sea surface temperature, change in frequency, intensity or tracks of cyclones, sea level rise may aggravate the potential risks to coastal zones.

In addition to the global warming-induced sea level rise, the Indian coast is also highly vulnerable to the natural disasters. The entire east coast of India, the Gujarat coast along the west coast and the islands of Lakshwadeep and Andaman and Nicobar face frequent cyclonic conditions which some times cause large scale destruction of life and property. The Super Cyclone had caused massive destruction along the coast of Orissa in 1999 and its impact was felt several kilometres inland. The tsunami, which occurred on 26th December 2004 was one of the most serious and unexpected natural catastrophes to occur along the Indian coast (Danielsen et al., 2005). The major destruction caused by this tsunami was to the life and property located along the coast of Andaman and Nicobar, Tamil Nadu, Pudicherry and Kerala. It would take several years to restore the damages caused by this natural catastrophe. While it is agreed that no human interference is possible to control such an event but precautionary measures such as coastal zone planning for locating coastal communities in safer areas, protecting and effectively managing the natural protecting ecosystems such as mangroves, coral reefs, shelter belt plantations, along with installation of early warning systems, timely evacuation and relief measures can minimize loss of life and property to a large extent (Danielsen et al., 2005). Despite the socio-economic importance of mangrove to humans, humans degrade mangroves for various developmental activities. Thus, the coastal environment becomes highly vulnerable to the natural catastrophes such as tsunami, cyclone, etc. For example, the Pichavaram mangrove of south east coast of India is degraded by various human activities (Fig. 3). Due to the strict conservation practices by the government agencies, the area of mangrove is increased significantly (Vasudevan, personal communication) which controlled the encroachment of 2004 Indian Ocean tsunami waves and saved the human population. Compared to other regions affected by tsunami, the damage caused is comparatively less in the Pichavaram region mainly due to the extensive mangrove which acted as a natural bio-shield (Kathiresan and Rajendran, 2005).

The socio-economic pressures drive changes in coastal ecosystems. The international programme on Land Ocean Interaction in the Coastal Zone (LOICZ) (for details refer www.loicz.org) has identified four areas of investigations. One of these addresses changes to coastal systems due to social and economic activities. Since coastal eco-tourism and related anthropogenic activities have intensified during the last two decades, there were adverse

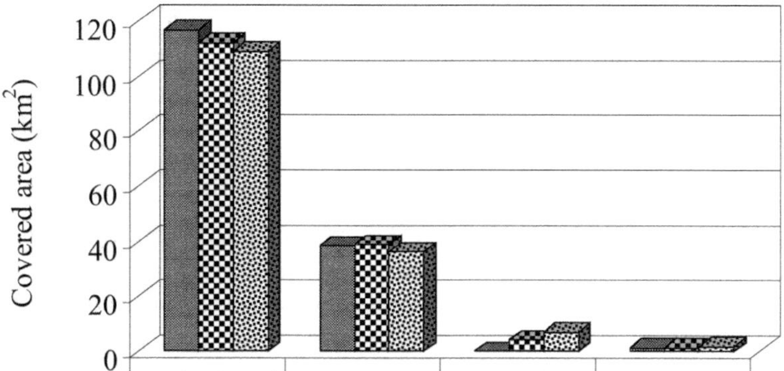

	Agricultural land	Mangrove area	Aquaculture land	Forest plantation
1970	116.55	38.41	0	1.2
1987	112	38.92	3.99	1.25
1998	108.79	36.23	6.99	1.8

Figure 3: Decadal change in the land use pattern in the Pichavaram mangrove catchment area (modified from Yeon et al., 2004).

impacts of tourism on coasts worldwide along with coastal pollution and adverse impacts of natural disasters.

For the purpose of protecting and conserving the coastal environment of India, the Ministry of Environment & Forests issued the Coastal Regulation Zone (CRZ) Notification dated 19.2.1991 under the Environment (Protection) Act, 1986. This notification regulates all developmental activities in the Coastal Regulation Zone area. The CRZ Notification has placed India amongst the select countries in the world to frame laws to legally protect sensitive coastal ecosystems, to formulate guidelines for coastal activities, and to demarcate areas for conservation (Mascarenhas, 1998).

In the current scenario of the global environmental change (i.e. global warming, sea level rise, climate change, pollution, etc.), the Indian coast offers an opportunity to study the use and effectiveness of coastal legislations with respect to human activity and eco-sensitiveness of coastal systems.

4. MAJOR INITIATIVES FOR PROTECTION OF MARINE AND COASTAL ENVIRONMENT IN INDIA

In order to protect its marine environment, the Government of India, even before 1992, had initiated a number of programmes. These acquired a new significance post-1992. This section highlights the major policies and programmes adopted.

From a coastal development-environment interface point, the most significant Indian policies are those contained in the Indian Constitution. Article 297 of the Indian Constitution serves a two-fold purpose: (i) it treats the territorial waters as part of the territory of India and (ii) it vests the maritime territory in the Union as against the federating coastal states (MOEF, 1992). Landward, the shore and the beaches are the property of the coastal state; land inland from that point is private property.

The Environment Protection Act (EPA) 1986 is the main statute to regulate the coastal zone of India. Under Section 5, it grants the Central government the power to close down any industry if found violating the law. The Act allows for public interest litigation for the purpose of protecting the environment. In 1991, a notification on the Coastal Regulatory Zone was issued under the 1986 EPA declaring the coastal stretches of seas, bays, estuaries, creeks, rivers and backwaters which are influenced by tidal action in the landward side up to 500 m from the high tide line and the land between the low tide line and high tide line as the coastal regulation zones. The Notification aims at protecting and improving the quality of the coastal environment. The Notification declares the limits of the Coastal Zone and classifies it into four categories for the purpose of regulation. CRZ I includes areas which are ecologically sensitive, areas of outstanding natural beauty, historical heritage or rich genetic diversity. CRZ II includes the areas that have already been developed up to or close to the shoreline. Areas that are relatively undisturbed are classified under CRZ

III. CRZ IV includes the coastal stretches in the Andaman and Nicobar, Lakshwadeep and other small islands except those designated as CRZ I, II and III.

The notification lays down certain prohibitions with some exceptions. Prohibited activities include setting up of new industries (except those which are directly related to the water front or which directly need foreshore facilities) and expansion of existing industries including fish marine and coastal environment. 216 processing units are manufacturing, handling storage or disposal of hazardous wastes and substances, discharge of untreated wastes and effluents and dumping of municipal wastes as landfills or otherwise. Withdrawal of groundwater within 200 metres of the High Tide Line (HTL) is prohibited with some exceptions. In most of these areas, an area of 200 metres from the high tide line (HTL) has been declared a no development zone. Several restrictions have been imposed for carrying out development in the area between 200 and 500 metres from the HTL. These measures have been adopted to protect fragile ecosystems which exist in the area and are vital for sustaining the ecological balance. Mangroves and coral reefs have been declared ecologically sensitive areas (CRZ I) under this notification and regular monitoring using satellite imagery is in progress. A state-wise Mangrove Committee has been formed for effective management of the mangrove ecosystem. Mining of corals and coral sands has been banned. The CRZ notification also offers protection to coastal communities such as traditional fishermen. Although the draft was open for public scrutiny before it was notified, pressure from various lobbies led to the formation of an apex committee, appointed to look into various representations. Subsequently, the first amendment was issued in August 1994 (MOEF, 1994) This order redefined the HTL, and proposed six amendments: distance from HTL for rivers, creeks and backwaters was reduced to 50 m; construction of basements and barbed wire fencing was granted; density of certain constructions could increase; and flattening of dunes and permanent structures for sports were banned. Simultaneously, all coastal states were bound to formulate coastal zone management (CZM) plans classifying coastal stretches as CRZs.

4.1 Implementation of the CRZ Notification

The notification states that the respective State governments should have identified, classified and recorded all the CRZ areas in the State Coastal Zone Management Plans[1], which was to be approved by the MOEF and monitored by the respective State Coastal Zone Management Authorities (SCZMA) which was constituted on 26th November 1998. A National Coastal Zone Management Authority (NCZMA) was set up to monitor and implementation of the CRZ

[1] Para I(ii) and Para 3(3)(i) of the CRZ Notification state that the Coastal Zone Management Plans should identify and classify the various CRZ areas.

Notification's provisions across the nation. In these CRZ areas, from the date of the CRZ Notification i.e 19th February 1991, certain restrictions would be imposed on various anthropogenic activities including the setting up and expansion of industries, operations or processes etc. This would include several activities that would possibly be taken up during the rehabilitation process. The responsibility of implementing the CRZ Notification rests with the State Governments and the MOEF (See Para 3 of the CRZ Notification). The notification outlines the activities that are to be permitted by the MOEF (See Para 2 and Para 3 of the CRZ Notification) and under which conditions. All other activities are to be regulated by the State governments and Union Territory administrations within the framework of these approved CZMPs.

4.2 Current Status of the CRZ

The implementation of the CRZ was by and large ignored by many state governments. Vested interests from various lobbies such as the tourism and industry have constantly sought to get rid of this notification. The CRZ Notification has been amended at least 19 times, and each dilution rendered the law more incapable. The 26th December 2004 Indian Ocean tsunami starkly demonstrated the impact of this gross neglect.

In order to restructure the CRZ, the MOEF has constituted the Swaminathan Committee under the chairmanship of Prof. M.S. Swaminathan in July 2004. Its mission was to enable the MoEF base its coastal regulations on strong scientific principles and to devise regulations that would meet the urgent need for coastal conservation and development/livelihood needs. The report of the committee was submitted to the MOEF in February 2005. The main recommendations of this report were: (a) a rationalization of management boundaries based on coastal vulnerability, (b) moving away from mere 'regulation' to the larger concept of 'integrated management' and replacing the categories of 'Coastal Regulation Zones' (CRZ I to IV) with 'Coastal Management Zones' (CMZ I to IV), based on ecological importance, coastal vulnerability, and other socio-cultural concerns. However, there has been some reshuffling of activities to be permitted under each of these categories, and (c) creating a new institutional structure specifically geared for coastal management, including the establishment of a National Institute for Sustainable Coastal Zone Management.

The primary objectives of the CZM Notification are: (i) to manage the sustainable development of the coastal ecosystems, (ii) provide sustainable livelihoods to the concerned resource users at the end, and (iii) to conserve the ecologically and culturally fragile and significant resources. The CZM Notification classified the coastal ecosystems into four classes as the Coastal Zone Management (CZM) I, II, III and IV zones and established different ways for managing these areas from various activities.

1. CMZ I: The CZM I areas are the 'Ecologically Sensitive Areas' (ESA) (See the CZM Notification 2007). These include mangroves, estuaries, coral reefs, mud banks etc.
2. CMZ II: CMZ II shall consist of areas, other than CMZ I and coastal waters, identified as "Areas of Particular Concern (APC)" such as economically important areas, high population density areas, and culturally/strategically important areas (See the CZM Notification 2007). The administrative boundaries of these "Areas of Particular Concern" would be boundaries of CMZ II. These are heritage areas, mining sites, port and harbours, power plants, and defense installations.
3. CMZ III: They shall consist of all other open areas including coastal waters, that is all areas excluding those classified as CMZ I, II and IV.
4. CMZ IV: They shall consist of inlands territories of Andaman and Nicobar, Lakshwadeep, and other offshore islands.

However, the CRZ Notification review mechanism by the Swaminathan Committee was criticized because it was devoid of the participation of public interest groups or coastal communities. Though the Swaminathan Committee was allegedly set up to impart better science into the law and devise regulations and policies for coastal conservation and better livelihood, it fell short on many counts.

4.3 Objections against the CZM Notification

The shrillest criticism of this committee has been that the regressive regulatory framework suggested in the final chapter and annexes of its report is completely adversative to the initial drawn principles of integrated coastal zone management (Sridhar et al., 2006). Although the Swaminathan Report provides an alternative model for coastal zone management, it does not really attempt the problems with the CRZ notification, which is presently in force. The following are the main objections against the CZM notification.

4.3.1 The CZM Notification – Conservation

All eco-sensitive areas which are included in the CZM I are not defined quantitatively (area, size) and qualitatively (structure, function). Therefore, it will not be feasible to draft any protective measures for these critical ecosystems since these are responsible for key ecosystem services and that the health of these habitats is directly linked to the productivity of the fisheries and livelihoods in these areas. Since, the fragile zones of the CZM I play a vital role in the global hydrological cycle and global biogeochemical nutrient cycles, these ecosystems should be characterized properly for their conservation. Therefore, a better protection and conservation to these areas should be required for more productive fishery and also ensure long-term livelihood security for fishing communities.

4.3.2 The CZM Notification – Sustainable Development

The term 'Sustainable Development' is not clearly defined in the CZM notification. Appendix III of the proposed CZM Notification contains an assorted listing of various 'areas of particular concern' that are identified as CZM II areas. In fact, the list of mentioned areas in this part is typically those which are currently subject to provisions in the CRZ notification. These are the areas prioritized for development i.e. special economic zones and tourist areas stand to gain immensely by their presence on this list as the CZM II areas will hardly be subject to regulations. The rapid increase in the industrial units along the coastal zone will pose a serious environmental problem by degrading the ecological health of the coastal ecosystems. The released pollutants (heavy metals and organic pollutants) will be biomagnified in the coastal biological resources and pose serious health problems over the coastal population which will certainly hinder the sustainable development activities.

4.3.3 The CZM Notification – Livelihoods

The list of activities that are listed in Appendix V of the CZM Notification 2007, allows the developmental activities on the seaward side of the setback line. These activities are not environmental friendly and disturb the ecological stability of the coastal environment. By allowing these activities in the coastal environment push back the coastal communities behind the setback line, the government is facilitating the transfer of ownership and access of beach fronts from the fisher communities to the commercially interested non-coastal agencies. These activities further accelerate the coastal erosion and will change the geomorphology for the high productivity breeding grounds (e.g. mangroves and estuaries) for the commercially important fishes and prawns. Hence, this proposal indeed promotes the commercial exploitation of the coastal zones and seems to be curse for the coastal societies. Thus, in the CZM notification, there is no particular focus and apprehension on the rights of the coastal communities.

However, based on the criticisms and comments made by various stakeholders, and recommendations of the four-member Committee under the Chairmanship of Prof M.S. Swaminathan constituted by the MOEF to recommend future steps on the draft Coastal Management Zone (CMZ) Notification, 2008, the MOEF Minister for State, has recently issued a statement to let the CMZ Notification, 2008 lapse and incorporate amendments as recommended in the existing CRZ Notification, 1991 for better coastal management (http://thehindu.com/2009/08/20/stories/2009082061580800.htm). Further, the Swaminathan Committee has advised the Government to consider necessary legal and scientific issues for the integrated coastal zone management (for more information refer Report of the Expert Committee on the draft Coastal Management Zone (CMZ) Notification dated July 16, 2009).

5. CONCLUSION

The Indian environment is highly productive and harbours several ecosystems. Presently, Indian coastal ecosystems are facing increasing human pressures e.g., over-exploitation of marine resources, dumping of industrial and toxic wastes, oil spills and leaks which have resulted in substantial damage to its ecosystems. The impact of global warming-induced sea level rise due to thermal expansion of near-surface ocean water has a great significance to India due to its extensive low-lying densely populated coastal zone. Sea level rise is likely to result in loss of land due to submergence of coastal areas, inland extension of saline intrusion and groundwater contamination and may have wide economic, cultural and ecological repercussions. The expected impacts of the global environmental change are increased storm frequency and severity as well as loss of wetlands and mangroves, and reef degradation. Coastal currents, water temperature, and marine populations will all be affected. Low-lying regions may be flooded and water supplies contaminated.

The CRZ notification 1991 was drafted with a clear vision to protect and conserve the ecologically sensitive coastal zone of India for the sustainable development. The rapid changes in the ecosystem structure and dynamics by various internal and external forces drive the administrative authorities to strengthen the coastal zone regulatory principles. This leads to emergence of the CZM notification. But, the draft CZM notification contains completely no monitoring mechanism to protect the very important and highly vulnerable Indian coastal environment. The modifications proposed in the CZM notification attract the setting up of the industrial units along the coastal zone and increase the coastal environmental pollution which is detrimental for the coastal population health. Similarly, the changing of the setback line limits encourages the commercial exploitation of the coastal biological resources and is having very adverse effect on the local traditional fishing communities. Thus, a large number of coastal populations have strongly opposed the implementation of the CZM notification which has led the Government to let the CZM notification lapse and to keep the CRZ notification as the basic framework. Further, the Swaminathan Committee has recommended the Government of India to set up a high level scientific and legal committee to propose necessary guidelines to strengthen the current coastal zone management guidelines and practices for sustainable development.

REFERENCES

Acharya, P. and Thakur, N.K. (1999). Aquatic resources management policy to cater to the needs of 21st century. In: Vision on Indian fisheries of 21st century. Abidi, S.A.H., Thakur, N.K., Biradar, R.S., Shenoy, L. (eds.) Proceedings of the national seminar CIFE, Mumbai, pp. 35-46.

Aggarwal, D. and Lal, M. (2000). Vulnerability of Indian coastline due to sea level rise. *In:* Mimura, N., Yokoi, H. (eds.) Proceedings of the APN/SURVAS/LOICZ Joint Conference on Coastal Impacts of Climate Change and Adaption in the Asia-Pacific.

Burke, L., Kura, Y., Kaseem, K., Revenga, C., Spalding, M. and McAllister, D. (2001). Pilot Analysis of Global Ecosystems: Coastal Ecosystems. World Resources Institute, Washington D.C. 93 p.

Church, J.A., Gregory, J.M., Huybrechts, Kuhn, M., Lambeck, K., Nhuan, M.T., Qin, D. and Woodworth, P.L. (2001). Changes in sea level. *In:* Climate Change 2001: The Scientific Basis. Contribution of Working Group I to the Third Assessment Report of the Intergovernmental Panel on Climate Change. Cambridge University Press, Cambridge, UK, pp. 639-693.

Danielsen, F., Sørensen, M.K., Olwing, M.F., Selvam, V., Parish, F., Burgess, N., Hiraisi, T., Karunagaran, V.M., Rasmussen, M.S., Hansen, L.B., Quarto, A. and Suryadiputra, N. (2005). The Asian tsunami: A protective role for coastal vegetation. *Science*, **310**, 643, doi:10.1126/science.1118387.

Devaraj, M. and Vivekanandan, E. (1999). Marine capture fisheries of India: Challenges and opportunities. *Current Science*, **76(3)**, 314-332.

FAO (2003). FAO Yearbook of Fishery Statistics. FAO Fishery Information, Data and Statistics Unit, Aquaculture production 2001. Vol. 92/2. Rome, FAO.

FSI (2001). State of Forest Report 2001, Forest Survey of India, pp. 36-39.

Kathiresan, K. and Rajendran, N. (2005). Coastal mangrove forests mitigated tsunami. *Estuarine Coastal Shelf Science*, **65(3)**, 601-606.

Krishnan, M., Birthal, P.S., Ponnusamy, K., Kumaran, M. and Singh H. (2000). Aquaculture in India: Retrospect and prospects. *In:* Krishnan, M. and Birthal, P.S. (eds.) Aquaculture development in India: Problems and Prospects. Workshop Proceedings organised by National Centre for Agricultural Economics and Policy Research (NCAP) with Central Institute of Brackishwater aquaculture (CIBA), New Delhi, pp. 11-31.

Mascarenhas, A. (1998). Coastal Sand Dune Ecosystems of Goa: Significance, Uses and Anthropogenic Impacts. *Current Science*, **43**.

MOA (Ministry of Agriculture) (1996). A hand book on Fisheries Statistics 1996, Ministry of Agriculture, 217 p.

MOEF (Ministry of Environment and Forests) (1992). National Conservation Strategy and Policy Statement on Environment and Development. New Delhi, 36 p.

MOEF (Ministry of Environment and Forests) (1994). The Gazette of India, Notification, S.O. no. 595 (E), 18 August 1994.

Pandian, T.J. (1999). Preface. *Current Science*, **76(3)**, 313.

Rao, T.A., Molur, S. and Walker, S. (1999). Mangroves of India: Report summary 1998, Publ. Zoo Outreach Organization ISSN 0971-6378, XIV (2), pp. 1-49.

Somvanshi, V.S. (1999). Indian fisheries in 21st Century: Some thoughts on marine fisheries resources management. *In:* Vision on Indian fisheries of 21st century, Abidi, S.A.H., Thakur, N.K., Biradar, R.S., Shenoy, L. (eds.) Mumbai: Proceedings of the national seminar CIFE, pp. 54-64.

Sridhar, A., Arthur, R., Goenka, D., Jairaj, B., Mohan, T., Rodriguez, S. and Shanker, K. (2006). Review of the Swaminathan Committee Report on the CRZ Notification, Draft submitted to UNDP, New Delhi. URL: http://www.undp.org.in/dmweb/tsunami; accessed 20th September 2007.

TERI (1996). The economic impact of one meter sea level rise on Indian coastline: Methods and case studies. Report submitted to the Ford Foundation.

UNEP (1989). Criteria for Assessing Vulnerability to Sea Level Rise: A Global inventory to High Risk Area. Delft Hydraulics, Delft, The Netherlands, 51 p.

Yeon, C.H., Lakshumanan, C. and Natesan, U. (2004). Coastal wetland and shoreline change mapping of Pichavaram, southeast coast of India using satellite data. Map India Conference 2004, Beijing, China.

4

Climate Resilient Coastal Zone Development in Bangladesh: Participatory Governance for Common Resources Management

Paramesh Nandy and M. Aminul Islam[1]
Bangladesh Forest Research Institute, P.O. Box 273, Chittagong, Bangladesh
[1]UNDP in Bangladesh, IDB Bhaban, Dhaka 1000, Bangladesh

1. INTRODUCTION

The current scenario in coastal zone of Bangladesh is characterized by a dispersed and ineffective manner of increasing/regenerating natural capital, which leads to increased competition and conflict over scarce resources. In context to this scenario, appropriate processes to adapt policy, governance arrangements and co-management practices for forest resource development along the coastline in response to climate change have not been developed or implemented. This is further aggravated, as there is a communication gap between local resource management initiatives and scientific communities engaged in natural resources management. There is also a lack of both financial and technical assistance for local community-based users to improve their adaptive capacities through targeted training and funding for innovative trial and demonstration which can make a breakthrough in poverty reduction. Poverty Reduction Strategy Paper of the country always focussed on the economic development and income generation activities but the issue of the empowerment of rural community efforts for resource generation hardly received due attention while the latter can be instrumental as a catalyst for the sustainability of poverty alleviation strategies of the country. The trend of economic development is not eco-friendly. Such a mismatch of productive interaction between social capital and natural capital is one of the critical factors for sustaining the vicious cycle of poverty. Natural resources of Bangladesh coastal belts are under considerable threat where high incidence of natural disasters coupled with population pressures highly contributes to the environmental deterioration. Accordingly, the question of sustainable development needs critical attention for ensuring support to the survival strategies of poverty stricken rural community efforts for the enhancement of their access to natural capital.

This article highlights on bottom-up innovative participatory management approaches for local and regional level coastal zone specific common property resource development. This initiative is intended to bring a win-win situation for enhancing climate resilient natural resource base and poverty reduction through a change by interacting arrangement among institution, technology, natural capital and target community. Further consideration made to mainstream the effective management of common pool resource generation involving poor communities through an appropriate strategy needs responsive planning and policy formulation.

2. COMMON PROPERTY REGIME: AN ENTRY POINT FOR POVERTY REDUCTION

In a densely populated and poverty stricken coastal zone, the importance of common property regime (CPR) in natural resource management policy making and sustainable development can hardly be over-emphasized. As things stand now, ownership, access and management of natural resources in coastal zone, especially relating to land, water and forestry lie at the heart of at least three fundamental problems in coastal area: (i) poverty, (ii) resource degradation and (iii) resource use conflict. Thus, the long-standing commons should be reinvented as important topics for serious research and continued policy analysis. In an age of private property regime and dominance of market forces, a major challenge is to revive the concept of CPR and make it widely acceptable to policy makers and other social forces that matter.

2.1 Participatory Governance of the Common Resources

Till today, creation of common wealth for common people and its common management has not been successfully considered as an important aspect of in-country poverty alleviation programmes. Its central importance is that it would encourage and increase the capability approach of rural stakeholders as well as contribute to the welfare economics of the country. This is particularly the case if large number of rural people can be involved in group-wise resource generation activities through immediate application of locally developed technologies by the public institutions of the country. In this article, poverty alleviation through resource regeneration and its participatory governance has been regarded as an effective measure for maximum use of community capability production and sharing with equity and justice. In this regard, formation of multiple groups of communities by type of occupation, by type of resource harvesting nature as well as management may contribute to multidimensional poverty alleviation programmes of Bangladesh. In this article, attempt has been made for a balanced and multidimensional poverty alleviation programme through community-based formation of numerous groups to be engaged in common resource generation under common management. Particular emphasis in this article has been given to the formation of resource user groups associated with climate change impact in the coastal areas.

2.2 Case Studies on Individual, Community and Systematic Capability Approach

Out of eight case studies conducted by the Empowerment of Coastal Fishing Communities (ECFC) for Livelihood Security (BGD/97/017) project supported by UNDP, four unique examples of enhancing capability approach after getting one time capital grant from Micro Capital Grant (MCG) of the project have been highlighted in this article.

Case 1: Rahmatullah (22 years), son of Osman of Shahaparirdip, a poor vulnerable coastal fisherman consisting 11 family members having no cultivable land had been maintaining his family (hand to mouth) by selling his labour to other fishing boat owners. He was the only earning member of his family and had to be at sea all day fishing. He received training on improved dry fish processing and started dry fish business with support of Tk. 5,100/- from MCG of ECFC project in August 2004. Within two years, he was able to pay back his loan with a running capital of Tk. 20,000/- after maintaining his eleven member family expenditures.

Case 2: Abul Fayez (32 years), son of Zafor Ahmed of North Moheshkhalipara, a poor coastal fisherman received training on cow fattening and started his business with the procurement of single oxen for Tk. 7,200/- with the support of Tk. 5,100/- from MCG of ECFC in October 2003. He reared it for three months and sold it for Tk. 10,750/- and was immediately able to pay back his loan.

Case 3 – Community based resource development: Poor fisher community mobilized themselves to take lease of a derelict big tank of the local Union council for fish-culture under the ECFC project in Teknaf Union. The project supported the community to develop skill in cultured fish resource development and its cooperative management. Community participants in this common resource development seem to be optimistic in getting bumper production as a result of collective strength. This successful endeavour opens up an opportunity to go for bigger venture to plan for adaptation to meet the disasters including climate change risk.

Case 4 – Participatory governance: Community based small-scale irrigation system during February to April in rice cultivation of Bangladesh is another example of successful participatory governance of natural resources. Rural farmers themselves made a provision of annual fee for the access to irrigation per unit cultivated area which was used to cover the cost of irrigation and maintenance of water pump machine given by the Bangladesh Agricultural Development Corporation (BADC). Since 1968, such community based small-scale irrigation system has been maintained efficiently till 1986 and it did not require any intervention of new technologies. It appeared as most successful and sustainable system to getting more output per unit area leading to poverty alleviation of poor farmers spontaneously. However, the provision for auctioning the water pump machines by BADC in 1986 has appeared worsening the situation. In most cases individual big farmers bought and became the owner of these machines. Irrigation charges have been raised from Tk. 625/- per hectare in 1968 to Tk. 5,000/- per hectare in 1987. Thus, the community approach became unsustainable and has been converted to individual business approach.

Figure 1 provides a horizontal and vertical linkage of capacity building. This diagram represents the integration between four important capacity building levels. The figure highlights the linkages between capacity at the individual level and the linkages; therefore it also needs to be considered as a capacity for relationships between organizations (network level) and the broader enabling environment and so on as and where needed.

The above four cases are excellent examples of poverty alleviation activities because every one either individual or in group is involved in one way or another which ultimately provided positive changes in the economic condition of individuals or groups of the society. However, the economy of the country appears as one of the vulnerable in the world due to high density of population

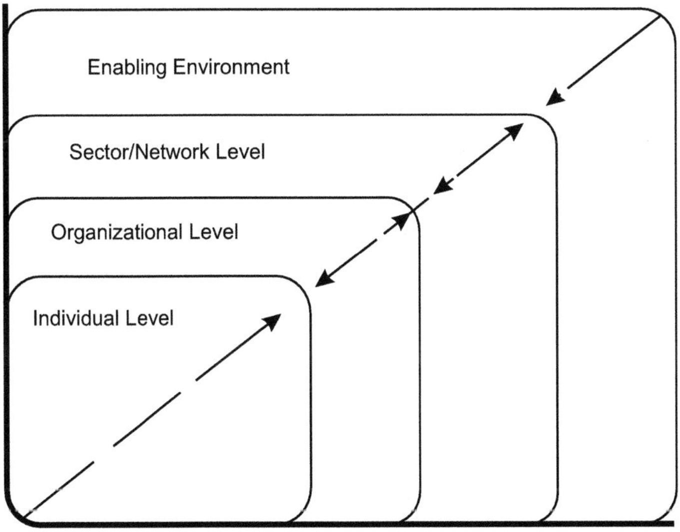

Figure 1: Horizontal and vertical linkage of capacity building.
Source: http://www.recoftc.org/site/index.php?id=376

and high incidence of natural disasters coupled with persistent political instability. Nearly half of its population is living below poverty line. What is urgently needed is to prioritize the area for natural resource regeneration through mass involvement of these rural people. To end this maximization of community capability approach and participatory governance of natural resources rather than individual approach could be the sustainable way of securing livelihoods of the people living in the areas of intense poverty. However, capacity building for such a resource generating initiative with equity and justice managed through a participatory management is the need of the time.

2.3 Discussion on Case Studies and Way Forward

Analyses of the above case studies apparently indicate that the development of alternative income generation opportunities have been introduced successfully in the coastal fishing communities. However, individual income generation activities cannot serve as an appropriate corridor of community empowerment for poverty reduction in the intense poverty areas like Bangladesh coastal areas where more than 20 million people are living below poverty line. It requires large-scale community led initiatives to accelerate the progress in poverty reduction. However, community led initiatives often are thought as an involvement of large group of homogeneous people while most of the rural communities are not homogeneous. In order to overcome this constraint, it is suggested to grouping of the rural communities by occupation, by resource harvesting nature, by resource type and use and so on as and where needed. Smaller multidimensional resource generating groups would be more effective than larger communities provided the participatory governance and decision making structure of each group can be put in place. Community risk assessment followed by the preparation of their own action plan and supported by the local government to implement priority options is found to be very effective.

With a view to involve the communities in resource generation activities in an organized group, it is suggested to identify the comparative geographical advantage and accordingly prioritize the area of resource generation in the targeted villages. After proper identification of priority areas and associated problems, the proposed strategy includes the formation of numerous groups for resources regenerations. It is recommended to constitute regional committee of multidisciplinary experts which will provide resource based group training at the expense of concerned public institutions as a part of transferring technologies developed by respective institutes.

3. COASTAL FOREST RESOURCES MANAGEMENT THROUGH SCIENTIFIC APPROACH

The coastal areas of Bangladesh experienced severe cyclone and tidal surges and hence planned to make vegetative shelterbelts throughout coastal regions.

At a glance, Bangladesh coastal forest today is absolutely a keora (*Sonneratia apetala*) forest with some patches of baen species (*Avicennia officinalis*) somewhere in the eastern coastal belt. This type of monoculture practice now encounters a number of problems viz. sudden onslaught of keora with stem borer infestations, non-existence of preferred mangrove species for regeneration and overall, there would be no vegetation after maturing/felling the existing keora trees. In response to climate change, this problem is further aggravated due to absence of other mangrove species endurable against natural disasters. The total area of man-made keora forests now comprises 151,000 hectares, two-third of which can now be involved in man-made afforestation with economically preferred mangrove species. However, for the creation of artificial coastal forests with preferred mangrove species, what is urgently needed is the selection of suitable coastal habitats for the favourable growth of individual mangrove species. The investigated species are the key mangrove species belonging to Sundarban natural forest. Bangladesh is blessed with the larger single tract of natural mangrove forest in the world. The southwestern coast of Bangladesh is greatly sheltered against natural disasters by the existence of Sundarban.

It is apparent from the investigations that ten mangrove species can be accommodated in most of the coastal areas as under-planting species in existing keora plantations. Thus, suitable coastal habitats/sites for the accommodation of ten mangrove species have been selected (Islam and Nandy, 2001) and found suitable for artificial plantation of these species in differently inundated coastal habitats (Table 1).

Application of above scientific approach will enrich and sustain coastal vegetation and also play a vital role to protect the lives and assets of millions of coastal dwellers from tsunamis, cyclones and other tidal surges. It will not only minimize the adverse impact of monoculture but also help to establish a planned man-made coastal forest with economically preferred mangrove species. It would appear as proven security against climate change impact as well as unique source of community-based resource generating activities for sustainable livelihood of coastal communities.

Table 1: Selection of suitable coastal habitats for favourable growth of selected species

Sl. No	*Frequency of Inundation (FI)*	*Suitable species*
1.	Sites with 3 months FI	*Cynometra ramiflora.*
2.	Sites with 3 and 6 months FI	*Ceriops decandra* and *Phoenix paludosa.*
3.	Sites with 3, 6 and 9 months FI	*Heritiera fomes, Excoecaria agallocha, Xylocarpus granatum, Lumnitzera racemossa* and *Bruguiera sexangula.*
4.	Sites with 3, 6, 9 and 12 months FI	*X. mekongensis* and *Aegiceras corniculatum.*

The same is applicable for mass propagation and plantation of palmyra palm (*Borassus flabellifer*) species. After devastating cyclone of 1991, which caused the death of 140,000 people in the coastal areas, this species has got priority and came into consideration as a disaster resistant tree component for coastal areas of Bangladesh. The spongy cotyledon of germinated seeds is attractive and highly tasty and hence seeds disappeared from planting spots. The main reason why the seedlings of this species are not available in commercial nurseries or why this species is not popular to the nursery men is the lack of easy technique for raising palm seedlings. An easy technique for raising palm seedlings from detached germtubes (Nandy et al., 1999), especially in polybags has already been developed by Bangladesh Forest Research Institute (BFRI). Thus, the problem of taking away germinated seeds and re-sowing of new seeds in the same spot can be overcome by raising polybag seedlings. It contributes to the national economy by saving the investment, securing the climate change impact through ensuring the sustainability of palm plantation in the fragile coastal belts. It also encouraged coastal rural people in raising commercial palm nurseries and thus contributes in alleviating the poverty through greater employment and income generation activities.

4. COASTAL LAND RESOURCES MANAGEMENT THROUGH PARTICIPATORY APPROACH

Bangladesh contains the largest river delta in the world. Its coastline is about 710 km long and broadly divided into three distinct regions viz. Eastern, Central and Western coastal regions. During pre- and post-monsoon period, cyclones are common in either part of these coastal regions. The low-lying areas, particularly the newly accreted (char) areas, are more vulnerable to floods and cyclones from Bay of Bengal.

Government land tenure policy narrates that the coastal lands are new formations and free from all rights and privileges. Newly accreted coastal lands are managed and controlled by the state owned Forest Department (FD) and solely used for forestry practices. Little consideration is provided on sustainable use of coastal land resources. These accreted lands are gradually raised and become suitable for human settlement in the mangrove plantations. Moreover, coastal land and aquatic resources provide numerous livelihood opportunities encouraging migrants from other places. In 1999, more than 8000 hectares of coastal plantations have been encroached in Noakhali Division alone (Canonizado, 1999) for which the participatory management approach appeared to be the best option for future sustainability. Regarding land resources management, Participatory Rural Appraisal (PRA) was conducted in each coastal region. It was revealed that the coastal dwellers have no tenural rights to use coastal lands and its resources. The limiting factor is the ownership of land which traditionally lies with the FD. Settlers deliberately keep a small portion of land in possession for securing their livelihoods which resulted in antagonistic relationship between local people and the FD. Antagonistic relationship between them encourages illegal activities by outsiders in the coastal forest resulting in major destruction of vegetative shelter belts.

Resettled population in the newly accreted coastal land opted for an interesting land-leasing schedule (Nandy et al., 2003) which includes initial leasing of land for five years with a provision of its renewal for 15 years. However, they can continue to settle in the allocated land with an arrangement of a deed called Semi-permanent Ownership for not less than 50 years. It has been recommended to introduce a practical Land Tenure Policy highlighting the importance of people's participation. It is apparent that the legal relationship, mutual trust and land ownership could be the essential part of a successful participatory management of coastal land resources.

5. CREATE MECHANISMS TO ENSURE STAKE-HOLDERS' PARTICIPATION IN BIODIVERSITY CONSERVATION AT LOCAL LEVEL

Decentralized management of biodiversity through local capacity building and empowerment has emerged as an important strategy throughout the developing world. Local level institutions have many advantages over central agencies in that they are familiar with the local biodiversity, its history and value. However, since many of the existing local institutions have disintegrated or been ignored in the decision-making process, there is an urgent need to strengthen and revitalize the existing ones and create new institutions wherever necessary. The national government should direct all efforts to build infrastructures and provide the necessary financial resources to the local institutions. There should also be provisions to build capacity in local-level institutions to maintain books/ database and develop management plans for biodiversity that has market value.

Specific actions

Action	*Key actors*	*Priority*
Build capacity of the local governments to manage biodiversity locally	MoEF and NGOs	High
Empower local-level institutions including religious ones to evaluate, monitor and regulate the harvest and trade in biodiversity	MoEF and NGOs	High
Establish cooperatives and regulated markets locally exclusively for biodiversity trade	MoEF, MoA, LGRD NGOs and CBOs	High
Build capacity in women to run the local level institutions	Ministry of Women Affairs, NGOs	High

MoEF is the Ministry of Environment and Forest
MoA is the Ministry of Agriculture
LGRD is the Local Government Engineering Department

6. ESTABLISHMENT OF COMMON WEALTH AND PARTICIPATORY MANAGEMENT

6.1 Common Wealth for Sustainable Management (CWSM) – With Special Reference to SPA

Earlier, it was not realized that the poverty alleviation programmes should be aimed at creating common wealth for common or joint management in a sustainable manner. Active participation of rural people can contribute substantially to the poverty alleviation of rural coastal dwellers through creating easy available improved resources where there should have entrance of the rural people for resource-use. Since 2000, BFRI has been establishing a series of SPA suitable for each coastal region of Bangladesh which may serve as a source of common resource pool. The seed sources with best trees owing to their superiorities over surrounding base populations are expected to have increased ability to withstand extreme conditions. Expected practical gain has been derived from in situ growth variations of these species. It has been shown that the use of selected tree seeds from SPA of eastern coastal belt may provide 24.1% more wood volume/ha, while it may increase up to 66.5% in keora plantations to be raised by using only the best tree seed sources of SPA (Nandy et al., 2004). Similarly, there would be 33.7 to 121.2% gain in central and 26.4 to 57.8% gain or increased yield/ha in western coastal belt (Nandy et al., 2001). The use of improved seed sources from the established SPA of baen species may provide 20.3 to 59.1% gain per hectare.

6.2 CWSM – With Special Reference to Shrimp Cultivation

Shrimp farming in the coastal areas is gaining popularity due to quick income generation. In 1984, the total area of shrimp farming was 51,834 ha which has grown to 108,279 ha in 1990. It is important to mention about the dramatic change of Bogda (*Penaeus monodon*) shrimp cultivation area which is highly practiced in Rampal, Paikgacha and adjacent areas of Sundarban. Bogda farming area has been increased from 80,015 ha in 1992 to 187,644 ha in 2005 and appeared to be commercially viable practice leading to land-use conflict in the above mentioned high saline zones.

In spite of the fact that the shrimp farming is an important and viable aspect of aquaculture in the coastal areas, it is associated with the major destruction of mangrove plantations. Its management in Bangladesh is unplanned and unscientific (Siddiqi, 2001). As for example, Chowdhury et al. (1990) reported that the maximum yield from shrimp farming was 180 kg/ha/yr in the Chokoria and 330 kg/ha/yr in the Cox's Bazar of eastern coastal belt while the same was 345 kg/ha/yr in the western parts of Bangladesh coasts. It indicates that unplanned shrimp cultivation by removing mangrove plantations is not only hazardous to the environment but also may substantially decline the availability of fish near coast line.

Sustainable Environment Management Programme of the Ministry of Environment and Forest with UNDP support has carried out a study to identify land suitability areas in south western coastal area to take the policy decision and guideline where and what type of crops including shrimp farming may be allowed by suitable zones. Accordingly, shrimp cultivation should be practiced and concentrated in well planned, selected, suitable and productive areas in order to minimize the land-use conflict and destruction of mangrove forests. In this regard, participatory approach by involving trained coastal farmers in large scale planned productive sites could contribute and accelerate the poverty alleviation of participated coastal dwellers. What is needed is to conduct thorough surveys to locate and quantify areas suitable for either component of aquaculture and involve trained group of farmers for the creation of their common resource for sustainable management and resource-use.

6.3 CWSM – With Special Reference to Coastal Embankments and Its Erosion Control

Bangladesh Water Development Board (BWDB) has constructed over 4000 km of coastal embankments along the Bay of Bengal and offshore islands to safeguard against inundation, intrusion of saline water to the farmers' field and devastation associated with repeated attacks to tidal surges and cyclonic storms. These earthen embankments are subjected to erosion which is acute in offshore islands. In order to protect these embankments from erosion, FD undertook extensive programme for quick afforestation on the shoulder of these embankments. Owing to the urgent necessity of establishing vegetative cover in both inner and outer slopes of the embankment, there was no opportunity to select the species tolerant to different degree of salinity or species suitable to either inner and outer slopes of the same embankment.

Site-suitable species have been selected (Nandy et al., 2002) for inner and outer slopes of these embankments, particularly for low, moderate, moderately high and extremely high saline zones of Bangladesh coastal regions (Table 2). The selection of adaptable species to specific coastal site(s) may have greater climate change impact and provide sustainability of coastal embankment plantations in Bangladesh.

Now a days, there is a dilemma about the erosion and stabilization of the embankment. Trees planted on the shoulder of the embankments are apprehended to cause erosion due to easy penetration of rain and tidal water into the embankments through root systems. Moreover, trees do not allow grass cover as undergrowth and enhance the erosion of surface soil from the slopes of the embankments. In view of this, trial plots with simultaneous planting of tree and vetiver grass (*Vetiveria zizanioides*) have been established on the slopes of the embankments in central and western coastal regions. Vetiver grass was

Table 2: Selection of saline tolerant mainland species for coastal embankments

Sl. No.	*Saline zones (SZ)*	*Selected species*	
		Most suitable	*Suitable*
1.	Low SZ	*Dalbergia sissoo*, *Acacia auriculiformis* and *Pithcellobium dulce*	*Acacia nilotica*, *Terminalia arjuna* and *Samanea saman*
2.	Moderate SZ	*Casuarina equisetifolia*, *P. dulce* and *T. arjuna*	*Leucaena leucocephala*, *D. sissoo*, *S. saman* and *A. auriculiformis*
3.	Moderately high SZ	*S. saman* and *L. leucocephala*	*C. equisetifolia*, *D. sissoo*, and *A. auriculiformis*
4.	Extremely high SZ	*A. nilotica* and *C. equisetifolia*	*A. auriculiformis*, *D. sissoo*, and *L. lecocephala*

found to stabilizing newly established embankments for which nursery and plantation techniques have been developed (Nandy, 2003). Being a low cost and easily adopted technique, introduction of *V. zizanioides* under the trees would provide better protection against soil erosion of coastal embankments.

Whatever technologies are developed, physical protection of the embankment depends on its close supervision by the adjacent farmers whose crop fields and assets are protected and secured. The embankments are found to disappear within a year or two of the start of erosion, which requires immediate repairing. Unfortunately, the mandate of BWDB has been restricted to major repairs only when embankment is close to failure or completely failed. Accordingly, the farmers could be the real and close supervisors of these embankments if they are involved in the group of beneficiaries from embankment plantations. Existing rules of FD provide 50% output (yield at maturity) of plantations to the interested stakeholders. Owing to their own necessity, they will take immediate measures to minimize the embankment erosion and provide security of its plantations. Appropriate technologies, as developed by BFRI, offer unique opportunity for participatory resource generation in the exposed embankments through the involvement of farmers group in raising plantations and its subsequent maintenance. And, this could be the right way for sustainable and participatory governance of common wealth.

6.4 CWSM – With Special Reference to Joint Coastal Forest Management

Forest department developed a new, collaborative approach to ensure more effective protection, conservation and sustainable utilization of designated Protected Areas with support from USAID under Nishorgo project. This new approach may be applied along the coastal forest resource development with community involvement under joint or co-management model adopted. Joint forest management lessons learnt in West Bengal for the conservation and management of forest protected areas contribute more effective biodiversity conservation as well as provide a higher level of economic benefit and alternative income generation opportunities for the surrounding communities.

In general, before initiating the activities of any development project in the coastal areas local farmers are never asked for how to maintain the existing resources, what their demands are, how to make the fair share of resources or how to address for sustainable resource management. As a result, most of these projects are found unsustainable or ended with minimum contribution to the national poverty alleviation programmes. With these in view, local farmers were organized to discuss among themselves regarding various issues and particular emphasis was given to the Joint Forest Management (JFM) in encroached afforested coastal areas of Bangladesh. Problems regarding participatory approach for productive use of allocated land and its resource control were prioritized at farmers level and validated (Nandy et al., 2003). Summary of farmers opinions for a number of provisions for each of following issues have been provided in Table 3.

This article discusses on the issues how to reduce the pressures for encroachment and conversion of the remaining mangrove forests by addressing the driving forces and root causes of the pressures on the coastal mangroves. In order to meet this challenge, it is suggested to ensure support for increased local participation in the collaborative or joint management of these forests and regeneration or enhance productive potential of the forest resources, in a manner that integrates opportunities for partnership governance. This approach also emphasized on sustainable use in association with poverty reduction in tandem with more effective conservation where local communities would be mobilized through co-management of public-private or community partnership.

7. CONCLUSIONS

This article reveals the fact that a paradigm shift from conventional income generation activities to resource generation activities all along the coastal zone opens up the threshold point to unlock the poverty trap through participatory governance of conservation of biodiversity and management of common property regime by the landless and marginal farmers who constitute majority of the population. Policy dimension toward creating enabling environment for bringing benefit beyond the boundary should consider the issues of climate resilient development.

The impact of inundation and severe storms on the natural forest of Sundarban is profound. Its key mangrove species had been surviving against climate change impacts for over a century. Its ten key species have been considered as endurable species against climate change impacts and successfully evaluated in the coastal areas of Bangladesh. Bangladesh coastal belt is most

Table 3: Perceptions of participating stakeholders

Sl. No	*JFM issues*	*Possible solutions*
1.	Farmers selection	Farmers unanimously agreed that the participating farmers should be landless having family and recognized by local authorities including Govt. and NGO.
2.	Allocation condition	Concord of opinion is that they are bound to reside and invest own labour/inputs in allocated plots. Considering their economic condition, they may earn money by using their additional labour outside.
3.	Allocation committee	Special Land Allocation Committee comprising members from Govt., NGO and farmers can be formed.
4.	Allocation size	Minimum allocation of land (1 ha) per family with valid land-leasing document may satisfy the farmers.
5.	Group size for JFM	They prefer 50 farmers in a group so that they can resist any attack from outside.
6.	Plot size for JFM	Farmers unanimously agreed that they need at least 30-40 m wide agri-plots in between shelter belts comprising six lines of trees per belt.
7.	Farmers right for resource use and resource replacement	Farmers assured not to cut the trees but to keep the allocated forest land always covered with mainland species provided the seedlings are available (it may be mentioned here that the mangroves are no more suitable in raised land while climate change impact necessitates to create vegetative protection belt).
8.	Share for common wealth to be created through JFM	Farmers agreed for 50:50 share of commercial produces at the end of maturity and also expect to get all dry leaves, lops, tops etc. after felling as remuneration for taking care of trees in shelter belts.
9.	Authority for community management	Farmers consider themselves as the main authority for their common wealth management while they unanimously agreed to form Community Development Committee by involving Govt., NGO and farmers.
10.	Responsibilities of such authority	Periodical monitoring of common wealth management as well as development of school, bazar, prayer house, community centre and even provide support in raising nurseries.

vulnerable to sea level rise. In this regard, the introduction of selected species is the best option to cope with different scenarios of sea level rise. Several techniques have been provided for eco-friendly maintenance of coastal embankments. In order to set the congenial environment for group-based resource generation activities of coastal dwellers, it has been recommended to combat poverty alleviation through the integration of both scientific and participatory approach.

PRA conducted throughout coastal belts of Bangladesh provides an interesting land leasing schedule, which can be incorporated at introducing coastal land tenure policy highlighting people's participation and their access for productive use of land to overcome land-use conflict. The same is applicable for shrimp cultivation, which should be practiced in the selected productive areas through the application of CPR in order to minimize land-use conflict and destruction of mangrove forests, while the CPR scheme will provide sustainability of poverty reduction for group farmers.

REFERENCES

Canonizado, J.A. (1999). Integrated Forest Management Plan of Noakhali C/A Division (1999-2008), FRMP TA Component. Mandala Agril. Dev. Corp./FD/MOEF, Dhaka.

Choudhury, A.M., Quadir, A.M. and Islam, J. (1990). Study of Chakaria Sundarbans using Remote Sensing Techniques. SPRRSO, Dhaka. 77 p.

Empowerment of Coastal Fishing Communities for Livelihood Security (BGD/97/017), Compendium of Micro-Capital Grant (MCG) Case Studies (2005). Cox's Bazar, Bangladesh (www.fao.org/Participation/english_web_new/content.../fisheries.doc).

Howlader, N.I. (1999). Forest Resources Management Project. Mid-Term Review. Ministry of Environment and Forest, Dhaka.

Islam, M.R. and Nandy, P. (2001). Site Suitability of Different Mangrove Species and Planting Techniques for Baen (*Avicennia officinalis*) in the Coastal Areas of Bangladesh. *In:* Siddiqi, N.A. and Baksha, M.W. (eds.). Mangrove Research and Development. Bangladesh Forest Research Institute, Chittagong. pp. 39-48.

Nandy, P. (2003). Technologies and Information Generated for Coastal Afforestation in Bangladesh. Plantation Trial Unit Division, Bangladesh Forest Research Institute. Barisal. 19 p.

Nandy, P., Alam, M.J. and Haider, M.R. (2004). Establishment of Mangrove Seed Production Area for *Sonneratia apetala*. *Journal of Tropical Forest Science*, **16(3)**, 363-368.

Nandy, P., Haider, M.R., Islam, M.R., Alam, M.J., Moula, M.G. and Habib, M.A. (2002). Growth Performance of 37 Species Raised in Embankments of Eastern, Central and Western Coastal Belts of Bangladesh. Research Bulletin. Coastal Greenbelt Project, Forest Department, Bangladesh. 41 p.

Nandy, P., Haider, M.R. and Moula, M.G. (2001). Establishment of Seed Production Area for Keora (*Sonneratia apetala* Buch.-Ham.) with Special Reference to the Central Coastal Belt. *In:* Siddiqi, N.A. and Baksha, M.W. (eds.). Mangrove Research and Development. Bangladesh Forest Research Institute, Chittagong. pp. 73-81.

Nandy, P., Moula, M.G., Isalm, M.R. and Habib, M.A. (1999). Techniques for Raising Seedlings from Detached Germtubes of *Borassus flabellifer* Linn. *Bangladesh Journal of Forest Science*, **28(2)**, 114-120.

Nandy, P., Paul, S.P., Moula, M.G., Islam, M.F., Basak, S.R., Miah, M.A.Q. and Mannan, M.A. (2003). Stakeholder Analysis through Participatory Rural Appraisal for Land-use and Joint Forest Management in Raised Coastal Areas. PRA Bulletin, PTU Division, Bangladesh Forest Research Institute, Barisal. 58 p.

Siddiqi, N.A. (2001). Mangrove Forestry in Bangladesh. Institute of Forestry & Environmental Sciences, University of Chittagong, Chittagong. pp.120-122.

5

Present Status, Challenges and Management of the Japanese Coastal Zone Environment

Manish Kumar and Rita Chauhan[1]

Department of Urban Engineering, University of Tokyo, Japan - 113-8656
[1]School of Environmental Sciences, Jawaharlal Nehru University
New Delhi, India - 110067

1. INTRODUCTION

Coastal regions are intensely dynamic areas, which are extremely important to humans. Coastal zones are transitional areas between sea and land—the coastline with coverage of the three spheres viz. air (atmosphere), water (hydrosphere), and land (lithosphere). It harbours unique, dynamic and irreplaceable ecosystems. At the same time, coastal zones are subject to intense use by humans for transportation activities, resources and energy procurement, industrial uses, and recreation. Furthermore, coastal zones are the first line of defence against inland disasters. They are buffer zones against the ravages of tsunamis, rough waves, flooding, and erosion. The coastal zones are linked to human beings through functional aspects: (i) provision of ecological services, (ii) disaster prevention, and (iii) human utilization. Each of these aspects is intricately linked. Consequently, it has become mandatory to monitor and manage these three facets of the coastal zone in an integrated manner to ensure that the human relationship to coastal zones remains harmonious.

Japan is an island country located in East Asia in the northwest Pacific Ocean. It ranks sixth amongst the world, with a population of 30,477,000 people living within 10 metres of the average sea level. The present chapter provides an overview of coastal conditions and trends, and reviews studies on vulnerability assessment to impacts of sea-level rise on the Japanese coastal zone. Moreover, the current status of the coastal environment in Japan is reviewed followed by the possible framework of integrated coastal zone management (ICZM) scheme in Japan.

2. FACTS ABOUT THE JAPANESE COASTLINE

Japan lies between 20° 24′ N and 45° 30′ N latitude with Pacific Ocean in the east and Sea of Japan in the west. It is composed of four main islands—Honshu, Shikoku, Kyushu and Hokkaido and 6800 other smaller islands, some of which are not inhabited (Bureau of Statistics, Ministry of Internal Affairs and Communications). Japan experiences a typical Asian temperate monsoon climatic region with relatively high sea temperatures and a warm climate. It is bathed by the warm Japan Current on the Pacific side and Tsushima Current on the Sea of Japan side of the country. The difference in tides at flood stage on the Pacific coast is approximately 1.5 metres, and approximately 0.2 metres on the Sea of Japan coast.

It has over 34,000 km of coastline and approximately 4.5 million sq. km of ocean within its territorial sea (Table 1). The term "Coastal Zone" was first coined in the third National Comprehensive Plan established in 1977. The definition of coastal zone was released by the Japanese Association of Coastal Studies as "coastal water can extend to the 12 nautical mile boundary of territorial waters and the terrestrial zone encompasses the administrative area of the municipalities that have a coastline, subject to consideration of the river system to the space".

The coastline comprises sandy coasts and rocky shores that are mixed together. The Pacific Ocean side of the central Tohoku district consists mainly of deeply indented coastlines, while the Sea of Japan side is rich in sandy coasts. The semi-closed Seto Inland Sea is located in the southern part of the main islands. About 12% of coastlines comprise sandy coast, while the 28% of coastline is covered by structure for shoreline protection (Fig. 1). The artificial seashore includes features such as coastal roads around the seashore and airport facilities along the coast. Due to highly developed region, the ratio of natural coast to entire coastline of Japan is only 53% (Shikda and Koarai, 1997). Until Edo period (1603-1867), Japanese coastal dwellers used the coast extensively for coastal fisheries and salt production. After the Meiji era (1869-1912), industrial development in the coastal zone started. The intensive port and coastal

Table 1: Dimensions for Japanese coast

Total land area	377,720 km^2
Length of coastline	34,536 km
Territorial sea	430,00 km^2
Contiguous zone	44.448 km
Economic zone area	4,470,000 km^2

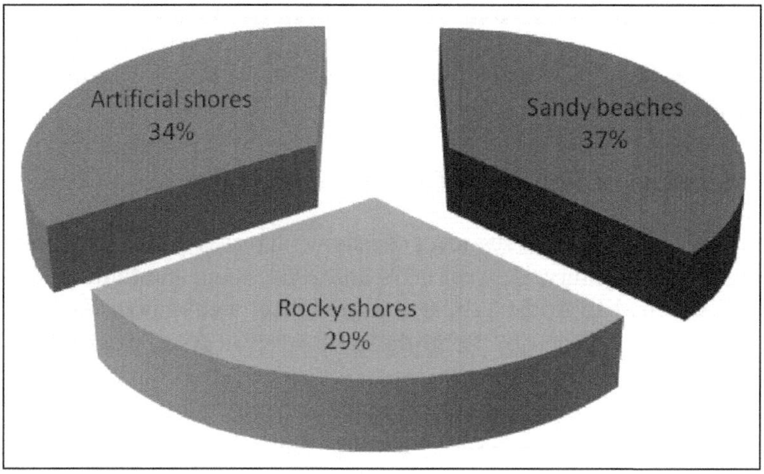

Figure 1: Distribution of physiographic features of Japan coastline.

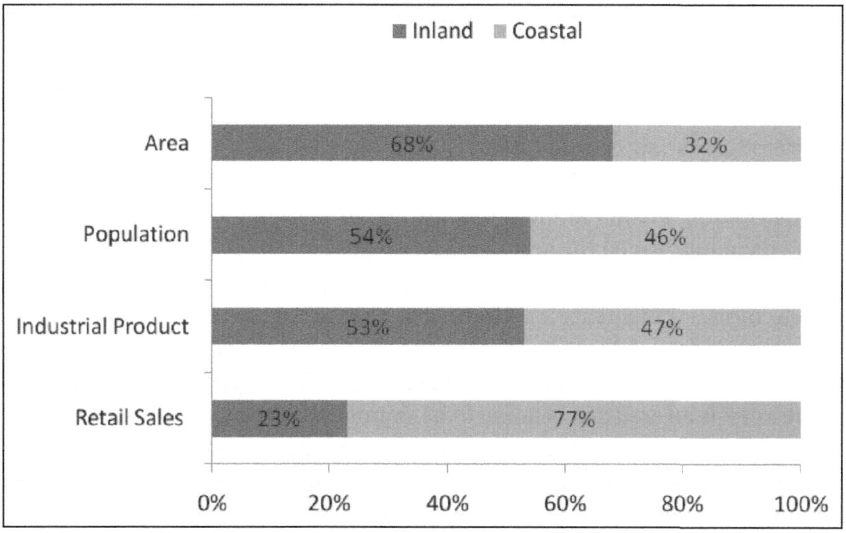

Figure 2: Relative contribution of coastal and inland sectors (Kojima, 2000).

area development under the Japan government encouraged the economic development under the policy of a wealthy nation and strong army. The majority of the population and economic activities in Japan are concentrated in coastal zones. As shown in Fig. 2, the area of coastal municipalities occupies only about 32% of the total area of 370,000 km^2. They hold about 46% of the total population of approximately 120 million. They produce about 47% of the industrial output, and amazingly 77% of the total expenditure for retail business or market goods is spent in the coastal municipalities. Demands on coastal and marine resources have been constantly increasing as coastal areas become more developed.

3. MAJOR CONCERNS OF JAPANESE COAST

The Japanese coast is facing following major threats, which are matter of great concern.

3.1 Sea Level Rise and Climate Change

Estimates of global mean sea-level rise show increases of tens of centimetres within the next century. Between 1994 and 2004, about one-third of the 1562 flood disasters, half of the 120,000 people killed, and 98 percent of the two million people affected by flood disasters were in Asia, marking Asia as particularly at risk from the detriments of water-related climate change. The current population movements are reported as being in the opposite direction, however, and countries including India and China, with its economic liberalization and growth-driving rapid urbanization and coastward population movements, are raising particular alarm among coastal environmentalists.

The population of Japan has been identified as being at great risk from rising sea levels and more intense cyclones linked to climate change (Kojima, 2000). In fact, the population density map of the country shows that many of Japan's largest urban areas, including Tokyo, Kawasaki, Yokohama, Nagoya, Osaka and Sapporo, have over 1000 people per km^2 living within the worrying 10-metre elevation coastal zone. The report cites the "recent expansion of international trade" and Japan's preference for ocean shipping over air freight as contributing factors to the pre-eminence of densely populated urban hubs along the country's coastline.

Japan has a lot of resources and has the capacity to not only protect, but also to move, settlements around to some degree, as long as it's over a long time period. Also, China, India, Bangladesh, Vietnam and Indonesia all have populations in the 10-metre zone greater than that of Japan. However, the 10-metre zone is an area that is going to be imminently flooded; thus the issues of sea level rise and coastal hazards need to be taken seriously in Japan. It is strongly recommended that movement towards, and growth of, the coast must be stopped in Japan; as they try to reclaim the land and somewhere in doing so it affects the natural ecology and facilitates more human life at stake from natural disaster.

3.2 Eco-hydrological Conditions

Healthy coastal ecosystems are vital to healthy coastal zones. However, coastal erosion and construction of coastal preservation structures are adversely affecting coastal ecosystems. Dry beaches, seaweed beds, and coral reefs are important elements of coastal ecosystems. Landfills and coastal preservation structures are causing coastal zones to become increasingly artificial. The basic fact is that as artificial coastline has increased, natural coastline has decreased.

An artificial coastal area is an area where structures have been erected in shallow waters. Semi-artificial coastal areas are those in which structures are erected but are either on the land side of the high tide shoreline or the sea side of the low tide shoreline. Despite a slowdown in the current speed of construction of landfill coastline and preservation structures, the increasing trend of artificiality of the coastline still continues.

Like Tokyo, many cities of Japan have combined sewer system to regulate the wastewater and rainfall. However, in Japan rainfall happens torrential which force direct discharge to bay area and river of wastewater. The effect of Combined Sewer Overflow (CSO) on the water and sediment quality has drawn much attention in recent times.

3.3 Anthropogenic Utilization and Their Impact

Coastal zone has been used for various human activities from the beginning of civilization. In ancient times they were used by hunters and gatherers as a place to harvest marine resources, today they are used as sites for airports, petroleum exploration, fuel storage and energy generation, industrial and commercial development, waste dumps, and recreation areas. Among the East Asian countries, Japan is having the highest rate of urbanisation (Fig. 3). Japan is covered with mountainous regions unsuitable for intense human utilization. Thus, its population and accompanying activities tend to be concentrated in flat areas, especially along the coast. The land area is just about the same, but the ratio of the population density in the drainage basin is 60:1 in Tokyo Bay's favour. This highlights the inescapable need in Japan to make efficient use of its coastal zones.

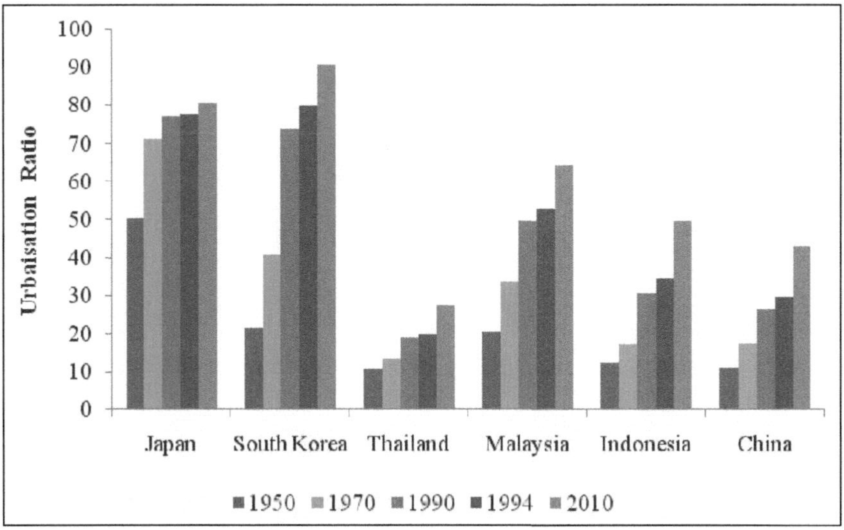

Figure 3: Urbanisation ratio in East Asian countries (United Nation, 1995).

3.4 Disaster Situation and Their Prevention

The Japanese coastal zone is prone to several types of coastal disasters, namely typhoons in the summer, wind waves in the winter, earthquakes and tsunamis. Sea-level rise is anticipated to increase, with dramatic impacts in those regions where the vulnerability to these events already exists. Figure 3 depicts the location of such disaster-prone areas, major earthquake epicentres and severely eroded beaches, as well as the path of major typhoons that have caused devastating storm surge damage (Takayama, 1997; Yoshikura 2000). The average annual number of damage cases accounts for more than 440, and the restoration expenditure exceeds 240 million US$.

The entire Japan is threatened by flooding and high waves during the typhoon season around September. In addition, the Sea of Japan side of Japan is buffeted by strong winds and rough seas in the winter. Storm conditions combined with high tides can cause especially severe damage. Figure 4 illustrates damage caused by flooding due to typhoons in Japan since 1900. High tides in the three major bays of Tokyo, Ise, and Osaka, as well as in the Ariake Bay, can be amplified by storm winds. A particularly damaging typhoon hit the Tokyo area in 1953. Maximum tide deflection was 3.9 metres in Tokyo Bay. Although it is not included in the figure, this typhoon leads to the drafting and enactment of the Coastal Act. The worst recorded typhoon-induced damage

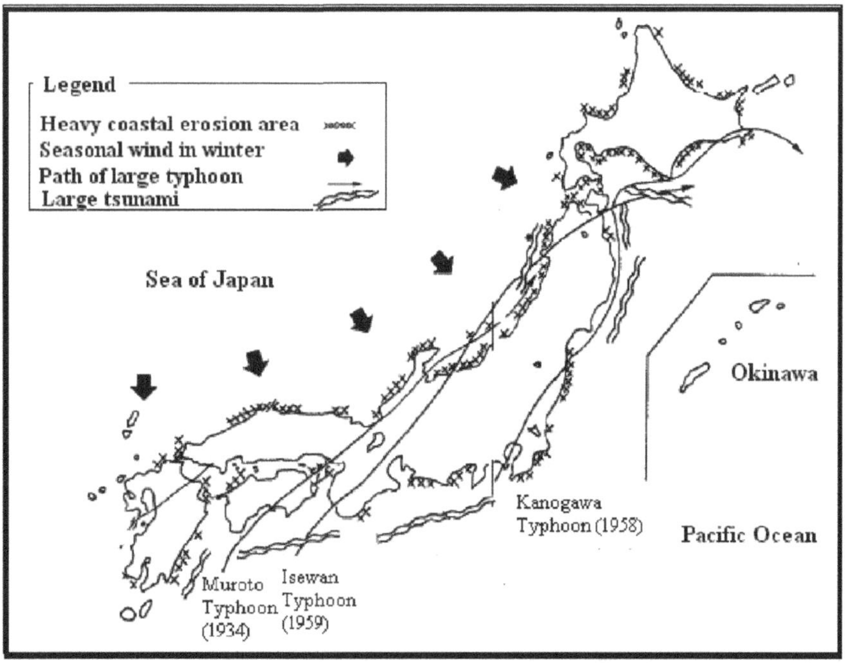

Figure 4: Layout of coastal area of Japan with their potential problem with the depiction of some major typhoons.

was experienced in the Ise Bay Typhoon of 1959 which caused a storm surge anomaly of 3.4 metres and resulted in more than 5000 deaths and damage to almost one million buildings. Damage caused by flooding has diminished in recent years, but it is unclear whether this is a result of efforts to protect the coastline, or the lack of severe typhoons in recent years. Thus, prevention of disasters related to typhoons, tsunamis, and erosion is one important functional aspect of coastal zones that must be taken into account in any Integrated Coastal Zone Management (ICZM) scheme developed in Japan.

4. COASTAL ZONE MANAGEMENT IN JAPAN

4.1 Legal Framework of Coastal Management

Japan's first coastal zone management scheme was embodied in the Coastal Act of 1953, with prevention of disasters as the prime mandate. The coastal zone is conglomeration of ports, harbours, fishing/aquacultural lands, and reclaimed agricultural lands. Each classified area is administered by different ministries such as the Ministry of Transportation, Ministry of Agriculture, Forest and Fisheries, and Ministry of Construction. The development and utilization of coastal zone are also governed by laws such as Harbor Act, Fishing Port Act, and Public Waterfront Landfill Act. The National Parks Act and the Seto Inland Sea Preservation Special Measures Act are set down to cover the conservation and protection of coastal habitats. These different laws result in differing responsibilities by different government agencies. There is, however, no law that serves to coordinate all the relevant aspects of each of these separate laws.

With the Earth Summit (UNCED) in 1992, the Chapter 17 of Agenda 21 highlighted the need for protection of marine and coastal environments, and gave the concept of integrated coastal management. The amendments to Coastal Act in 1999, the enactment of fishery basic laws and amendment to other marine and coastal related laws were few steps undertaken by Japanese government. Japan is still seeking a way to develop a comprehensive framework for Integrated Coastal Zone Management (ICZM) (Mimura et al., 2008).

4.2 Integrated Coastal Zone Management—Concept

Basic concepts on ICZM are to create or restore aesthetic safe and sound coastal areas for present and succeeding generations, balance competing demands of environments safety and various land and water uses in coastal areas and foster partnerships among various concerned local governments, agencies, groups and individuals (Kojima et al., 1999). Integrated Coastal Zone Management (ICZM) has become vital in order to solve balance conflicting demands from various concerned agencies, groups and individuals. Also there are some needs for adoption of integrated coastal zone management; the law reform from the viewpoint of environmental conservation is demanding the coastal management system of integration.

The National Land Agency of Japan provides technical information as in "the Guideline for the Integrated Coastal Management Plan". The boundaries of the coastal zone, subject to an ICZM plan, will be identified to treat an entire encompassed area as a natural system unit strongly influenced by each other in light of geographical shapes, water and soil movements. The zone encompasses a certain part of coastal areas extending in two directions: (a) a coastline direction and (b) an inland seaward direction. The local authorities will provide background directions for boundaries, which will be based on the portions of the whole country coastline which the guidelines show an inland seaward direction boundary extends in accordance with the characteristics of local areas and planning elements.

The key features of integrated coastal zone management are:

- To encourage participation and cooperation of various concerned groups such as the national government, local administrations, private sectors, Non Profit Organizations (NPO), fishermen and local communities
- To provide wide overviewed implementation giving full consideration to the whole of bays, inland seas and river
- To provide long-term viewed implementation setting a future vision of coastal areas under an analysis of natural cycles
- To provide a continuing implementation based on the results of continuous monitor and analysis of the execution of an ICZM plan

A detailed study of the features of coastal areas including the characteristics of the natural disasters, social economic interests' culture and history in coastal areas are being undertaken to define problems of coastal area (Kishida, 2000). The ICZM plan will include the boundaries of a coastal zone (to be indicated clearly on map), period of reauthorization of the plan (approximately every year), basic policy (basic policies for projects and activities in the coastal zone in light of the future visions), definitions of issues and basic directions of comprehensive management programmes for solving and managing the issues, organizations for the implementation of an ICZM plan and measures to evaluate the execution of the ongoing plan. Integrated Coastal Zone Management Commission shall be created consisting of representatives of each of the prefecture governments.

4.3 ICZM Practices in Japan

Japan has established no legal system for mitigation to maintain the integrity of the coastal zone. The United States, on the other hand, has systematized a mitigation system. The definition of mitigation used in the United States was set down by the Council of Environmental Quality. Also a definition, which focusses on what are called avoidance, minimization, and compensatory courses of mitigation action, was set down in a 1990 memorandum agreed to by the US Army Corps of Engineers and the Environmental Protection Agency. Artificial changes in the coastal zone which accompany activities such as development must, in the context of mitigation, be done in such a way that they do not destabilize or adversely affect the environmental infrastructure, for example, the topography, soil quality, water quality, or air quality. When planning and executing projects related to human utilization, disaster prevention, and ecological protection, a selection from among the possible plans must be made from the standpoint of maintaining the soundness of the environmental infrastructure.

Although no codified system of mitigation exists in Japan, restoration of the environment, creation of new environments and concern for the environment has been present (Isobe, 1994). These practices can be incorporated in the developing concept of ICZM in Japan. The maintenance of coastal sand drifts provides another example of ICZM. A sand bypass was employed to maintain what are perhaps Japan's most famous coastal sand drifts—a formation known as Ama-no-Hashidate (Bridge of Heaven) in Kyoto prefecture. Projects have been embarked upon to restore and create new beaches by beach nourishment (Habara et al., 1996). Examples include the artificial shore created in the Kasai Waterfront Park, a project aimed at creating natural habitat. In this project, one of two artificial islands that were created is used for recreation while the other is reserved exclusively as a natural sanctuary where people are not allowed to enter.

Methods for creating a base for seaweed to adhere using blocks that have been placed at the bottom of the sea and methods for transplanting seaweed have been employed to create sites for new seaweed beds (Ishizaki, 1989). A 60,000 square metre seaweed mound was created to replace a seaweed site that was destroyed by the construction of the Ikata power plant in Ehime prefecture. Efforts have been made to shape seawalls and sunken piers to make them into appropriate habitats. It was found that aquatic plants and animals were able to attach themselves easier to a slanted stacked stone seawall, which was built for the Kansai International Airport in Osaka Bay. Monitoring confirmed the adherence of seaweed to the structures and a return of reef fish, which were not present prior to construction. As another example, a block shaped, gently slanted seawall on the Sumiyoshi coast in Miyazaki prefecture was constructed so that sea turtles could lay their eggs on the beach. Compared to conventional gently slanted seawalls, the block shaped seawall resulted in an increase in the ratio of eggs.

In addition to the examples given above, there are numerous other examples of restoration of the environment and creation of new environments. Significant progress has been in the technology required to put ICZM into practice. Japan has very little marshland in comparison to the United States, thus, compared to the U.S., Japan will focus on creating sand beaches, tidal flats, and seaweed sites.

Local coastal zone management is another important best ICZM practices, which has been developed quite recently. It involves the integration at local area with reconciling the sectoral approaches. One such practice is community-based fisheries management, which is underway in Japan for past several decades (White, 1998). This management approach includes territorial use rights in fishing for common fishing grounds. The functional aspects are: (i) ecological aspects of coastal fisheries resources for conservation; (ii) economic aspect that fishers maximize their profit in catching particular target species and selling them to the market and (iii) the social and cultural background secures social equity and equity among local fishers and people in a fishing community. Its mechanism includes an adjustment function to avoid conflicts, and to enhance social unity through achieving consensus on local rules and their enforcement.

4.4 Challenges for Integrated Coastal Zone Management

The lack of an integrated management act for the coastal zone in Japan prevents sharing a clear definition. It is necessary to have a clear definition in order to develop the appropriate administrative arrangements. The following are few other constraints in sustainable management of Japanese coastal zone:

1. Environmental degradation – damage from coastal and marine litter add pressure to coastal environment and its aesthetic value (Japan Environment Action Network, 1999).
2. Coastal utilization, particularly non-industrial use, has been growing at an alarming rate (Harada, 1994).
3. About 72% of coastline is managed by a single authority, the National Land and Transportation Ministry, the rest of the coastal zone is under divisional management.
4. Conflicts arising out due to coastal uses or to balance and manage industrial and non-industrial use.
5. Currently there is no legal entitlement for ICZM plans, which may affect the enforcement of plans through legislative action.

5. CONCLUSIONS

The population of Japan has been at great risk from rising sea levels and more intense cyclones linked to climate change. The development in the post-war era has resulted in Japan's coastline becoming an artificial coastline. Coastal Zone Management in an integrated manner is absolutely essential to solve Japan's development-related coastal zone problems. Therefore, to achieve true ICZM, establishment of new laws and reforms in administrative bodies is desirable. The prowess of Japan in technological development can be harnessed to integrate separate technologies into a coherent mitigation-oriented system of technologies that can enhance the coastal zone environmental foundation.

REFERENCES

Environment Agency of Japan (1980, 1984, 1992). Basic Survey on the Preservation of the Natural Environment. Sea Region Survey Report, p. 339, p. 203 and p. 20, respectively.

Environmental Agency of Japan (1992). A basic study on natural environmental preservation. Report of Marine Area Study, 339 p.

Habara, Y., Takahama, S. and Imamura, H. (1996). Functional Evaluation of Artificial Dry Beaches Created for Mitigation. *Coastal Engineering*, **43**, 1161-1165.

Harada, M. (1994). Towards a renaissance of leisure in Japan. *Leisure Studies*, **13**, 277-287.

Ishizaki, Y. (1989). On the Creation of Substitute Seaweed Sites and the Adherence Conditions of Seaweeds in the Number 3 Reactor at the Ikata Atomic Energy Plant. *Inland Sea Science*, **3**, 33-37.

Isobe, M. (1994). Creation of Coastal Environments. Asakura Shoten, 208 p.

Kishida, H. (2000). Start of a new coastal management system. *Ecology and Civil Engineering*, **3**, 65-75.

Kitajima, S., Ito, T., Mimura, N., Tsutsui, Y. and Izumi, K. (1993). Impacts of Sea-level Rise and Cost Estimate of Countermeasures in Japan. *In:* Mclean, R. and N. Mimura (eds.), Vulnerability Assessment to Sea-level Rise and Coastal Zone Management. Proceedings of the IPCC Eastern Hemisphere Workshop, pp. 115-123.

Kojima, H., Gomya, M. and Mimura, N. (1999). Development of coastal zone management, its direction and countries status. *J Japanese Assoc Coastal Zone Studies*, **11**, 1-20.

Kojima, H. (2000). Vulnerability and adaption to sea level rise in Japan. Proceedings of APN/SURVAS/LOICZ joint conference on coastal impacts of climate change and adaption in the Asia-Pacific region, pp. 81-88.

Matsui, T., Tateishi, H., Isobe, M., Watanabe, A., Mimura, M. and Shibazaki, A. (1992). The vulnerability assessment of sea coastal zone in Japan under the sea-level rise. Proceedings of Coastal Engineering. *JSCE*, **39**, 1031-1035 (in Japanese).

Mimura, N. (ed.) (2008). Asia-Pacific Coasts and Their Management. Springer Publication, The Netherlands, 365 p.

Mimura, N. (1994). Responses of coastal landform to sea-level rise—focusing on sandy beaches and river mouths. Pre-prints of the 3rd Symposium on Global Environment, Committee on Global Environment. *JSCE*, 97-102 (in Japanese).

National Research Council (1990). Managing Coastal Erosion. National Academy Press. 182 p.

National Research Council (1995). Beach Nourishment and Protection. National Academy Press, 334 p.

Ports and Harbors Bureau, Ministry of Transport (1998). Coastal Protection in Japan, 24 p.

Shikida, A. and Koarai, M. (1997). Statistical Analysis of artificial modification of natural coastline in Japan since 1960. *J Japanese Assoc Coastal Zone Studies*, **9**, 17-25.

Takayama, T. (1997). Characteristics of Tsunami Disaster and Counter-measures against Tsunamis in Japan. The fourth Japan-China (Taipei) joint seminar on natural hazard mitigation.

Tamizu, Tatsuyuki (1992). Coastal Erosion Counter-measures Taking into Consideration Laying of Eggs by the Red Sea Turtle. Techno-Ocean '92.

Tanaka et al. (1993). Shoreline changes of Japan by comparing topographic maps. Proceedings of Coastal Engineering. *JSCE*, **40**, 416-420 (in Japanese).

Tanaka, Shigenobu, Koarai, Mamoru and Fukazawa, Mitsuru (1993). Changes in the National Coastline Based on a Comparison of Topographic Maps. *Papers on Coastal Engineering*, **40**, 416-417.

Uda, T., Ito, H. and Ohtani Y. (1992). Sea-level changes in Japan: 1955 to 1989. Proceedings of Coastal Engineering. *JSCE*, **39**, 1021-1025 (in Japanese).

United Nations (1995). World urbanization prospects: the 1994 revision.

UNEP (1996). Handbook on Methods for Climate Change Impact Assessment and Adaptation Strategies, 5-3.

White, A.T. (1998). Coastal Management in Asia: Are donor-assisted programs sustainable and beneficial? CRMP document no. 43-CRM, 1-12.

Yanagi, T. and Akaki, T. (1993). Variation of the mean sea-level around Japan. *Umi no Kenkyu*, **2**, 423-430 (in Japanese).

Yoshikura, K. (2000). On port-related hazards of 1999. *Wave and Beach*. **145**, 32-38 (in Japanese).

6

Integrating Hydrologic and Hydrodynamic Models for Decision Support Systems and Management of Coastal Zones and Estuaries

Prabin Paul and Vladimir Cvetkovic

Department of Land & Water Resources Engineering
Royal Institute of Technology, Brinellvägen 32, SE-100 44
Stockholm, Sweden

1. INTRODUCTION

An estuary may be defined as a semi-enclosed coastal basin, or system of basins, which has limited water exchange with the open sea, and which is measurably diluted with freshwater discharge from the land. Often the dynamics of an estuary are also strongly affected by tidal forcing. An exact understanding of all the processes occurring, and their application to make reliable predictions in real estuarine system, remains as a challenge.

River discharge, wind influence, temperature changes, climatic changes, seasonal variations and tidal motion are the main parameters that act as forcing and influence the estuarine balance (Dyer, 1991). The interaction is important from an ecological perspective because the estuarine circulation is the most efficient mechanism for flushing river-borne pollutants out to the open sea. The greatest horizontal mass flux occurs at maximum stratification and minimum turbulent mixing. Turbulent mixing increases the tracer residence time in the estuary (Nunes Vaz et al., 1989).

Most studies on Region of Freshwater Influence (ROFI) have been done in temperate regions, yet in global terms, tropical and subtropical ROFIs are much more important for delivering fresh water to the oceans (Souza and

Simpson, 1997). The strong seasonal variation that characterizes the outflow patterns, stratification and sediment dispersal has received relatively little scientific attention. Studies based on the climatic changes and its effect on the circulation and mixing in the estuaries are yet to be addressed, making us understand the extent of salinity intrusion, sedimentation, pollutant disposal by flushing and also helping in the prediction of seasonal changes in the biological process.

The study conducted on the multi-layer structure of the estuarine water identified four water masses according to temperature and salinity (Cho et al., 2004). The observations made clear that the temperature and salinity of the different water masses varied considerably with seasons, depending on the temperature of the incoming fresh and sea water. Runoff is sensitive to changes in extreme rainfall, as demonstrated by Chew et. al. (1991). The climate change projections indicate increases in extreme rainfall, where the mean annual rainfall is projected to decrease.

The changing climatic scenario has resulted in the rise in the sea level as a result of the global warming, many studies show. This rise in the sea level was later explained by the Intergovernmental Panel of Climate Change (IPCC). UNEP has identified India among the 27 countries that are most vulnerable to sea level rise. According to the trend analysis of the rise in Mean Sea Level, the trend at Cochin is found to be 125 mm/year (Dwivedi, 2005). This will result in the change in coastal zones, especially the wetlands around the lake, which results in land use changes in the region. A 42.4 km^2 loss of wetland is predicted by a 0.1 m rise in sea level (Dwivedi, 2005).

Most of the reported modelling studies on the estuarine dynamics in specific coastal areas focus either on a particular factor (the density dependence, tidal variations, temperature change, inflow variation, land use changes, etc.) or the effects of different management strategies on saltwater intrusion. Typically, all the specific studies consider a single factor affecting the mixing as changing with timescales, with all other major factors independent of time. Most importantly, hydrological forcing in terms of freshwater discharge is in most cases treated in a simplified manner, not accounting for spatially distributed effects in the surrounding catchments.

The objectives of this study are: 1. To study runoff changes associated with seasonal and climatic variations (precipitation, temperature, land use changes, etc.) using a high resolution GIS based hydrology model. 2. To develop a hydrodynamic model (using a general purpose FEM solver) that can simulate the dynamics in estuaries, subject to spatially distributed forcing under extreme conditions of hydrological (precipitation) and thermal (temperature) fluctuations over the discharging catchments. 3. To study the factors which are inter-dependent, upon integration and to expose the sensitivities of the coupled model.

This article describes the development and application of hydrology-hydrodynamic model built for the Vembanad Lake System and mainly deals with the results from the hydrology model, simulated over a period of 10 years.

2. STUDY AREA

We have identified a specific location, which satisfies most conditions relevant for this study, and is in many ways typical for tropical regions of the world. The Vembanad Backwater system is the largest estuary in South India into which six rivers (Periyar, Muvattupuzha, Manimala, Achencoil, Meenachil and Pamba) are directly draining and is representative of other systems in the humid tropics. There is a significant variation in the spatial and temporal distribution of rainfall, temperature and wind in the wetland system. The temperature of the region varies with time and season, from 15°C in the winter to 38°C in the summer. The average daytime temperature rises as high up as 35°C and in the night falls to 20°C due to the influence of sea-winds. The seasonal variation is quite extreme in the area. The average rainfall in the area is 3200 mm, with more rainfall in the highlands in comparison to lowlands. The precipitation pattern of the area reveals the heavy downpour during the rainy season, with much less or no rain during other seasons. The rivers draining into the system are monsoon fed with least discharge during the rest of the year. Finally, there is significant topographic gradient toward the estuaries, strongly influencing the freshwater inflow (as a key part of the forcing) from the surrounding catchments.

3. MATERIALS AND METHODS

In most of the integrated hydrology-hydrodynamic models a constant density condition is assumed (Langevin et al., 2005). In an estuary where more than one river meets, different basins acquire different densities in response to uneven cooling/heating, freshwater supply, and inflow of saline water from the sea, etc. Density gradients thus arise over the relatively narrow straits connecting the basins. Modelling the evolution of water properties in such a system obviously requires that all inflows and outflows be specified.

A hydrology model can give a reasonable inflow/runoff estimate of the region, with precipitation, evapotranspiration, temperature and infiltration put together with the geographical features of the regions like slope, land use, soil texture etc. Here a GIS based mass balance model is developed to simulate the runoff from the catchments associated with the lake system.

It is assumed that, for the coastal and estuarine flow, the vertical velocity component is relatively small in comparison with the horizontal components. Hence the continuity and Navier-Stokes equations can be integrated over the depth and solved numerically to give the depth averaged velocity field (Martin & McCutcheon, 1998). Based on Wang et al. (2003), estuarine and oceanic models normally assume groundwater exchanges as negligible or that the exchanges can be represented as a simple source term. The advection-diffusion module is coupled to the flow equations with an equation of state for the salt balance in the lake/estuary making the model density-dependent reflecting the real case scenario.

Figure 1 shows the methodology outline of the proposed model, which can be used to simulate the hydrodynamics, transport, and water quality of water bodies, such as estuaries and coastal seas. Time-dependent water level, flow and constituent-concentration data are required for model calibration and verification.

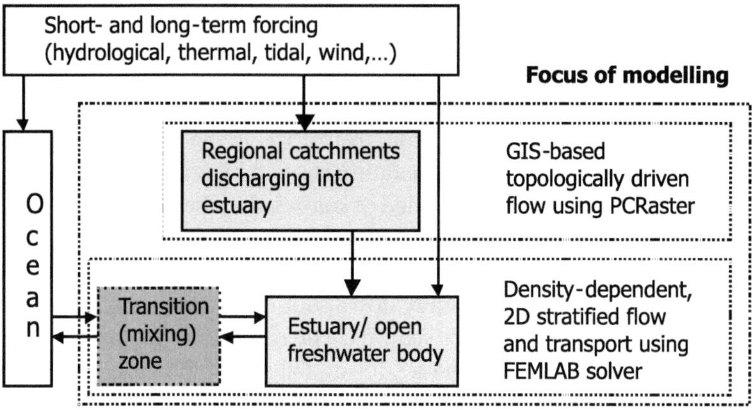

Figure 1: Schematic illustration of the proposed methodology.

The key components of our methodology may be summarised as follows:

Mixing/flow dynamics
Spatially distributed hydrologic forcing
Calibration, validation/verification by a case study

3.1 Mixing/flow Dynamics

Basic principles of dynamics in an estuary can be expressed by the equations of motion of water and the equations of continuity of water and mass of salt (Dyer, 1973). The equations to be solved are for the momentum/flow, salt concentration, and energy/temperature. All these equations can be expressed in a general balance form with different flux terms, sink/source terms and different boundary conditions. Several constitutive-type of relationships also need to be adopted. The model is designed to take care of the flow and constituent transport in vertically well-mixed estuaries, tidal embankments and inland waterways. The assumption is that velocity accelerations in the vertical direction are negligible thereby permitting a 2D numerical approximation of the governing equations in the horizontal plane.

The model was designed and developed specifically to compute time-dependent variable-density fluid flows in bodies of water whose depths, though varying, are shallow compared with their horizontal dimensions. The two-dimensional model numerically integrates the governing equations of mass and momentum conservation developed by neglecting vertical accelerations and replacing horizontal velocity components by their respective vertically averaged values. The equations of mass and momentum conservation are solved implicitly in alternating direction fashion in conjunction with vertically integrated transport equations for heat, salt, and dissolved constituents. In addition to accounting for local and convective accelerations, differential pressure gradients, horizontal density variations and forces due to bed and friction slopes, the model rigorously accounts for factors of importance in broad bodies of water such as Coriolis acceleration, wind stress effect, and horizontal momentum diffusion. Once all the coupled balance equations are formulated, they need to be solved by numerical methods. We use a recently developed, general purpose, so called "multi-physics" solver—Comsol—adapted to solving balance equations for physical systems. We use the convection-diffusion form of the mass-balance transport equation included in the FEM solver, which accounts for the generation, decay, and interaction of constituents.

The partial-differential equations used in this model to express horizontal mass and momentum conservation as defined in Leendertse (1987), are

$$\frac{\partial \zeta}{\partial t} + \frac{\partial (HU)}{\partial x} + \frac{\partial (HV)}{\partial y} = 0, \tag{1}$$

$$\frac{\partial U}{\partial t} + U\frac{\partial U}{\partial x} + V\frac{\partial U}{\partial y} - fV + g\frac{\partial \zeta}{\partial x} + \frac{gH}{2\rho}\frac{\partial \rho}{\partial x} +$$
$$gU\frac{(U^2+V^2)^{1/2}}{C^2 H} - \frac{\theta \rho_a W^2 \sin\psi}{\rho H} - k\left(\frac{\partial^2 U}{\partial x^2} + \frac{\partial^2 U}{\partial y^2}\right) = 0 \text{ and} \tag{2}$$

$$\frac{\partial V}{\partial t} + U\frac{\partial V}{\partial x} + V\frac{\partial V}{\partial y} + fU + g\frac{\partial \zeta}{\partial y} + \frac{gH}{2\rho}\frac{\partial \rho}{\partial y} +$$
$$gV\frac{(U^2+V^2)^{1/2}}{C^2 H} - \frac{\theta \rho_a W^2 \cos\psi}{\rho H} - k\left(\frac{\partial^2 V}{\partial x^2} + \frac{\partial^2 V}{\partial y^2}\right) = 0 \tag{3}$$

where C = Chezy resistance coefficient; f = Coriolis parameter; g = acceleration due to gravity; h = distance from horizontal reference plane to channel bottom (positive downward); H = temporal flow depth ($h + \zeta$); k = horizontal exchange coefficient; U = vertically averaged velocity component in x direction; V = vertically averaged velocity component in y direction; W = wind speed; ζ = water-surface elevation relative to horizontal reference plane; θ = wind stress coefficient; ρ = density of water; ρ_a = density of air; ψ = angle between wind direction and the positive y direction; and x,y,t = Cartesian coordinates and time.

In the solution, the Chezy (C) coefficient is treated as a dependent variable of the temporal flow depth and the bottom roughness, expressed by the Manning coefficient (n). The Chezy coefficient is related to the Manning coefficient according to:

$$C = \frac{\lambda}{n} H^{1/6} \tag{4}$$

The temperature and salinity are linked to fluid density, which includes dependence on salt concentration, through an equation of state; this can be expressed in different ways. Here, an equation of state for salt balance is included to account for pressure-gradient effects in the momentum equation; thus, the hydrodynamic and transport computations are directly coupled.

Boundary conditions were identified at all boundaries to permit a solution of the governing equations. This includes the forcing agents (wind, water level and salinity and temperature gradients) as well as boundary values on the turbulent kinetic energy and its dissipation, bottom friction and continuity equations. The initial boundary conditions are arrived by analyzing the data sets.

The boundary conditions specified are of three types: (1) Dirichlet—the value of the dependent variable (elevation or flux) is specified, (2) Neumann—the value of the flux is specified, and (3) Robin—a linear combination of the first two. These three types describe the physical situations of known-elevation, known-flux, or stage-discharge relations, respectively (the latter are often referred to as radiation boundary conditions). In finite element method, a Dirichlet condition means that the value of the dependent variable is known on the boundary; it is referred to as an essential condition. A specified-flux condition enters the right hand side vector in the set of discrete equations and is referred to as a natural boundary condition. At the land boundaries, no normal flow and no slip boundary conditions are used. At the open boundary sea level height is specified. Wind forcing is assumed to be spatially uniform over the entire model domain. For the momentum/flow, turbulence is typically parameterized using the eddy viscosity concept. The momentum balance equation for the flow is complemented by the conservation of mass. The bottom friction is calculated based on the bathymetry data provided in the model.

Once all the coupled balance equations are formulated, they are solved using the multi-physics solver. The mathematical link between the mixing dynamics in the estuary as formulated by the coupled balance equations, and the forcing due to seasonal and climatic variations in precipitation and temperature, is provided by the boundary conditions. The most important aspect of the forcing is the freshwater input into the estuary from the surrounding catchments, in turn controlled primarily by precipitation, temperature and topography over areas that are spatially distributed and are significantly larger than the estuary.

3.2 Spatially Distributed Hydrologic Forcing

Freshwater input into the estuary is resolved using a GIS-based, spatially distributed hydrological model. For our purpose, we use PCRaster (an open-source code, www.geog.uu.nl/pcraster) to analyze the digital maps and the precipitation and climatic values to predict the discharge from different river basins to the estuary (one example of an application of PCRaster is for the

Norrström basin, given in Darraque et al. (2005)). PCRaster uses spatio-temporal operators with intrinsic functionality especially designed for construction of spatio-temporal models. The central concept of PCRaster is spatial discretization of the landscape, resulting in 2D cells. Each cell can be regarded as a set of attributes defining its properties, but can receive and transmit information to and from neighbouring cells. This representation of the landscape is often referred to as 2.5 D: the lateral directions in a landscape are represented by a set of neighbouring cells making up a map. Relations in vertical directions, for instance between soil layers, are implemented using several attributes stored in each cell. GIS operations or operations used in modelling can be regarded as functions that induce a change in the properties of the cells on the basis of the relations within cells (between attributes on one cell location) or between cells. Key output variable for our purpose is flow rate controlled by topography, which form a drainage network ultimately discharging into the estuary. The spatial-temporal temperature variations (seasonal and due to elevation differences within the catchments) are duly accounted for.

The water model consists in determining the factors that are seen as determinant to describe the water fluxes (De Wit et al., 2000):

- The long-term average total runoff or precipitation surplus Q (mm.yr^{-1}):
  $Q = P - E_a$
  P = long term average annual runoff (mm.yr^{-1})
  E_a = long term average actual evapotranspiration (mm.yr^{-1})
- The denomination "long term" always refers to an annual average made over several years (here ten years).
- The groundwater recharge index (groundwater recharge/total runoff) Q_{gw}/Q.
  Q_{gw} = long term average total groundwater recharge (mm.yr^{-1})
- The groundwater residence time RT_{gw} (yr): average time that a water particle needs to flow from the groundwater table to the point of outflow to the surface water.

The implementation of the forcing for the estuary includes specifying freshwater discharge at given (digitalized) locations along boundaries in Comsol, as obtained from the PCRaster. The sub-domain settings in Comsol describe the physics on a model's main domain, which is divided into sub-domains. While specifying boundary conditions, Comsol can differentiate between exterior and interior boundaries: an *exterior boundary* is an outer boundary of the modelling domain, whereas an *interior boundary* is a dividing interface between two sub-domains in the modelling domain; both of these boundaries will be used in the modelling.

3.3 Calibration, Validation/Verification with a Case Study

The model developed by combining dynamics in the estuary with spatially distributed, GIS-based hydrological and thermal forcing, has to be calibrated and validated/verified with observation data from a particular area, and the stated hypotheses tested. Selecting the area for the verification of the model is very important in the study. The area selected should be prone to extreme hydrological forcing; climatic, seasonal, temperature, wind and tidal variations.

The different types of data required for model implementation and verification are bathymetry data, water levels, currents, temperature and salinity, and winds. In order to calculate the inflow of fresh water from the rivers draining into the estuary, it is necessary to get the physical/hydrological/structural features of the river basins associated with it. These include the digital maps, river discharge velocities and precipitation. The digital map sets include the drainage network, soil maps and land use map. The digital map sets were processed in a GIS tool with the precipitation and river discharge values, as a base for hydrological modelling to get the fresh water inflow from the rivers to the estuary, using PCRaster.

Bathymetry gives the depth at every node point and the bottom profile of the modelling region, which was used to produce a depth contour covering the modelling area and defining relevant layers. Continuous observations of water levels at fixed stations, representative of the entire area, were used to understand the effect of tides on the water level. Currents/Tidal data are required to know the variations with time and season. It gives the horizontal current distribution at many levels as a complement to the time series of vertical current profile between two fixed stations. Temperature and salinity measurements at regular intervals at all representative points especially at the estuarine openings, interior points where the freshwater inflow from the rivers, and at important points in the estuary necessary to understand the vertical variation in temperature and salinity at different time scales, were taken. Surface wind is another important factor, which influences the mixing in estuary that will be estimated.

Similar data sets for the previous years were used for the calibration of the model. The data selected for calibration should be complete and accurate in all respects and with homogenous conditions. Such data sets for all the representative periods should be acquired for calibration of the model. This calibrated model will be used to verify the primary data to be collected from the modelling area. The data collection is a real task where reliability, consistency and quantity have to be considered. Some of the data are available with different agencies.

4. RESULTS AND DISCUSSIONS

Prior to performing the simulation of estuarine dynamics with the integrated model, the hydrology model was independently developed and calibrated to the extent possible. After the surface-water model was developed, thus simulated runoff stages were applied as boundary conditions at the river inlet boundaries in the lake model. This stepwise approach had three advantages. First, it was relatively easy to identify and correct input and runtime errors for the individual models before they were integrated. Second, it is a tedious task to combine models run at two different backend platforms, one in GIS and the other in FEM. And third, the surface-water model gave the results in a few minutes, whereas the integrated model will require over 24 h to run.

Despite data scarcity and consistency, which is a common problem in the developing countries, the hydrology model produced reasonable simulation results at the monthly time steps with parameters calibrated against measured discharge data (Fig. 2). The model performance significantly depends on the quality of input data such as rainfall, temperature and evapotranspiration. The effects of groundwater infiltration was evaluated by performing simulations without the process and comparing results with the base case simulation, with interactions with groundwater. It compares worse with field data than the base case integrated model. The model was very sensitive to the groundwater recharge index calculated based on the land use, soil characteristics, etc. It was also observed that change in land use had significant impact on the modelled discharge at the outlet.

The model runoff results reflected the variations in the hydrological factors for all the six catchments. Figure 3 shows the variation of the rainfall and temperature of the biggest catchment, Periyar, draining into the Vembanand Lake system. The variation in the discharge to the lake system can significantly affect its dynamics including the estuary. The flushing capacity of the lake decreases with the decrease in the discharge. Salinity intrusion due to the tidal action will increase changing the ecological processes in the lake.

Only minor discharge adjustments at the watershed outlet boundaries were required as part of the calibration process. Limitations were periodically encountered using the explicit, time-lagged approach to couple the hydrology and hydrodynamic models. For some sensitivity simulations with very large input rates, convergence could not be achieved during solution of the shallow water equation. Evaluation of the convergence problems indicated that very high discharge rates caused numerical oscillations in the implicit solution. Future

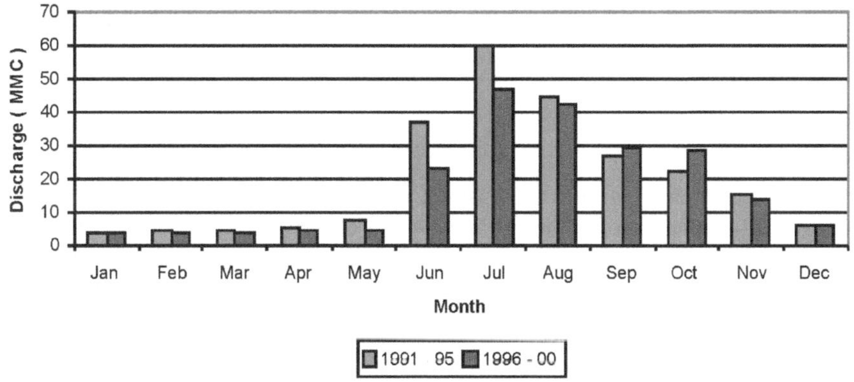

Figure 2: Calibrated model monthly discharge of Periyar River.

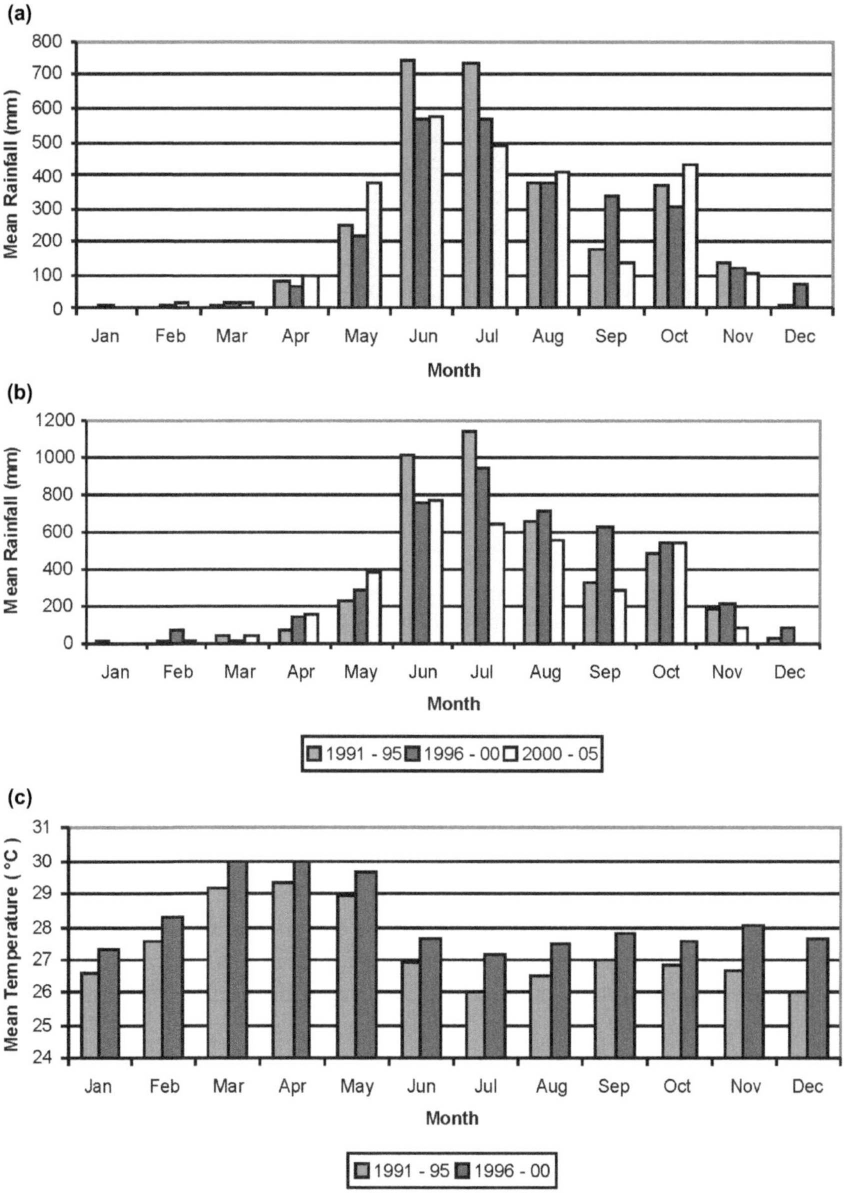

Figure 3: a) Monthly mean rainfall pattern at the highland for the Periyar catchment **b)** Monthly mean rain for the lowland region of Periyar catchment **c)** Monthly mean temperature variation over the entire catchment. (1991-95: calibration period, 1996-00: validation period, 2000-05: recent data to show the variation, but not included in the model)

efforts using the integrated model should follow the example of Fairbanks et al. (2001) and focus on determining the relation between accuracy and efficiency for different coupling approaches and time step lengths.

The freshwater flow into the Vembanad Lake occurs primarily through the wetland channels broken up from the main river channel at the wetland-watershed boundary, rather than as a single output channel for a catchment. This flow pattern was neglected in the model to reduce the high complexity of the hydrodynamic model and also due to the lack of data for the individual channels. Confidence in the predictive capability of the integrated model is due largely to the accuracy and long-term record length discharge data.

5. CONCLUSIONS

The article described a FEM approach for simulating integrated surfacewater flow and flow dynamics in estuaries. The approach combined the GIS based PC-Raster code for run-off modelling and the two-dimensional hydrodynamic flow and solute-transport model solved using the FEM solver, Comsol Multiphysics.

The integrated model was applied to the Vemband Lake system for the period 1991–2000 and to evaluate the effects of selected hydrologic and hydrodynamic processes. Field data indicate a strong seasonal pattern in lake and coastal wetland salinities, which are highly affected by the changing catchments runoff necessitating the need of an integrated hydrology-hydrodynamic model to study real time spatial and temporal scenario in the lake/estuary.

Future modifications to the water-management system in major catchments due to the proposed dam constructions will also alter the freshwater deliveries to the Lake. Based on the performance of the model to match the seasonal flushing pattern, the model should be able to predict the effects of these altered water deliveries on hydrodynamics of the lake-estuary system. For the hydrology model, groundwater infiltration was found significant when the water infiltration was switched off in the model; the calibration was achieved only when the groundwater component was brought back into the model. On the other hand, the disadvantage of this hydrodynamic model is that it doesn't take into consideration the ground-water interactions in the lake-estuary system and also the wetland system associated with it, which are important factors to be considered in a realistic system.

In general, comparisons between simulated and observed flow and mixing patterns in lake-estuary system indicate that important system processes and behaviour are represented by the model, and within the model's limitations, it could reasonably simulate the major flow characteristics to predict the effects changing hydrology and other climatic scenarios which affect the hydrodynamics of the system. The general approach described here would also be applicable to other coastal regions where restoration or contaminant transport issues are of concern. The integrated code showed numerical instability due to high sensitivity to boundary and initial conditions which could be corrected changing the boundary conditions and the physical constants.

REFERENCES

Chew, C.Y., Moore, L.W. and Smith, R.H. (1991). Hydrologic Simulation of Tennessee's North Reelfoot Creek Watershed. *Research Journal of Water Pollution Control Federation*, **63**, 10-16.

Cho, Y.-K., Park, L.-H., Cho, C., Lee, I.T., Park, K.-Y. and Oh, C.-W. (2004). Multilayer structure in the Youngsan Estuary, Korea. *Estuarine, Coastal and Shelf Science*, **61**, 325-329.

Darracq, A., Greffe, F., Hannerz, F., Destouni, G. and Cvetkovic, V. (2005). Nutrient transport scenarios in a changing Stockholm and Mälaren valley region, Sweden. *Water Science & Technology*, **51**, 31-38.

de Wit, M., Meinardi, C., Wendland, F. and Kunkel, R. (2000). Modelling water fluxes for the analysis of diffuse pollution at the river basin scale. *Hydrological Processes,* **14**, 1707-1723.

Dwivedi, D.N. and Sharma, V.K. (2005). Analysis of sea level rise and its impact on coastal wetlands of India. Proceedings of the 14th Biennial Coastal Zone Conference, New Orleans, Louisiana.

Dyer, K. (1973). Estuaries: a physical introduction. John Wiley & Sons, London.

Dyer, K.R. (1991). Circulation and mixing in stratified estuaries. *Journal of Marine Chemistry*, **32**, 111-120.

Edinger, J.E., Buchak, E.M. and Kolluru, V.S. (1998). Modeling flushing and mixing in deep estuary. *Water, Air and Soil Pollution*, **102**, 345-353.

Fairbanks, J., Panday, S. and Huyakorn, P.S. (2001). Comparisons of linked and fully coupled approaches to simulating conjunctive surface/subsurface flow and their interactions. Proceedings of MODFLOW 2001 and Other Modeling Odysseys, 356-361.

Fischer, H.B., List, E.J., Koh, R.C.Y., Imberger, J. and Brooks, N.H. (1979). Mixing in inland and coastal waters. Academic Press, Inc., New York.

Langevin, C., Swain, E. and Wolfert, M. (2005). Simulation of integrated surface-water/ground-water flow and salinity for a coastal wetland and adjacent estuary. *Journal of Hydrology*, **314**, 212-234.

Leendertse, J.J. (1987). Aspects of SIMSYS2D, a system for two-dimensional flow computation. Santa Monica, CA, Rand Corporation Report, R-3572-USGS.

Lindsay, P. (1996). Temporal and spatial variations in density and velocity fields of the water of Gareloch, Scotland. *Journal of Environmental Geology*, **28(4)**, 201-212.

MacCready, P. (1999). Estuarine Adjustment to changes in river flow and tidal mixing. *Journal of Physical Oceanography*, **29**, 708-726.

Martin, J.L. and McCutcheon, S.C. (1998). Hydrodynamics and Transport for Water Quality Modeling. Lewis Publishers, New York.

Naudin, J.J., Cauwet, F., Chretiennot-Dinet, M. -J., Deniaux, B., Devenon, J. –L. and Pauc., H. (1997). River discharge and wind influence upon particulate transfer at the land–ocean interaction: Case study of the Rhone River plume. *Estuarine, Coastal and Shelf Science*, **45**, 303-316.

Nunes Vaz, R.A., Lennon, G.W. and de Silva Samarasinghe, J.R. (1989). The negative role of turbulence in estuarine mass transport. *Estuarine, Coastal and Shelf Science*, **28**, 361-377.

Rogberg, P. and Cvetkovic, V. (2003a). New scaling law for the decay exponent of bimolecular reactions in unbounded transitional flows. *Physical Review Letters*, **90**, 028303-1.

Rogberg, P. and Cvetkovic, V. (2003b). Kolmogorov-Sinai entropy for the A+B≥P reaction in transitional flows. *Journal of Chemical Physics*, **120(14)**, 6423-6429.

Song, Y., Semazzi, F.H.M., Xie, L. and Ogallo, L.J. (2004). A coupled regional climate model for the Lake Victoria basin of East Africa. *International Journal of Climatology*, **24**, 57-75.

Souza, A.J. and Simpson, J. H. (1997). Controls on stratification in the Rhine ROFI system. *Journal of Marine Systems*, **12(1-4)**, 311-323.

Stenström, P. (2003). Exchange flows in estuaries and archipelagos. PhD Thesis (Trita - LWR PHD 1005), Department Land & Water Resources Engineering, Royal Institute of Technology, Sweden.

Svensson, J., Lindhal, S., Liungman, O. and Svensson, U. (1996). Calibration and validation of and Öresund application of the 3-D numerical model Phoenics. Swedish Meteorological and Hydrological Institute, Sweden.

Tomczak, M. (2002). Definition of estuaries; empirical estuary classification. http://www.es.flinders.edu.au/~mattom/

Turrell, W.R., Brown, J. and Simpson, J.H. (1996). Salt intrusion and secondary flow in a shallow, well-mixed estuary. *Estuarine, Coastal and Shelf Science*, **42**, 153-169.

UNEP (United Nations Environment Programme), 1989. Environmental Data Report - 2nd Edition. Oxford, UK: Basil Blackwell.

Velasco, G.G.D. and Winant, C.D. (2004). Wind and density-driven circulation in a well mixed inverse estuary. *Journal of Physical Oceanography*, **34**, 1103-1116.

Wang, J.D., Luo, J. and Ault, J.S. (2003). Flows, salinity, and some implications for larval transport in South Biscayne Bay, FL. *Bulletin of Marine Science*, **73(3)**, 695-723.

7

Creation of System "Delta-Sea" as a Basis of Ecosystem Approach to the Management of Large Aral Sea's Coastal Zone

Yusup Khai Rysbekov
Department of Management, Scientific Information Centre
Interstate Coordination Water Commission of Central Asia
11, Karasu-4, Tashkent, 100187, Republic of Uzbekistan

1. INTRODUCTION

Aral Sea region is referred to be the ecological disaster zone of global scale (Gerasimov et al., 1983; Mikhlin, 1988, 2007). Intensive irrigation development in Amu Darya and Syr Darya rivers' basins (RBs) in 1970-1980s have caused sharp decrease of sea level, and hope for water reception from outside (from Siberian rivers) was not justified. (Amu Darya and Syr Darya are two great rivers of Central Asia, which run into the Aral Sea.) In 1911-1960 average multiyear river flow of the Aral Sea Basin equalled about 117 km^3/year, including Amu Darya RB – 80 km^3, and Syr Darya RB – 37 km^3. During this period, sea received the water in volume 56 km^3/year, on average, including 42 km^3 from Amu Darya River and 14 km^3 from Syr Darya River. In the subsequent period a sharp decrease of inflow to the sea was observed. So, in 1961-1970 inflow to the sea was reduced to 30.0 km^3/year, 1971-1980 – 16.7 km^3, 1981-1990 - 3.45 km^3, 1991-1999 – 7.55 km^3/year, on average. In 1980-2001 years Amu Darya and Syr Darya rivers' flow did not reach Aral Sea. Accordingly, sea level has decreased from 53 m (beginning of 1960s) up to 41 m (1985-1986), and 30 m (2001). Therefore, since 1960 the Aral Sea coastal line has receded to 130 km. In 1980s Aral Sea was divided into the Large and

Small Aral Seas (LAS and SAS). Now the LAS was divided in two parts i.e. Western and East Aral Seas (WAS and EAS). Dynamics of the basic parameters of the Aral Sea for 1960-2000 is given in Tables 1 and 2.

According to the expert analysis, more than 70% of present Aral Sea level lowering are caused by the anthropogenic impact, and rest of these changes are implication of climatic factors (natural aridity). Formation of vast saline desert with area almost 3.6 Mio ha is major consequence of the sea shrinkage that has complicated social, political and ecological situation in this region. By estimations, damage from wetlands' degradation makes more than 144 Mio $ per year. Last year many recommendations for Amu Darya and Syr Darya rivers delta wetlands' restoration and management on preservation of SAS, WAS and EAS and their coastal zones were prepared. Now, after realization of a line of necessary measures by Government of Kazakhstan (Kokaral dam was constructed, which has separated SAS from LAS, and blocked water receipt to the EAS), experts estimate a management of SAS and Syr Darya delta as relatively satisfactory (Fletcher, 2007; etc.). At the same time, considerably more difficult situation developed in the Amu Darya river delta and concerning preservation of WAS and/or EAS.

Table 1: Dynamics of the basic parameters of the Aral Sea for 1960-1985

Year	*Water level, m*	*Water area, thousand km^2*	*Water volume, km^3*
1960	53.40	69.79	1056.1
1965	52.30	62.38	972.5
1970	51.43	58.92	941.2
1975	49.01	54.67	802.7
1980	45.75	49.21	631.8
1985	41.94	43.08	444.6

Table 2: Dynamics of the basic parameters of the Large and Small Aral Seas for 1986-2000

Year	*Large Aral Sea*			*Small Aral Sea*		
	Water level, m	*Water area, thousand km^2*	*Water volume, km^3*	*Water level, m*	*Water area, thousand km^2*	*Water volume, km^3*
1986	41.02	38.56	380.6	40.90	2.83	22.47
1990	38.24	33.67	280.4	40.50	2.75	21.84
1995	36.50	30.04	217.3	40.50	2.75	21.84
1996	35.48	28.54	195.6	40.50	2.75	21.84
1997	34.80	26.91	173.4	41.20	2.91	22.67
1998	34.21	25.75	168.4	42.50	3.24	27.03
1999	33.98	24.12	147.6	36.80	2.09	12.03
2000	33.50	22.93	139.5	39.80	2.62	19.26
2006-07	30-33	-	-	42	3.3	-

2. SOUTHERN PREARALIE: DEVELOPING SITUATION

Study area enters into sub-region known under name "Southern Prearalie", which occupies territory of Amu Darya delta on site from Mezhdurechensk reservoir in south up to coast of Aral Sea in north and from Usturt plateau in west up to Kyzylkym desert in east (Nachtnebel et al., 2007); as a whole Amu Darya delta and contiguous to WAS and EAS areas, most part of which is in Uzbekistan (Karakalpakstan and Khorezm). Till beginning of 1960s the Aral Sea was ranked as the fourth lake over the world after the Caspian Lake, the Great Lakes (North America) and Lake Victoria (Africa). In beginning of 1960s the Aral Sea had area more than 64,000 km^2 (on other data 68,000 km^2); now area of three "Aral Seas" – SAS, WAS and EAS – together makes less than 20,000 km^2. SAS has area more than 3000 km^2, WAS about 5000 km^2 and EAS 10,800 km^2 with its gulf (Nachtnebel et al., 2006, 2007; Shutter et al., 2006). As a result, a strong degradation of the Amu Darya and Syr Darya river deltas' wetlands was observed; water reduction, water mineralization increase, etc. (Bortnik et al., 1999; Ellis, 1990; Rasakov, 1996). Lakes' area in the Amu Darya delta varied from 120 km^2 (1997, wet year) up to 26.0 km^2 (2000, dry year), in the Syr Darya delta from 450 km^2 (1982) up to 262 km^2 (2000). EAS receives water from Amu Darya river after satisfying needs of its delta, which includes wetland systems and numerous lakes. In turn, WAS receives water only in case, when water is enough, and it overflows to the EAS. It is necessary to notice that Amu Darya river delta does not always receive water in quantities sufficient for its sustainable existence. So, in 2004 and 2005, there was rather a lot of water, and situation in the Amu Darya delta was estimated as relatively satisfactory. However, in 2006 situation has become difficult because of sharp reduction of river flow to the river delta. The picture of the Aral Sea's degradation is given in Fig. 1.

As a result, during last quarter of 20th century, a management of coastal zone of Sea was strongly complicated, especially in the Amu Darya delta.

Figure 1: Degradation of the Aral Sea.

Here the former ecosystems are destroyed, many lakes have dried up, on place of dried up bogs and area between delta and coastal zone of WAS and EAS salt desert was formed, flora and fauna were sharply reduced, local climate has changed (Semenov et al., 1999). In these conditions a search and development of acceptable ways for restoration and steady management of zone from delta up to the Sea is especially important.

2.1 Basic Results of Researches Carried Out and Suggestions for Aral Sea's Survival

Analysis of results of previous projects, including some projects developed in the Soviet period in 1980s (Greenberg, 2006; Nachtnebel et al., 2006, 2007; Shiklomanov, 1989; Shivareva, 1996; Shutter et al., 2006) shows that the uniform opinion on ways of WAS and EAS saving does not exists. So, in the end of 1980s, when the task of revival of Aral Sea and Southern Priaralie was put, three zones for rehabilitation were selected:

First zone: Amu Darya delta in limits from irrigation area in the south and up to former Aral Sea level in north.

Second zone: Drained sea area from its former level (53.0 m) and up to level (29.0-30.0 m), at which it is supposed to stabilize Aral Sea level in the future.

Third zone: Aral Sea itself.

Decision on problems of each abovestated zones required development and performance of the appropriate measures, for which purposes were:

For first zone: arrangement of Amu Darya delta for restoration of its historically ecological mode and creating the conditions for normal ability of the population's to live.

For second zone: Arrangement of drained Sea areas for a mitigation of negative consequences from Sea receding via creation of a combination of reservoirs and wood-protective zones.

For third zone: Preservation of bio-potentiality of the Sea zone and the negative consequences from reduction of its existing area.

However not all stipulated measures were executed. Collapse of the USSR is called by some experts as one of the reasons of such situation.

2.2 Eternal Russian Question: What to do?

Other projects also have not given a concrete answer to question on steady management of Amu Darya delta, its wetlands and coastal zone of Aral Sea on territory of Southern Priaralie. According to appropriate analysis (Nachtnebel et al., 2006; Rysbekov, 2006; Shutter et al., 2006) and other scientific sources, the suggested ways for preservation of water ecosystems in coastal zone of Aral Sea are reduced to the following basic variants (other variants are not left in this framework listed below.)

1. Reduction of total area of Aral Sea (there are variants);
2. Reduction of area of Aral Sea, but with NAS liquidation;
3. Separation of NAS and WAS;
4. Separation of NAS and WAS, at reduction of WAS area;
5. Preservation of NAS and WAS, at separating of EAS;
6. Preservation of reduced LAS and NAS;
7. Preservation of reduced LAS, but NAS is salty reservoir;
8. Preservation of reduced EAS, but WAS and NAS are salty reservoirs;
9. Preservation of reduced WAS and NAS, but EAS is salty reservoir;
10. WAS exists, EAS is completely separated;
11. EAS exists, WAS is completely separated;
12. EAS and WAS continue to exist in natural condition, without human intervention.
13. Water of Amu Darya is directed to WAS.

Analysis shows that variant 13 is most expedient, attractive and possible to realization, and this variant can be accepted for a basis for preservation of WAS. But this idea, which was offered by Academy of Sciences of Uzbekistan, is not considered by authors of the Report under the Project INTAS (Nachtnebel et al., 2006; Shutter et al., 2006) as correct. At the same time, in developing situation it is necessary to recognize and to prove the decision on saving one of Large Aral Seas (WAS or EAS), and it is possible only at connection of the Amu Darya delta with one of abovenamed seas, and creation of "Delta Sea" ecosystem (Rysbekov, 2006, 2008). In this context, creation of the "Amu Darya delta – WAS" ecosystem is more expedient. If the problem of WAS preservation by this way is not solved, water ecosystems of the Amu Darya delta will be unstable and will change their parameters depending on river flow receipt. Summarizing abovenamed theoretical ways of preservation of the Aral Sea (or its part), it is obvious that they are reduced to the following three variants concerning WAS and EAS:

1. EAS is cut, WAS exists.
2. WAS is cut, EAS exists.
3. WAS and EAS exist in natural condition (the anthropogenic intervention is absent).

There are two theoretical variants of intervention for the current situation's change to the best party, if the purpose of preservation of one of seas will be put: 1. WAS preservation; and 2. EAS preservation.

According to carried out scientific and experimental researches, there are also two hypothetical routes of artificial channel's construction to the WAS, namely:

1. Through Sudochie Lake – Southern channel, and
2. Through zone of Rybachie and Muynak Lakes – Northern channel.

Analysis shows that a Southern channel's construction is preferable from the point of view of technical practicability, smaller expenses of financial assets and future efficiency for nature protection in this area. As is marked above, now in the considered zone there are three separate ecosystems, each of which is difficult to manage:

Ecosystem 1: Amu Darya river delta.

Ecosystem 2: Dried seabed from former Aral Sea shoreline at level of 53 m to locations with isobaths at level 29-30 m, where the sea level should be stabilized in the future.

Ecosystem 3: WAS and EAS itself.

Our project idea consists in substantiation of the WAS preservation and sustainable management by artificial created ecosystem "Delta Sea", which will include also a number of wetlands and lakes in Amu Darya river delta. For this purpose it is necessary to change approaches to management of the WAS coastal zone and Amu Darya river delta zones. Management of Amu Darya delta will be more effective at its connection with WAS, which has more attractive conditions, in comparison with EAS:

1. WAS is deeper than EAS;
2. Realization of the technical measures (construction of dams, etc.) will be cheaper; and
3. Channel from Amu Darya delta will be constructed on the route, which includes Sudochie Lake, where water will flow in the WAS itself. Sudochie Lake of the Ramsar Convention is a very important monument as well.

It is supposed to determine for creation of flowing ecosystem "Delta WAS":

- Minimal river flow, which is necessary for functioning of new ecosystem. These water volumes should not be used for other needs (irrigation, etc.) in upstream located zones.
- Optimum parity of water and land areas in future delta (optimum areas of wetlands and lakes). Accordingly, it will be necessary to refuse from some wetlands (Akushpa, Tayly, etc.), which have few chances for survival, and parameters of which do not promote sustainable management by them (small area and depth, high water mineralization, etc.).
- Optimum water level in WAS, on which it is necessary to save it. It can be determined after calculation and establishment of necessary water limits for new artificial ecosystem.
- Volumes of lost biodiversity restoration (fish, muskrat, bird, other alive essences, and vegetative communities) and economic activity in the Amu Darya delta and WAS. It will depend on quantity and quality of water in the Amu Darya wetlands and WAS.
- Necessary measures on maintenance of ecological stability in the Amu Darya delta. As a rule, it is operational measures.

- Necessary volumes of cultivation of woods and other vegetation in the deserted territory, which has stayed outside direct influence of the WAS coastal zone (area, vegetative cultures' kinds, etc.) etc.

3. CONCLUSION

As it is shown above and follows from simple logic, creation of uniform flowing ecosystem "Amu Darya Delta – Western Aral Sea" is represented most acceptable. At the same time, this idea conflicts with the opinion which has strongly become in public consciousness – saving of one more of three Aral Seas is impossible. Realization presented in this article will allow solve a noble task of restoration of ecological stability in this zone. Management in this zone will remain unstable without creation of flowing ecosystem "Amu Darya Delta - Western Aral Sea". While Amu Darya delta will receive water by "residual principle", a problem of restoration of the Southern Priaralie's ecosystems cannot be decided. Now Western Aral Sea receives only collector-drainage water through Sudochie Lakes' system, but these water volumes are insignificant. At the same time, Eastern Aral Sea receives much more water than Western Aral Sea. Practically all superfluous water of Amu Darya river flow down to the Eastern Aral Sea via Mezhdurechensk reservoir and Dumalak Lakes' system.

In each separate year water surpluses from the small Aral Sea are dumped to the Western Aral Sea also. If Western Aral Sea will receive water flowing down to the Eastern Aral Sea, it is possible to create system offered in the present article. In the future it will be possible to make accounts for preservation and Eastern Aral Sea, which can become the final water receiver from other two Aral Seas – Western and Small Aral Seas. Preliminary analysis concerning development of such idea is present (Rysbekov et al., 2008). But it is already a theme of separate scientific research and analysis.

REFERENCES

Bortnik, V., Kuksa, V. and Cycarin, A. (1999). Modern condition and possible future of the Aral Sea. *News of Academy of Sciences of USSR*. Geographical series, **4**, 62-68.

Ellis, W. (1990). A Soviet Sea Lies Dying. *National Geographic* (Feb. 1990), 73-93.

Fletcher, M. (2007). The return of the Sea. *The Times*. http://www.timesonline.co.uk/tol/news/world/europe/article1975079.ece

Gerasimov, I., Kuznetsov, N., Kes, A. and Gorodetskaya, M. (1983). Problem of Aral Sea and anthropogenic desertification in Priaralie. *Desert Development Problems*, **6**, 22-33.

Greenberg, I. (2006). As a Sea Rises, So Do Hopes for Fish, Jobs and Riches. *The New York Times*. http://www.nytimes.com/2006/04/06/world/asia/06aral.html

Micklin, Ph. (1988). Dessication of the Aral Sea: A water management disaster in the Soviet Union. *Science*, **241**, 1170-1176.

Micklin, Ph. (2007). The Aral Sea Disaster. *Annual Review of Earth and Planetary Sciences*, **35(4)**, 47-72.

Nachtnebel, H., Dukhovny, V., Sorokin, A., Tuchin, Tashmukhamedov, B. and Mirabdullayev, I. (2007). The Aral Sea – Past, Present and Future. Scientific-Information Centre of Interstate Coordination Water Commission of Central Asia. Tashkent, pp. 26-37.

Nachtnebel, H., Schutter, J., Dukhovny, V., Sorokin, A., Roschenko, Y., Kadirova, R., Tuchin, A., Umarov, P., Uhalin, U., Beloglazov, A., Korshak, E., Temlyanceva, E., Tashmukhamedov, B., Mirabdullayev, I., Vasilyev, O., Kvon, V., Kuzin, V., Semchukov, A. and Martyanov, V. (2006). Report on the Project INTAS - 0511 REBASOWS: The rehabilitation of the ecosystem and bioproductivity of the Aral Sea under conditions of water scarcity. Scientific-Information Centre of Interstate Coordination Water Commission of Central Asia. Vienna/Tashkent, 75 p.

Rasakov, R. and Kosnasarov, K. (1996). Dust and salt transfer from exposed bed of the Aral Sea and measures to decrease its environmental impact. *NATO ASI*. Series 2, Environment, Springer, **12**, 95-102.

Rysbekov, Yu. (2006). Substantiation of the necessity of creation of artificial ecosystem "Amu Darya Delta – Western Aral Sea". Proceedings of the Gisig Italy Eco-Imagine International Conference. Genoa, Italy, pp. 60-63.

Rysbekov, Yu. and Tillaiev, B. (2008). About problem of the Aral Sea and possible ways of its decision in thesis form or attempt of the concrete answer on eternal Russian question. Collection of the proceedings "Problems of the Aral sea and Priaralie" Scientific-Information Centre of Interstate Coordination Water Commission of Central Asia. Tashkent, Uzbekistan, pp. 26-37.

Semenov, O. and Shapov, A. (1990). Sand-salt storms in Priaralie. Hydrometeorological problems of Priaralie. Hydrometeoizdat, Leningrad, pp. 132-216.

Shiklomanov, I. (1989). Influence of economic activity on river flow. Hydrometeoizdat, Leningrad, p. 334.

Shivareva, S. and Smerdov, B. (1996). About modeling of the Aral Sea levels and its parts on prospect. *Hydrometeorology and Ecology*, **3**, 96-115.

Shutter, J. and Dukhovny, V. (2006). Integrated Water Resources Management in the Aral Sea Basin with the Purpose of Filling of the Southern Prearalie's water surfaces. Scientific-Information Centre of ICWC of Central Asia. Tashkent, Uzbekistan, 84 p.

SECTION 2

Coastal Zone Water Resources (Quantity and Quality): Challenges for Sustainability

8

Critical Evaluation of the Recent Development and Trends in Submarine Groundwater Discharge Research in Asia

Manish Kumar*, AL. Ramanathan[1], B.R. Neupane[2], Dinh Van Tu and Seok-hwi Kim

Dept. of Environmental Engineering, Kunsan National University, Korea
[1]School of Environmental Science, Jawaharlal Nehru University
New Delhi, India
[2]UNESCO, New Delhi, India

1. INTRODUCTION

In majority of the arid and dry regions of the world and in monsoon-dependent countries in Asia, groundwater is a major freshwater resource for drinking and other uses. The coastal regions, which support maximum density of population, mainly depend on the ground water. Due to the population growth and the fact that about 50% of the world population now already live in coastal regions, the groundwater issues in the coastal areas are increasingly becoming crucial (UNWWDR, 2009). Over-exploitation of groundwater in these areas can potentially lead to saltwater intrusion, land subsidence, permanent damage to the ability of an aquifer to store and transmit water, and reduced discharges to rivers, streams, and critical aquatic habitat areas (Fig. 1). Further, coastal groundwater plays an important role in nutrient flux to the ocean. The influence of submarine groundwater discharge (SGD) on the coastal water quality, their biogeochemical process and their ecology are very significant in most of the coastal regions. Investigations of interactions between groundwater and coastal

*Corresponding Author

seawater have been restricted mainly to the case of water movement from sea to the land, i.e. saltwater intrusion (Segol and Pinder, 1976; Reilly and Goodman, 1987) while submarine groundwater discharge (SGD) considers the water output from a basin-scale hydrological cycle, representing an input into the ocean (Fig. 2).

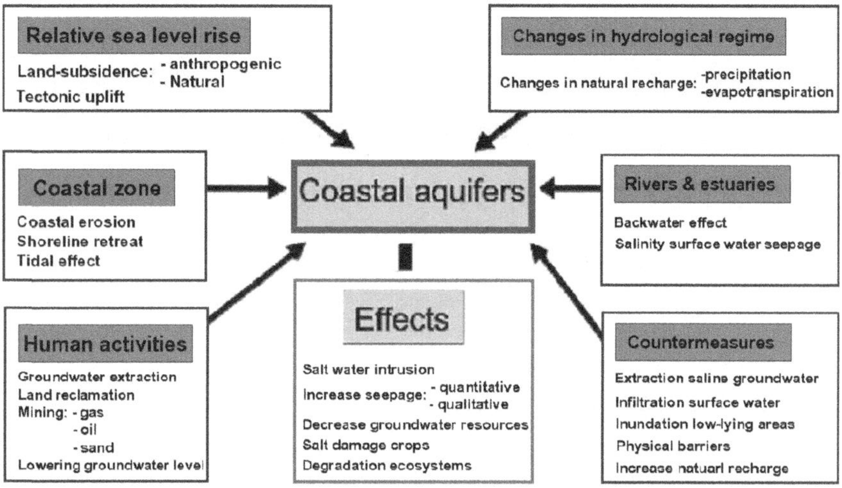

Figure 1: A compilation of processes affecting coastal groundwater.

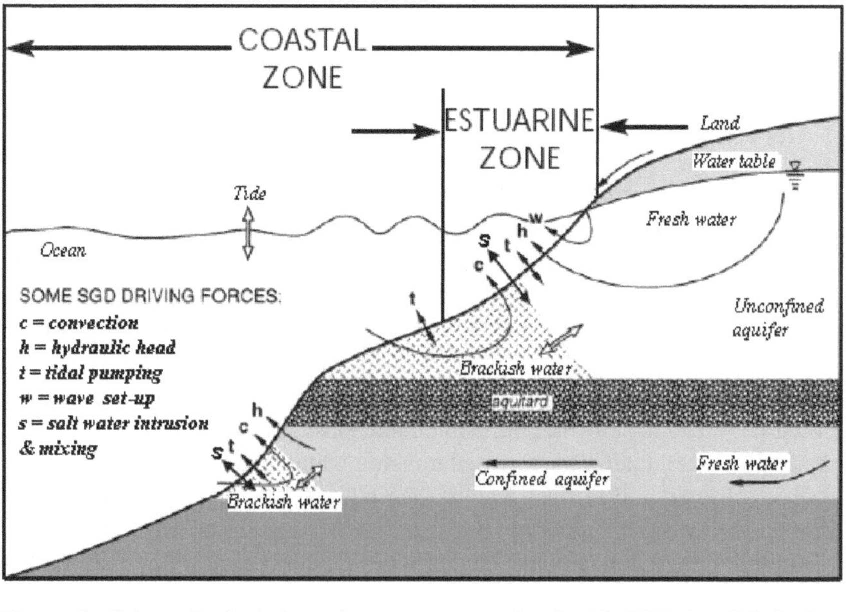

Figure 2: Schematic depiction of processes associated with SGD (modified after Taniguchi et al., 2002). Arrows indicate fluid movement.

Discharge of groundwater from the seafloor is known at least since the days of the Romans. Sonrel (1868) reported the discharge of freshwater from submarine springs and speculated on their use and risks for sailors. Since then several studies have considered the importance of fluid discharge from sediments for nutrient budgets of coastal environments, formation of offshore plankton blooms, hydrological cycles, or the release of trace elements and gases such as radon from the seafloor (Kohout, 1966; Zektser et al., 1973; Dzhamalov et al., 1977; Johannes, 1980; Cathles et al., 1987; Valiela et al., 1990; Zektser and Loaiciga, 1993). A peak in awareness of the global importance of SGD was noticed after 1996 paper by W.S. Moore in *Nature*, and much effort has subsequently been devoted to develop new tracer techniques and methods for identifying and quantifying submarine groundwater discharge (Moore, 1996; Monastersky, 1996; Cable et al., 1997; Burnett, 1999; Wilson, 2005). The study of SGD is valuable for understanding the availability and quality of coastal groundwater and its control on many coastal processes that span the disciplines of geology, geomorphology, geochemistry, biology, hydrology, and ecology. Specific examples of research areas in SGD studies can help to solve interdisciplinary problems which include: (i) assessment of the redox-controlled and microbially controlled delivery of nutrients and trace elements in SGD (linking the fields of geochemistry and coastal hydrogeology); (ii) evaluation of coastal-ecosystem change in response to variations in SGD (linking the fields of biology, ecology, and biogeochemistry to coastal hydrogeology and meteorology); (iii) examination of shore-faces breached by submarine groundwater discharge and the influence that SGD associated with paleo-channels has on erosional hotspots (linking the fields of coastal geology and geomorphology to coastal hydrology); and (iv) developing the capability to forecast processes and events associated with SGD (for example, to predict lag times between the initiation of wastewater discharge into coastal aquifers and the onset of impacts on coastal ecosystems by nutrients derived from SGD).

River discharge is a major pathway for discharge from land to the ocean in the global water cycle. Rivers are highly visible open channels that discharge water from great distances inland, and their contribution to the oceans are easily quantifiable. SGD per unit length of coastline could be very significant as a discharge process, because the length of coastline where SGD occurs is very great, and will occur whether or not rivers are present. Global estimates of SGD vary widely (approximately 0.2–10% of river flow); (Garrels and MacKenzie, 1971; COSOD II, 1987). Geochemical cycles of some major and minor elements (such as nitrate, phosphate) may be strongly influenced either by the direct discharge of fresh groundwater into the sea or by chemical reactions that occur during the recirculation of seawater through a shallow coastal aquifer system (COSOD II, 1987; Buddemeier, 1996). Although it is difficult to detect groundwater seepage through sediments, this diffusive input may occur over broad areas and deliver potentially significant amounts of fresh water and dissolved components to the world's coastal oceans. Thus, in spite of the recognition that many land-sea interfaces of the world are characterized by "leaky continental margins" (IOC-INF, 2000), it is still not known exactly how important SGD is in terms for overall marine hydro-geochemical budget of major ions and nutrients.

In the last two decades, SGD has emerged as a potentially significant pathway of dissolved substances and diffuse pollution into the coastal zone, at least in some regions of the world (Johannes, 1980; Burnett et al., 2001; Monastersky, 1996; Burnett, 1999). Recent research has focussed on characterizing, differentiating (terrestrial versus recirculated sources), and quantifying sources of SGD to the coastal ocean (Cable and Martin, 2008; Martin et al., 2006, 2007; McCoy et al., 2007a; Thompson et al., 2007; Wilson, 2005). Studies conducted in several regions of the world suggest that the terrestrially derived, freshwater component of SGD may or may not be a significant component of total SGD; however, geochemical changes associated with seawater (and freshwater) as it cycles through pore spaces in coastal aquifers and back across the sediment-water interface may supply appreciable geochemical fluxes (Lautier, 1998; Cable and Martin, 2008). Although, reports concerning fluid discharge in coastal areas, e.g., off Italy, Greece, Crete, Japan, Israel, Lebanon, Florida, and the Baltic proper, underlined the worldwide occurrence of this transport pathway from land to the ocean (Kohout, 1966; Zektser, 1996; Taniguchi et al., 2002), the current status of SGD research in the most populated continent i.e. the Asia is still ambiguous. Therefore, the present article first aims at providing a crisp review on the forces behind SGD process and the importance of coastal zone management, followed by critical evaluation of current status of SGD research in the Asian perspective. The Asian perspective has been dealt specifically by comparing the research carried out in Japan, Korea, China, Bangladesh and India.

2. PHENOMENOLOGICAL DESCRIPTION AND DRIVERS

Submarine groundwater discharge (SGD) is a process complementary to saltwater intrusion, which is defined as the discharging flow out of the aquifer across the sea floor. If mixing between fresh groundwater and seawater wedge is weak, it can be assumed that the interface between the two areas is sharp (Fig. 2). In such cases, during the winter period, SGD consists of fresh groundwater, which flows to the sea above the saltwater wedge, and the interface is located more landward than in summer. This is due to the relatively low groundwater levels occurring in fall. In summer, the interface moves seaward due to the rise of the groundwater level occurring in spring, while, simultaneously to fresh groundwater discharge, seawater outflow takes place from the area of saltwater wedge. The seawater outflow can be greater than the fresh groundwater discharge (Michael et al., 2005). In most cases, the interface between the freshwater and the saltwater is not sharp. Due to dispersion and molecular diffusion a freshwater-seawater mixing zone is built between the two areas (Fig. 2). The mixing causes the salt water in the deeper part of the aquifer to become lighter, rise up and discharge back to the sea. In this case, SGD consists of a mixture of fresh groundwater and re-circulated seawater. The re-circulation of seawater, in the near shore part of the aquifer, is enhanced by further mechanisms acting from the marine side such as tidal pumping and wave set-up. The variation of SGD in time exhibits components of small time scale (minutes, hours, days), which are due to the wave action and tides, as well as components of large time scale, which are due to the seasonal moving of the mixing zone and large-scale sea level variations (Fig. 2).

Concerning the contribution of the fresh groundwater flow and the seawater re-circulation to the total SGD, it is known that: (a) the terrestrially-derived SGD ranges from 6 to 10% of the surface waters discharging into the ocean (Burnett et al., 2003); (b) in cases that re-circulation is only due to density effects and mixing, recirculated seawater can be up to 70% of the SGD (Smith, 2004; Kaleris, 2006); and (c) in cases that re-circulation is due to wave setup and tide, the recirculated sea water may constitute up to 96% of the SGD (Li et al., 1999). SGD can take place not only near the shore but also at great distances. The latter occurs when confined aquifers, underneath the continental shelf, have their outcrop on the sea bottom and a large distance from the shore. SGD is usually distributed over large areas. Concentrated outflows at the sea bottom occur in cases where the outcrops of alluvial aquifers are of limited area. Such outflow locations on the sea bottom sometimes form pockmarks in the fine-grained sediments of the sea floor. Particularly strong, concentrated, SGD occur in case of karstic aquifers discharging into the sea (karstic submarine springs). Considering the different regional settings, different modes of fluid flow from sediments can be distinguished: (1) focussed flow along fractures in karst and rocky areas, (2) dispersed flow through soft sediments, and (3) recirculation of seawater through sediments. The compositions of fluids range from nearly pure freshwater, as is the case in karst areas, to the seepage of saline water, as reported for regions off Florida (Cable et al., 1997). From the perspective of a sustainable water management, the groundwater renewal rate is crucial number. It indicates how much water can be recovered from the aquifer, avoiding over pumping and decrease of the groundwater table.

3. ESTIMATION OF THE MAGNITUDE OF SGD

In addition to a large number of individual studies addressing the problem of SGD estimation, an initiative was launched by the International Atomic Energy Agency (IAEA) and UNESCO in 2000 to assess the importance of SGD and the effectiveness of prediction methodologies for coastal zone management. The results of the projects undertaken in the framework of this initiative have been presented in SCOR-LOICZ (2004) as well as in Burnett et al. (2006). A further comprehensive report concerning current knowledge about SGD has been presented by Gallardo and Marui (2006). One of the outcomes of this recent interest in SGD has been the establishment of a working group of scientists to define more accurately and completely how submarine groundwater discharge influences chemical and biological processes in the coastal ocean (Burnett, 1999; Kontar and Zektser, 1999). This group (SCOR Working Group 112, 'Magnitude of Submarine Groundwater Discharge and Its Influence on Coastal Oceanographic Processes') is co-sponsored by the Scientific Committee on Oceanic Research (SCOR) and the Land-Ocean Interaction in the Coastal Zone (LOICZ) project, a core activity of the International Geosphere Biosphere Program (IGBP). As a consequence, SGD assessment inter-comparison exercises have been organized at several 'flagship' coastal sites. These experiments, which are currently in progress, are being performed in order to compare directly several independent methodologies to evaluate the groundwater discharge. The methods to estimate SGD can be broadly distinguished in various ways: (a) Investigations from the land side, (b) Investigations from the sea side and (c) Indirect indicators.

3.1 Investigations from the Land Side

This group involves the hydrogeological methods, with the most important being the water balance approach and numerical models. In water balance approach, the estimation of SGD is based on the water balance equation at the basin scale. For extended periods (i.e. years), over which the change of water storage is negligible, precipitation, evapo-transpiration and surface discharge must be accurately known in order to reliably estimate SGD. The water balance approach only provides the fresh groundwater component of SGD as the total volume of groundwater discharging to the sea during the selected time period. Thus, the results of the water balance approach are not comparable with locally measured values or with results of numerical models, which provide the distribution of SGD over the discharge area. An example of the water balance approach is the work of Sekuliè and Vertaènik (1996), who estimated SGD in the Adriatic Sea.

Numerical models used in SGD studies are of different complexity. Groundwater models neglecting density effects, as for instance MODFLOW (McDonalt and Harbaugh., 1988), have been widely used. They provide only the freshwater component of SGD and can appropriately simulate the flow in the part of the coastal aquifer upstream of the saltwater intrusion area. SGD in such simulations is the groundwater flux at the seaward model boundary, provided that there is no groundwater extraction in the area downstream of this boundary up to the coast. The collection of the data needed for the model calibration requires a considerable effort and it is time consuming particularly for the parameters varying in time such as water levels and fluxes. However, in areas in which such data are available, numerical models represent an efficient tool for the estimation of the freshwater part of SGD.

In the last decades numerical models considering density effects (an extensive review of such models has been presented in Bear et al., 1999) have been applied in SGD studies. These models include the near-shore part of the aquifer as well as parts of it lying below the sea bottom; where density effects are relevant, provided both the fresh groundwater discharge as well as the discharge of the seawater is re-circulated in the aquifer. Their application is reasonable only if data is available for the near-shore part of the aquifer. However, particularly for the part of the aquifer below the sea floor, the collection of such data is difficult. Therefore applications of such models, as those presented by Smith and Turner (2001), Kaleris (2002), Langevin (2003) and Destouni and Prieto (2003), are limited.

3.2 Investigations from the Sea Side

The most important methods for investigations of SGD from the sea side are seepage meter measuring techniques and the use of geochemical tracers such as radium isotopes and radon. Both tracers are enriched in groundwater relative to seawater. An overview of different types of seepage meters used to measure SGD at the bottom of the sea, as well as conclusions concerning their applicability in the field, have been presented by Taniguchi et al. (2003). Seepage meters provide local SGD values because of the variability of the hydraulic properties of the sea bottom material and the relatively small area (usually $\ll 1$ m^2), over which SGD is measured. Taking into consideration the variability of SGD in time, it can be concluded that in order to obtain representative values of SGD for a shore, which can be a few kilometres long, a large number of measurements at different locations and in different times must be performed. The results of seepage meter applications and comparisons with other methods have been presented by Burnett et al. (2006).

Radon is produced by the decay of radioactive isotopes in the sediments. Therefore, it is strongly enriched both in the freshwater part as well as in the re-circulated part of SGD relative to coastal marine waters (Oberdorfer, 2003). Thus, these two parts of SGD can hardly be distinguished if its value is estimated from radon measurements. To what extent a near-shore water sample, used to measure radon activity, provides representative values for a larger part of the near-shore water body, depends on the degree of mixing of SGD with the surface water. The case study on Ubatuba coast by Povinec et al. (2008) also has focussed on the radioactive (^{3}H, ^{222}Rn, ^{223}Ra, ^{224}Ra, ^{226}Ra, and ^{228}Ra) and stable (D and ^{18}O) isotopes measurements in seawater, direct seepage measurements using manual and automated seepage meters, pore water investigations using different tracers and piezometric techniques, and geoelectric surveys probing the coast.

Radium is absorbed onto the surface of the geologic materials and is immobile in fresh groundwater. When saltwater intrudes the aquifer and replaces freshwater in the pores, radium is desorbed from sediment and becomes mobile

(Moore, 1996). Thus, radium is transferred to the near-shore seawater primarily by the re-circulated seawater actively involved in saltwater intrusion (Oberdorfer, 2003). Since this water discharges into the sea mixed with the fresh groundwater, it is not possible, as in case of radon, to distinguish the portions of fresh groundwater and seawater in SGD.

An overview of the technologies developed in the last years for the measurement of radon and radium activities, advantages and disadvantages of the methods as well as applications for the estimation of SGD, have been presented by Burnett et al. (2006). Conceptual model of use of continuous radon measurements for estimating submarine groundwater discharge in a coastal zone (Fig. 3) was well established by Burnett and Dulaiova, 2003.

3.3 Indirect Indicators of SGD

The application of the methods described above, involves considerable financial, labour and/or computational effort. Thus, before such methods are applied in coastal zone management studies, it is important to determine if the magnitude of SGD is relevant. Results of investigations presented in the literature can hardly help in such assessments as they vary widely. High values (above 100 cm/day) as well as low values (below 5 cm/day) have been observed (Burnett et al., 2006). A way to determine the importance of SGD in an area, is to use indirect indicators (SCOR-LOIZ, 2004), for example: (i) the groundwater escaping from the sea bottom might be coloured by tiny gas bubbles; (ii) surrounding sediment might be stained red by the oxidation of iron; (iii) there

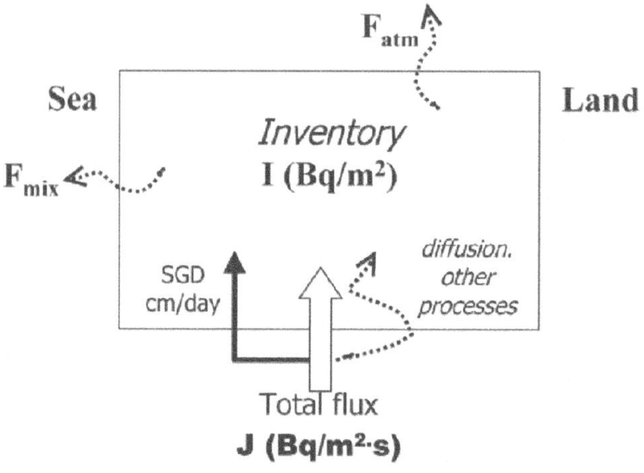

Figure 3: Conceptual model of use of continuous radon measurements for estimating submarine groundwater discharge in a coastal zone. The inventory refers to the total amount of excess ^{222}Rn per unit area. Decay is not considered because the fluxes are evaluated on a very short (1–2 hours) time scale relative to the half-life of ^{222}Rn (Reproduced from Burnett and Dulaiova, 2003).

are cold-water anomalies in the open water during the summer and warm-water anomalies in winter as well as salinity anomalies; and (iv) the levels of radium, radon, methane, hydrogen sulphide or carbon dioxide might be elevated.

3.4 Mixing Near the Sea Shore

The concentration of pollutants, transferred in the near-shore sea area by SGD, depends on the intensity of mixing in this area, which is due to hydrodynamic circulation, transport and dispersion. The estimation of the pollutant concentration in the near-shore sea using numerical models, which simulate the aforementioned processes, is computationally intensive and requires data, which in most cases is not available. A more convenient approach is to use cell models (well mixed reservoir models).

An application of a cell model is given by Uchiyama et al. (2000), who investigated the nutrient concentration in the near-shore sea due to SGD. They found that the long-term pollutant concentration depends on the relationship between the pollutant fluxes due to SGD and the exchange flux between the near-shore sea and the open ocean. For the conditions of the investigated case, these authors found that the volumetric exchange flux between the near-shore sea and the open ocean is three orders of magnitude larger than the SGD flux. Thus, the long term pollutant concentrations, resulting under the assumption that the pollution of the open ocean water is negligible, were found to be of minor importance for the near-shore sea environment.

For estimations of the exchange of water between the near-shore sea (surf zone) and the offshore zone, a relatively simple model presented by Inman et al. (1971) can be used. The data required are the wave and beach characteristics (breaker height, breaker angle, beach slope in the surf zone, beach slope offshore). For open beaches, typical wave and beach characteristics (Inman et al., 1971), it is probable that the water exchange between the surf zone and the offshore strongly dilutes pollutants transported in the surf zone by SGD, provided that SGD is of the order of magnitude determined in the existing studies, which are reported in Burnett et al. (2006) and Gallardo & Marui (2006). However, in cases of closed bays or near-shore seas with weak wave action and circulation, it can be shown using cell models, that the long-term pollutant concentration in the near-shore environment, which is due to the impact of SGD, is not negligible. In such cases managerial measures, which could reduce the pollutant load of groundwater, are required.

4. IMPORTANCE OF SGD IN ASIAN OUTLOOK

SGD has been recognized in the last decades as a mechanism for transporting land-derived pollutants to the sea (Moore, 1996). Nutrients, organics, metals and pathogens, which are dissolved in terrestrial groundwater, can induce chemical and biological effects in the near-shore sea area, where groundwater discharges. These effects can be enhanced by the re-circulated seawater as it reacts with the aquifer sediment and can be enriched by the pollutants. The pollutant concentration in the near-shore water and its impact on the chemistry and biology of this area depends not only on the fluxes but on the intensity of mixing and the exchange with the open ocean as well. Dissolved chemical discharge into the sea with groundwater depends primarily on submarine groundwater runoff (Dzhamalov and Safronova, 2000). The distribution of groundwater total dissolved solids (TDS) over the watershed areas involved in the calculation showed that it varied mostly from 0.2 to 0.8 g/l (Dzhamalov and Safronova, 2000). Especially, in Africa, Australia and Asia continents, SGD is characterized by the high total TDS and accordingly high concentrations of chemical elements (Ca, Mg, Na+K, HCO_3, SO_4, and Cl) (Table 1). Total subsurface dissolved solids discharge and export of individual macrocomponents (Ca, Mg, Na+K, HCO_3, SO_4, and Cl) into the sea is estimated to be the highest (~290 million ton/year) in Asia (Fig. 4). Considering the total submarine water and dissolved solids discharge into the seas and oceans, the average TDS of groundwater discharging into the sea from the top hydrodynamic zone was found around 0.6 g/l (Zektser et al., 1984). It should be emphasized that the calculated value of TDS (0.6 g/l) is in good agreement with the measurements made in the top hydrodynamic zone (Table 1). This allows this value of TDS to be used in estimating the overall characteristics of chemical element discharge into the sea with groundwater.

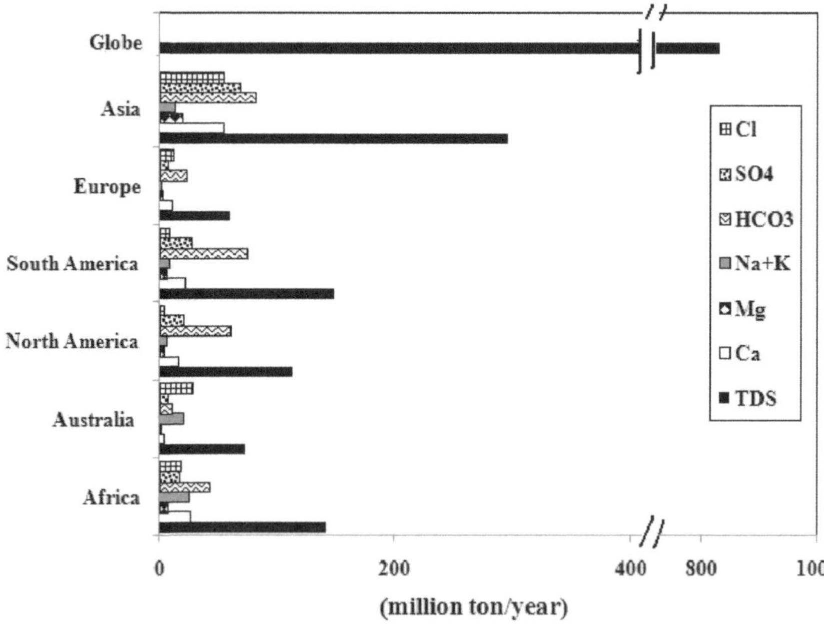

Figure 4: A continent-wise comparison of total subsurface dissolved solids discharge and export of individual macrocomponents (million t/yr) (modified from Dzhamalov and Safronova, 2000).

Table 1: Macrocomponent concentrations (mg/l), and TDS (g/l) of submarine groundwater (reproduced from Dzhamalov and Safronova, 2000)

Continent	*TDS*	*Ca*	*Mg*	*Na+K*	*HCO_3*	*SO_4*	*Cl*
Africa	0.2-10	40-420	10-120	30-4200	100-900	20-1200	20-4000
Australia	0.4-12	40-360	20-120	80-4200	200-750	30-1500	100-5000
North America	0.2-1.2	40-120	10-40	20-350	160-450	30-150	30-400
South America	0.2-0.9	45-170	12-60	10-45	140-250	35-250	15-170
Europe	0.1-1.2	20-120	10-40	20-350	90-450	20-150	10-400
Asia	0.4-10	40-260	20-120	60-3200	200-750	80-1500	60-4000
Global (mean)	0.6						

Thus, coastal zone managers, in order to determine whether SGD in an area of interest is of actual or probable importance, need to estimate: (a) the magnitude of SGD and (b) the intensity of mixing in the near shore sea area and the exchange between the near-shore sea and the open ocean.

5. CURRENT STATUS OF SGD RESEARCH IN ASIA

Locations of specific SGD estimates showed that many independent studies have been performed on the east coast of the United States, Europe and Oceania (Burnett et al., 2002; Taniguchi et al., 2002). Fewer studies have been done on the west coast of the US, South America, and Hawaii. However, in Asia SGD research, led by Japan and Korea, is still in its juvenile stage as evident from the lack of any quantitative data from Vietnam, Cambodia, Laos, or China, though indications of groundwater discharge have been reported for India, Thailand and Bangladesh. Here, we outline some of the research work carried out in Asian countries to evaluate the status of SGD research and development in the continent.

5.1 Japan

As mentioned before, the SGD research in Asia is mainly led by two coastal countries Japan and Korea. Japanese researchers start reporting their research findings in the last decade of 20th century while estimating nutrient transport by groundwater to near-shore zones. Taniguchi et al. (1999) analyzed the groundwater seepage rate into Lake Biwa, Japan, to evaluate the capture zone of groundwater entering a surface water body. Transient numerical simulations were made using a two-dimensional (2-D) unsaturated-saturated model with three-layered sediments. They concluded that calculated values agreed well with observed groundwater seepage rates when the thickness of the aquifer was estimated to be 110 m. This model also agreed with the capture zone results estimated by stable isotope data ($\delta^{18}O$ and deuterium). The results of this study showed that aquifer thickness and hydraulic conductivity values were the most important factors for reliable estimates of groundwater seepage rates by theoretical and numerical analysis.

Taniguchi et al. (2002) continuously recorded seepage flux rates in Osaka Bay, Japan, from May to August 2001 and analyzed these data via the Fast Fourier Transfer (FFT) method to discern the dominant periods of variation (Fig. 5). The study showed that there is both a semi-diurnal to diurnal tidal relationship to SGD and a semi-monthly variation in flow reflecting the neap-spring lunar cycle. Superimposed on this predictable behaviour in tidally driven response, are variations in terrestrial hydrologic parameters (water table height, etc.). This result demonstrates the overlapping nature between terrestrial and marine SGD forcing components.

Submarine groundwater discharge (SGD) rates were measured continuously by automated seepage meters to evaluate the process of groundwater discharge to the ocean in the coastal zone of Suruga Bay, Japan (Taniguchi et al., 2005). The ratio of terrestrial fresh SGD to total SGD was estimated to be at most 9% by continuous measurements of electrical conductivity of SGD. Semidiurnal changes of SGD due to tidal effects and an inverse relation between SGD and barometric pressure were observed. Power spectrum density analyses of SGD, sea level, and groundwater level show that SGD near shore correlated to groundwater level changes and SGD offshore correlated to sea level changes. SGD rates near the mouth of the Abe River are smaller than those elsewhere, possibly showing the effect of the river on SGD. The ratio of terrestrial groundwater discharge to the total discharge to the ocean was estimated to be 14.7% using a water balance method.

Taniguchi et al. (2007) have noted that a semi-diurnal variation of SGD with average conductivity is 38.8 mS/cm, the highest conductivity of SGD

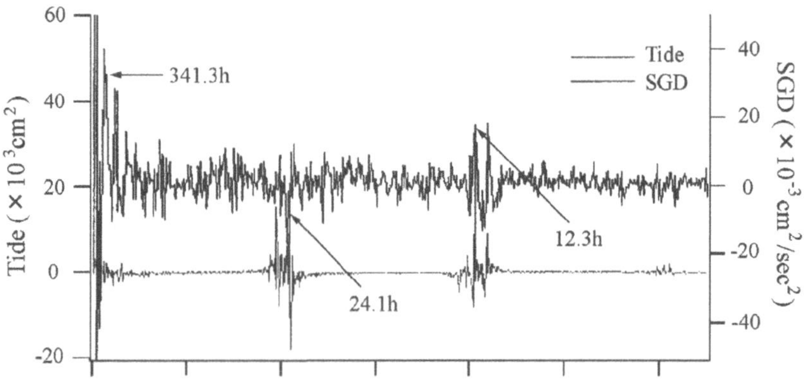

Figure 5: Time series analysis (FFT method) of long-term SGD measurements and tides in Osaka Bay, Japan, from May 29 to August 23, 2001. The main SGD frequencies correspond to semi-diurnal (12.3 h), diurnal (24.1 h), and bi-weekly (341.3 h) lunar cycles (reproduced from Taniguchi, 2002).

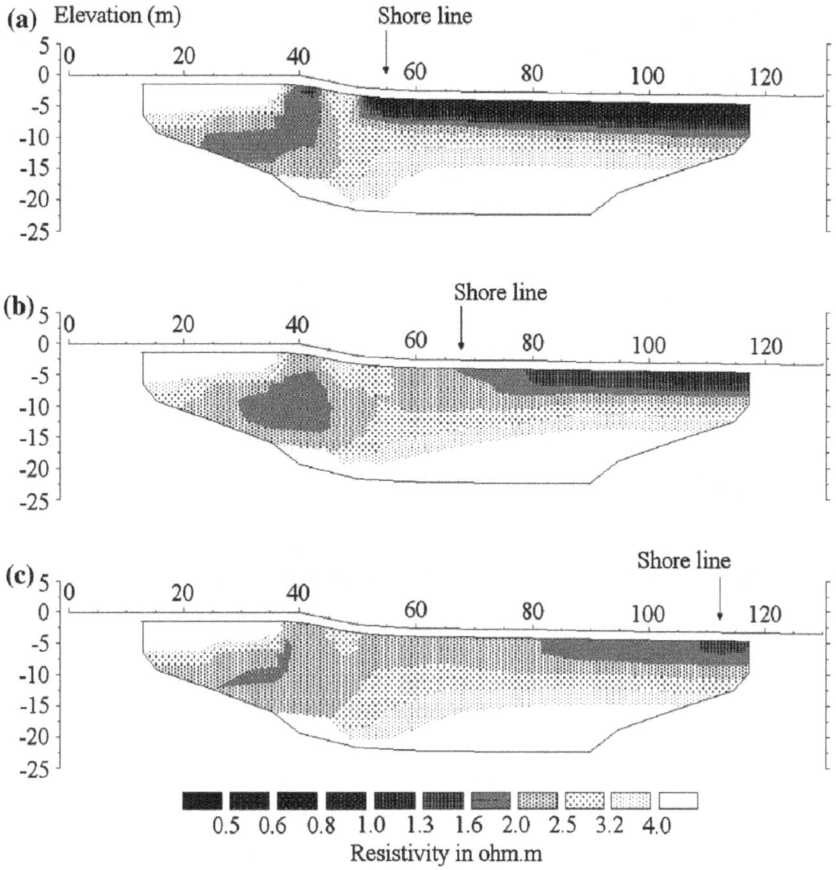

Figure 6: Temporal variations of resistivity of pore water along the coast of Sriracha at (a) high tide (09:45), (b) intermediate tide (15:45), and (c) low tide (23:45) on January 27, 2004. Darker shades indicate saltier water (low resistivity), while lighter hues denote fresher, high-resistivity solutions (reproduced from Taniguchi et al., 2007).

was found at low-high tides. Figure 6 shows high conductivity at high SGD, and diurnal changes of the electric conductivity of SGD with ranges of variations of EC of 0.5 mS/cm. The conductivity of ambient seawater and freshwater in the bore was 51.6 mS/cm and 0.066 mS/cm, respectively. Besides, borehole temperature data near the coast was also used for estimation of SGD into Tokyo Bay, Japan (Taniguchi et al., 1998) and a saltwater-freshwater interface in Toyama Bay, Japan (Taniguchi, 2000).

5.2 Korea

In Korea, SGD research found momentum by the beginning of 21st century when Kim and Hwang, (2002) have reported longer-term (weeks to months) tidally modulated cycles in seepage based on continuous measurements of the groundwater tracers radon, methane and automated seepage meter observations. The study showed that there is not only a semi-diurnal to diurnal tidal relationship to SGD but also a semi-monthly variation in flow reflecting the neap-spring lunar cycle. Superimposed on this predictable behaviour in tidally driven response, are variations in terrestrial hydrologic parameters (water table height, etc.). This terrestrial influence has been demonstrated by groundwater discharge which was more limited in the dry season when the aquifer was not recharging.

Kim et al. (2005) have estimated the magnitude of submarine groundwater discharge (SGD) into the Yellow Sea, which is one of the largest continental shelves in the world, using ^{226}Ra and ^{228}Ra isotopes. On the basis of ^{228}Ra and ^{226}Ra mass-balance models, they estimated the advective flux of ^{226}Ra through SGD to be ~270 × 10^{12} dpm yr^{-1} in the Yellow Sea. Using this ^{226}Ra flux and the measured ^{226}Ra activity in coastal groundwater, the submarine discharge of groundwater (mostly brackish groundwater) was calculated to be at least 40% of the river-water input (~2.3 × 10^{11} m^3 yr^{-1}). Then, on the basis of the SGD and the concentration of Si in coastal groundwater, they estimated the flux of Si through SGD to be 20–100% of that associated with river discharge (~23 × 10^9 mol yr^{-1}). This large SGD is likely to be due to the high tidal range (up to 10 m) and the wide distribution of sandy sediments in the Yellow Sea, favourable for the recirculation of seawater through bottom sediments and rocks. This result from such a large area implies that the Si flux through SGD may be significant on a global scale.

Hwang et al. (2005) have measured the concentrations of nutrients and radium isotopes (^{223}Ra, ^{224}Ra, and ^{226}Ra) in surface seawater and coastal groundwater in Yeoja Bay (in the southern sea of Korea) to estimate submarine groundwater discharge (SGD) and associated nutrient fluxes. In general, the radium and nutrient concentrations in brackish groundwater were an order of magnitude higher than those in ambient bay water or stream water. They determined the water residence time and SGD in the bay using the simultaneous equations for ^{226}Ra, ^{223}Ra, and Si mass balances. The mean residence time of bay water was about seven days in Yeoja Bay. The inputs of submarine groundwater to the surface layer (0–3 m) were estimated to be approximately 2.6 × 10^7 m^3 day^{-1}. The nutrient fluxes driven by SGD were approximately 26, 0.11, and 26 mmol m^{-2} day^{-1} for dissolved inorganic nitrogen (DIN), phosphorus (DIP), and silicate (DISi), respectively. These fluxes of nutrients through SGD were much higher than those through stream flow and/or diffusion from bottom sediments in this bay. This excess nutrient input from coastal groundwater is the most likely cause of harmful algae blooms occurring in the open sea areas out from the bay.

5.3 India

Peninsular India, with a coastline of about 7000 km, faces water scarcity in many coastal regions mainly due to the quick surface runoff through the rivers and streams during the monsoon seasons. Another possible cause may be the significant water loss in the coastal regions as SGD which has not attracted much attention of the scientific community until very recently. In the Indian scenario, most of the studies conducted in the coastal regions are concentrated on water quality aspects including seawater intrusion in aquifers. A few studies were conducted in the Bengal basin using barium, ^{226}Ra and ^{87}Sr/^{86}Sr indicate substantial groundwater discharge into the Bay of Bengal through the Ganga-Brahmaputra river systems (Moore 1997; Basu et al., 2001). Very recently a hydrogeological and groundwater modelling study has been conducted in the shallow aquifers of Kerala to identify the existence of SGD and to estimate its rate in the coastal waters of Vizhinjam, Thiruvananthapuram, by the in situ monitoring of dissolved ^{222}Rn in water (Jacob et al., 2009). The SGD in this region is found to be a combination of fresh groundwater and recirculated seawater which is governed by the hydraulic gradient in the adjacent aquifer and varying tidal conditions in coastal waters. A first-hand estimate of SGD rate is computed from the ^{222}Rn mass balance in the coastal waters and is found to be 10.9 ± 6.1 cm/day.

Another study was carried out on the groundwater potential, quality and subsurface groundwater discharge and flux for the management of the coastal aquifers of Cuddalore area in India (Fig. 7). The study was mainly focussed on the subsurface flux and measuring the SGD in coastal area and developed a model for future forecasting of the changes by using geophysical methods. The geophysical survey and data interpretation using by IPWIN2 software provided very encouraging results. They observed high nitrate concentrations in natural waters flowing in the subterranean environment (Chidambaram and Ramanathan, 2008).

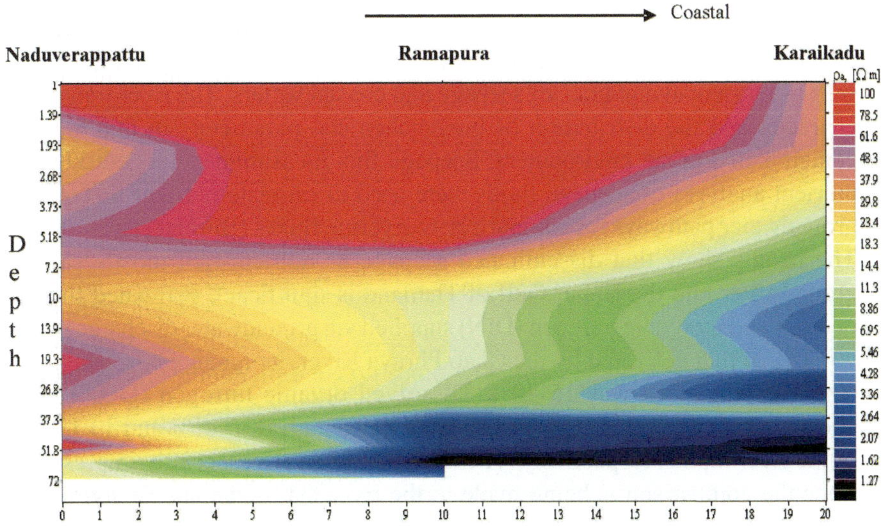

Figure 7: Resistivity survey in Cuddalore coast.

5.4 Bangladesh

Moore (1997) has documented that during low discharge of the Ganges-Brahmaputra River (Bangladesh), fluxes of ^{226}Ra and Ba to the northern Bay of Bengal are comparable to expected river-derived fluxes during peak river discharge. A large non-riverine source of Ra and Ba is required to explain the high fluxes during low river discharge. The author suggested that this source was submarine discharge of groundwater containing high concentrations of ^{226}Ra and Ba. Total annual fluxes of ^{226}Ra and Ba to the ocean from this system were significantly greater than estimated previously.

5.5 Thailand

Taniguchi et al. (2007) have investigated submarine groundwater discharge and salt water-fresh water interactions at two locations along the shoreline of the Upper Gulf of Thailand to evaluate mechanisms of water and material transport into the coastal zone. At the coastal sites investigated in the Gulf of Thailand, they observed that the maximum conductivities and flow rates occurred offshore, that is, higher than areas located closer to the coast. This result is similar to their observations from southern Japan (Taniguchi et al., 2006). The electric conductivities of seepage fluids measured directly inside benthic chambers had significant negative correlations to the remotely measured resistivities of pore water in the subseabed. Thus, resistivity measurements may be used to evaluate the spatial distribution of subseabed water quality (total dissolved salt) and combined use of seepage and resistivity measurements provides an approved understanding of coastal groundwater dynamics.

Nutrient and standard oceanographic surveys were conducted during the wet and dry seasons along the Chao Phraya River, Estuary and out into the Upper Gulf of Thailand. Additional measurements in selected near-shore regions of the Gulf included manual and automatic seepage meter deployments, as well as nutrient evaluations of seepage and coastal waters. The river transects characterized the distribution of biogeochemical parameters in this highly contaminated urban environment. Seepage flux measurements together with nutrient analyses of seepage fluids were used to estimate nutrient fluxes via groundwater pathways for comparison to riverine fluxes (Burnett et al., 2007). The results show that disseminated seepage of nutrient-rich mostly saline groundwater into the Upper Gulf of Thailand is significant. Estimated fluxes of dissolved inorganic nitrogen (DIN) supplied via groundwater discharge were 40-50% of that delivered by the Chao Phraya River, inorganic phosphate was 60-70%, and silica was 15-40%. Dissolved organic nitrogen (DON) and phosphorus (DOP) groundwater fluxes were also high at 30-40% and 30-130% of the river inputs, respectively. These observations are especially impressive since the comparison is being made to the river that is the largest source of fresh water into the Gulf of Thailand and flows directly through the megacity of Bangkok with high nutrient loadings from industrial and domestic sources.

5.6 China

Spatial and temporal variations of SGD have been evaluated by automated seepage meters from the Yellow River delta to 7 km offshore in the Bohai Sea, China (Taniguchi et al., 2008). They identified three zones from the coast to offshore based on different relationships between tidal and SGD changes. The results indicate that the point of maximum SGD shifted 2 km offshore from September 2004 to September 2006. This spatial change is thought to be caused by sediment deposition near the coast. Integrating submarine fresh groundwater discharge (SFGD) along the coastline of the Yellow River delta using measured values of SFGD per unit length of shoreline shows that discharge of fresh groundwater along the entire delta would be equivalent to 4.5 to 7.0% of the river discharge.

Peterson et al. (2008) used salinity and pH as indicators of the terrestrial and recirculated seawater components of discharging groundwater and radium isotopes to quantify offshore transport rates. They then used an hourly time series of multiple radium isotopes (^{224}Ra, ^{223}Ra, and ^{226}Ra) to quantify SGD rates and also used ^{222}Rn and seepage meters to independently quantify SGD rates as a comparison to the radium results. Offshore transport rates were found to range from 3.3 to 4.7 cm s^{-1}. Modelled time series radium activities indicated average SGD rates ranging from 4.5 to 13.9 cm d^{-1} in September 2006 and from 5.2 to 11.8 cm d^{-1} in July 2007. Temporal trends associated with the radium approach agree with SGD patterns revealed by automated seepage meters deployed nearby, but the absolute fluxes are about 70% lower than those determined by the seepage meters. Modelled SGD rates based on ^{222}Rn (mean = 13.8 cm d^{-1} in 2006 and 8.4 cm d^{-1} in 2007) agree with those determined by the radium analysis. Differences in derived SGD rates between the different radium isotopes (^{226}Ra highest; ^{224}Ra lowest) are likely results of uncertainties in the background activities and our limited selection of appropriate groundwater/pore water end-member values. Scaling the results to the entire Yellow River delta, they find SGD fluxes (and corresponding nitrate fluxes) 2-3 times that of the Yellow River.

6. FUTURE PERSPECTIVES

The importance of SGD is recognized in the Asian context and available methods to understand SGD are briefly reviewed in the present article. The important researches of SGD in different countries in Asia have been illustrated by some abovementioned studies. Even though, there were some articles related to SGD which appeared commonly for Japan and to a lesser extent for Korea, there are no quantitative SDG data that have been reported from Vietnam, Cambodia, Laos, Mongolia, Africa, and China, although there are a few reports that document submarine springs. Although, there were some methods that have been applied for investigating the SGD into coastal area, yet there are many limitations imposed on the SGD research and its development in Asia mainly due to the lack of concerns of decision makers and scientists relating to this phenomenon.

The involvement of professionals with SGD studies, shows that researchers from Asia are not particularly well represented in this field. Another review article (Gallardo and Marui 2006), which demonstrates the participation of hydrogeologists, in particular, has been negligible. SGD comprises three words, two of which (groundwater and discharge) are the main domains of hydrogeologists. It is a fact that number of research papers published from Asia especially from India and China pertaining to groundwater is highly significant. Therefore, while appreciating the undertaken research work, no doubt is left that the participation of hydrogeologists of Asia in SGD studies has been less than what it should have been.

Measurement of SGD requires sea-borne equipment that oceanographers and marine scientists are more familiar with. This obstacle, however, is gradually vanishing because of the ease of operation of sophisticated modern equipment. Professional organizations and scientific bodies involved with SGD research projects include Intergovernmental Oceanographic Commission (IOC), Scientific Committee on Oceanic Research (SCOR), Land-Ocean Interactions in the Coastal Zone (LOICZ) and International Association of Physical Sciences of Oceans (IAPSO). Again, here, one could easily see that these bodies, in general, have an ocean-marine root with minimal connection with land hydrology or hydrogeology. As a milestone, the International Atomic Energy Association (IAEA) and UNESCO initiated a five-year plan in 2000 to study the importance of SGDs and the methodologies used to study SGDs in the management of the coastal zone. This comprehensive research project (CRP) was carried out by: the Isotope Hydrology Section of IAEA in Vienna, Austria; the Marine Environmental Laboratory in Monaco; the Intergovernmental Oceanographic Commission (IOC), and the International Hydrological Program (IHP), along with nine laboratories in eight countries including four Asian countries (Brazil, India, Italy, Japan, Russia, Turkey, Slovenia and USA). Still SGD research is underdeveloped in Asia. Moreover, a couple of groundwater commissions like International Association of Hydrogeologists (IAH), have not yet focussed on this subject.

In addition, decision makers, managers and researchers in Asia need to be aware of SGD because the unawareness of SGD and fresh groundwater in the offshore realm can lead to misunderstanding of the hydrogeology of coastal aquifers. In this study we have already shown that maximum loads of subsurface suspended solids are being transported to sea in Asia (nearly five times higher than Europe), which is likely to have a great impact on the salinization of a coastal well field instantly attributed to seawater encroachment. However, if relatively fresh groundwater is present below the seafloor, this explanation is incorrect and a different cause is to be held responsible for the salinization process. Therefore, even if hydrogeologists are not directly involved in quantification of SGD itself, they need to be more aware of it than is presently the case in Asia. The field of SGD has to be progressed rapidly in Asia over the next few years and the advance research should focus on quantitative and qualitative characterization of SGD sources. Regional management plan like integrated coastal zone management plan or coastal zone management plan for SGD and coastal aquifers has to be evolved for its better management. Thus there is a need for the scientific community to develop a keen interest in SGD research. Arguably, it appears that SGD and associated nutrient and toxic fluxes to coastal oceans are still not perceived as an issue in the developing countries of Asia and much is needed to transfer both the technology as well as to convey the benefits of understanding them from the developed world.

7. SUMMARY

From the above discussions and case studies it can be inferred that study of SGD, although proven for the coastal water resource security in Asia is still at infancy and much requires to be done to gain ground in the current level of knowledge. SGD must be recognized as a ubiquitous phenomenon that coastal managers should consider before sketching out any water resource management plan. There is a need to include SGD in the curriculum of groundwater hydrology and hydrogeology subjects of the Geology and Earth Science departments of universities. Topics like definition and description of SGD, sampling and measurement techniques, the impact of SGD on the receiving waters and the coastal processes, factors controlling SGD quantity, the impact of seawater level fluctuations on the quantity of SGD, SGD as a water resources management issue and case studies worldwide should be taught as part of this subject. Ironically, seawater intrusion, the other side of the SGD coin, has been a core research subject in Asia. However, SGD has not found its rightful place in Asian context and there is a justified need to enhance a thorough research in this important field for integrated coastal zone management.

REFERENCES

Basu, A.R., Jacobsen, S.B., Poreda, R.J., Dowling, C.B. and Aggarwal, P.K. (2001). Large groundwater strontium flux to the oceans from the Bengal Basin and the marine strontium isotope record. *Science*, **293**, 1470-1474.

Bear, J., Cheng, A.H.-D., Herrera, I., Sorek, S. and Ouazar, D. (1999). Seawater Intrusion in Coastal Aquifers: Concepts, Methods, and Practices, Kluwer Academic Publishers, Dordrecht, Holland.

Buddemeier, R.W. (ed.) (1996). Groundwater Discharge in the Coastal Zone: Proceedings of an International Symposium. LOICZ/R&S/96-8, IV+179 pp. LOICZ, Texel, the Netherlands.

Burnett, W.C. (1999). Offshore springs and seeps are focus of working group. *EOS*, **80,** 13-15.

Burnett, W.C., Taniguchi, M. and Oberdorfer, J. (2001). Measurement and significance of the direct discharge of groundwater into the coastal zone. *J Sea Res*, **46**, 109-116.

Burnett, W.C., Chanton, J., Christoff, J., Kontar, E., Krupa, S. and Lambert, M. (2002). Assessing methodologies for measuring groundwater discharge to the ocean, *EOS*, **83**, 117-123.

Burnett, W.C., Bokuniewicz, H., Huettel, M., Moore, W.S. and Taniguchi, M. (2003). Groundwater and pore water inputs to the coastal zone. *Biogeochemistry*, **66(1-2)**, 3-33.

Burnett, W.C. and Dulaiova, H. (2003). Estimating the dynamics of groundwater input into the coastal zone via continuous radon-222 measurements. *J. Environ. Radioactivity*, **69**, 21-35.

Burnett, W.C., Wattayakorn, G., Taniguchi, M., Dulaiova, H., Sojisuporn, P. and Rungsupa, S. (2007). Groundwater-derived nutrient inputs to the Upper Gulf of Thailand. *Continental Shelf Research*, **27(2)**, 176-190.

Burnett, W.C., Aggarwal, P.K., Aureli, A., Bokuniewicz, H., Cable, J.E., Charette, M.A., Kontar, E., Krupa, S., Kulkarni, K.M., Loveless, A., Moore, W.S., Oberdorfer, J.A., Oliveira, J., Ozyurt, N., Povinec, P., Privitera, A.M.G. and Rajar, R. (2006). Quantifying submarine groundwater discharge in the coastal zone via multiple methods. *Science of the Total Environment*, **367**, 498-543.

Cable, J.E., Burnett, W.C. and Chanton, J.P. (1997). Magnitudes and variations of groundwater seepage into shallow waters of the Gulf of Mexico. *Biogeochemistry*, **38**, 189-205.

Cable, J.E. and Martin, J.B. (2007). In situ evaluation of near shore marine and fresh pore water transport into Flamengo Bay, Brazil. *Estuarine, Coastal and Shelf Science*, **76**, 473-483.

COSOD II (1987). Fluid circulation in the crust and the global geochemical budget. Report of the Second Conference on Scientific Ocean Drilling, Strasbourg, France, 6-8 July.

Cathles et al. (1987). Fluid circulation in the crust and global geochemical budget, Report of the second conference on scientific ocean drilling.

Chidambaram, S. and Ramanathan, AL. (2008). Comparison of subsurface ground water discharge in east and west coast of India (DST- ILTP project Progress report 2007-2008, Govt of India, 32 p, Unpublished).

Destouni, G. and Prieto, C. (2003). On the possibility for generic modelling of submarine groundwater discharge. *Biogeochemistry*, **66**, 171-186.

Dzhamalov, R.G., Zektser, I.S. and Meskheteli, A.V. (1977). Groundwater discharge into the seas and world ocean, Nauka (Science), Moscow, 94 p. (in Russian).

Dzhamalov, R.G. and Safronova, T.I. (2000). On estimating chemical discharge into the world ocean with groundwater. *Water Resources*, **29(6)**, 626-631.

Gallardo, A.H. and Marui, A. (2006). Submarine groundwater discharge: an outlook of recent advances and current knowledge. *Geo-Mar Lett*, **26**, 102-113.

Garrels, R.M. and MacKenzie, F.T. (1971). Evolution of Sedimentary Rocks, Norton & Co., New York.

Hwang, D.W., Kim, G., Lee, Y-W. and Yang, H-S. (2005). Estimating submarine inputs of groundwater and nutrients to a coastal bay using radium isotopes. *Marine Chemistry*, **96**, 61-71.

IOC/INF (2000). Assessment and management implications of submarine groundwater discharge into the coastal zone. IOC/INF- 1140 Paris, 2 June 2000.

Inman, D.L., Tai, R.J. and Nordstrom, C.E. (1971). Mixing in the surf zone. *J. Geophys. Res.*, **76**, 3493-3514.

Jacob, N., Babu, D.S., Suresh, and Shivanna, K. (2009). Radon as an indicator of submarine.

Johannes, R.E. (1980). The ecological significance of the submarine discharge of groundwater. *Mar. Ecol. Prog. Ser.*, **3**, 365-373.

Kaleris, V. et al. (2002). Modelling submarine groundwater discharge: an example from the western Baltic Sea. *J. Hydrology*, **265,** 76-99.

Kaleris, V. (2006). Submarine groundwater discharge: Effects of hydrogeology and of near-shore surface water bodies. *J. Hydrology*, **325**, 96-117.

Kim, G. and Hwang, D.W. (2002). Tidal pumping of groundwater into the coastal ocean revealed from submarine ^{222}Rn and CH_4 monitoring. *Geophys Res Lett*, **29**, 10.1029/2002GL015093.

Kim, G., Ryu, J.-W., Yang, H-S. and Yun, S-T. (2005). Submarine groundwater discharge (SGD) into the Yellow Sea revealed by ^{228}Ra and ^{226}Ra isotopes: Implications for global silicate fluxes. *Earth and Planetary Science Letters*, **237**, 156-166.

Kohout, F.A. (1966). Submarine springs: A neglected phenomenon of coastal hydrology. *Hydrology*, **26**, 391-413.

Kontar, E.A. and Zektser, I.S. (1999). Submarine discharge and its effect on oceanic processes in the coastal zone. *Water Resources*, **26**, 512.

Langevin, G.D. (2003). Simulation of Submarine Ground Water Discharge to a Marine Estuary: Biscayne Bay, Fl. *Ground Water*, **41(6)**, 758-771.

Lautier, J.C. (1998). Hydrogeologic Assessment of the Proposed Deepening of the Wilmington Harbor Shipping Channel, New Hanover and Brunswick Counties, North Carolina. NC Dept. of Environment, Health, and Natural Resources, Division of Water Resources, Raleigh, NC.

Li, L., Barry, D.A., Stagnitti, F. and Parlange, J-Y. (1999). Submarine groundwater discharge and associated chemical input to a coastal sea. *Water Resources Research*, **35**, 3253-3259.

Martin, J., Cable, J., Smith, C., Roy, M. and Cherrier, J. (2007). Magnitudes of submarine groundwater discharge from marine and terrestrial sources: Indian River Lagoon, Florida. *Water Resources Research*, **43**, W05440, doi: 10.1029/2006WR005266.

Martin, J.B., Cable, J.E., Jaeger, J., Hartl, K. and Smith, C.G. (2006). Thermal and chemical evidence for rapid water exchange across the sediment-water interface by bioirrigation in the Indian River Lagoon, Florida. Limnol. *Oceanogr.*, **53(3)**, 1332-1341.

McCoy, C.A., Corbett, D.R., Cable, J.E. and Spruill, R.K. (2007a). Hydrogeological characterization and quantification of submarine groundwater discharge in the southeast Coastal Plain of North Carolina. *J. Hydrol.*, **339**, 159-171.

McDonalt, M.G. and Harbaugh, A.W. (1988). A modular three-dimensional finite-differences groundwater flow model. US Geol Surv Open-File Report 83-875, 528 p.

Michael, H.A., Mulligan, A.E. and Harvey, C.F. (2005). Seasonal oscillations in water exchange between aquifers and the coastal ocean. *Nature*, **436**, 1145-1148.

Moore, W.S. (1996). Large groundwater inputs to coastal waters revealed by ^{226}Ra enrichment. *Nature*, **380**, 612-614.

Moore, W.S. (1997). High fluxes of radium and barium from the mouth of the Ganges-Brahmaputra River during low river discharge suggest a large groundwater source. *Earth Planet Sci Lett*, **150**, 141-150.

Morastersky, R. (1996). Seep and ye shall find: Hidden water flow. *Science News*, **149** (20 April), 245.

Oberdorfer, J.A. (2003). Hydrogeologic modeling of submarine groundwater discharge: comparison to other quantitative methods. *Biogeochemistry*, **66**, 159-169.

Peterson, R.N., Burnett, W.C., Taniguchi, M., Jianyao, C., Santos, I.R. and Ishitobi, T. (2008). Radon and radium isotope assessment of submarine groundwater discharge in the Yellow River delta, China. *Journal of Geophysical Research*, ISSN 0148-0227, 113, noC9, [Note(s): C09021.1-C09021.14] (3/4 p.)

Reilly, T.E. and Goodman, A.S. (1987). Analysis of saltwater upcoming beneath a pumping well. *Journal of Hydrology*, **89**, 169-204.

SCOR-LOIZ (2004). Submarine groundwater discharge management implications, measurements and effects. IHP-VI Series on Groundwater 5. IOC Manuals and Guides 44. UNESCO, Paris.

Segol, G. and Pinder, G.F. (1976). Transient simulation of saltwater intrusion in southeastern Florida. *Water Resources Research*, **12**, 65-70.

Sekuliè, B. and Vertaènik, A. (1996). Balance of Average Annual Fresh Water Inflow into the Adriatic Sea. *Water Resource Development*, **12(1)**, 89-97.

Smith, A.J. (2004). Mixed convection and density-dependent seawater circulation in coastal aquifers. *Water Resou Res*, **40** W08309. doi:10.1029/2003WR002977.

Smith, A.J. and Turner, J.V. (2001). Density-dependent surface water-groundwater interaction and nutrient discharge in the Swan-Canning Estuary. *Hydrological Proc*, **15**, 2595-2616.

Taniguchi, M., Sakura, Y. and Ishii, T. (1998). Estimations of saltwater-fresh water interfaces and groundwater discharge rates in coastal zones from borehole temperature data. Proceeding of Japanese Association of Groundwater Hydrology Meeting, Tokyo, October (1998), pp. 86-89.

Taniguchi, M., Inouchi, K., Tase, N. and Shimada, J. (1999). Combination of tracer and numerical simulations to evaluate the groundwater capture zone, IAHS Publ 258, pp. 207-213.

Taniguchi. M. (2000). Evaluation of the saltwater-groundwater interface from borehole temperature in a coastal region. *Geophys Res Lett*, **27**, 713-716.

Taniguchi, M., Burnett, W.C., Cable, J.E. and Turner, J.V. (2002). Investigations of submarine groundwater discharge. *Hydrol Process*, **16**, 2115-2129.

Taniguchi, M. et al. (2003). Spatial and temporal distribution of submarine groundwater discharge rates obtained from various types of seepage meters at a site in the northeastern Gulf of Mexico. *Biogeochemistry*, **66**, 35-53.

Taniguchi, M., Ishitobi, T. and Saeki, K.-I. (2005). Evaluation of Time-Space Distributions of Submarine Ground Water Discharge. *Ground Water*, **43(35)**, 336-342.

Taniguchi, M., Ishitobi, T., Burnett, W.C. and Wattayakorn, G. (2007). Evaluating Ground Water-Sea Water Interactions via Resistivity and Seepage Meters. *Ground Water*, **45(6)**, 729-735.

Taniguchi, M., Stieglitz, T. and Ishitobi, T. (2008). Temporal variability of water quality of submarine groundwater discharge in Ubatuba, Brazil. *Estuarine, Coastal and Shelf Science*, 484-492.

Taniguchi, M., Ishitobi, T., Chen, J., Onodera, S., Miyaoka, K., Burnett, W.C., Peterson, R., Liu, G. and Fukushima, Y. (2008). Submarine groundwater discharge from the Yellow River Delta to the Bohai Sea, China, *J. Geophys. Res.*, **113**, C06025, doi:10.1029/2007JC004498.

Thompson, C., Smith, L. and Maji, R. (2007). Hydrogeological modeling of submarine groundwater discharge along the continental shelf of Louisiana. *J. Geophys. Res.*, **12(C3)**, C03014.

Uchiyama, Y., Nadaoka, K., Rölke, P., Adachi, K. and Yagi, H. (2000). Submarine groundwater discharge into the sea and associated nutrient transport in a sandy beach. *Water Resour. Res.*, **36(6)**, 1467-1479.

UNWWDR (2009). Water in a Changing World, World Water Development Report 3, World Water Assessment Programme UNESCO.

Valiela, I., Costa, J., Foreman, K., Teal, J.M., Howes, B. and Aubrey, D. (1990). Transport of groundwater-borne nutrients from watersheds and their effects on coastal waters. *Biodegradation*, **10(3)**, 177-197.

Wilson, A.M. (2005). Fresh and saline groundwater discharge to the ocean: A regional perspective. *Water Resour. Res.*, **41**, W02016.

Zektser, I.S., Ivanov, V.A. and Meskheteli, A.V. (1973). The problem of direct groundwater discharge to the seas. *J. Hydrol.*, **20**, 1-36.

Zektser, I.S., Dzhamalov, R.G. and Meskheteli, A.V. (1984). Podzemnyi vodoobmen sushi i moray (Subsurface Water Exchange between the Continent and the Sea), Leningrad: Gidrometeoizdat, 207s.

Zektser, I.S. and Loaiciga, H. (1993). Groundwater fluxes in the global hydrological cycle: Past, present and future. *J. of Hydrology*, **144**, 405-427.

Zektser, I.S. (1996). Groundwater discharge into the seas and oceans: state of the art. *In:* Groundwater Discharge in the Coastal Zone, Buddemeier, R.W. (ed.). LOICZ/Russian Academy of Sciences: Texel, Netherlands/Moscow; pp. 122-123.

9

Influence of Climate Factors on the Groundwater Resources of Coastal Tamilnadu

S. Chidambaram, K. Trumalesh[1], AL. Ramanathan[2], G. Johnsonbabu, P. Paramaguru, U. Karmegam, R. Manivannan, K. Srinivasamoorthy and S. Manikandan

Department of Earth Sciences, Annamalai Universty, Annamalai Nagar – 608002

[1]Scientist D, Isotope Hydrology Application Division, BARC, Mumbai – 400085

[2]School of Environmental Sciences, Jawaharlal Nehru University New Delhi – 110 067

1. INTRODUCTION

Groundwater bodies which were recharged by precipitation represent a potential source of information on past climate conditions. The isotopic and elemental compositions of groundwater influenced by climate conditions at the time of recharge may serve as indicators of climate change. The relationship between climate and mean annual stable isotope contents of precipitation (Dansgaard, 1964; Rozanski et al., 1992) provides significant insights into paleoclimatic conditions. Therefore, selected confined aquifers can be used as continental paleoclimatic archives (Fontes et al., 1993). It is essential to appropriate aquifers to avoid the various processes and to have a good sampling strategy to make climatic interpretation. Such studies have been carried out, for example, in Africa (Dray et al., 1983; Andrews et al., 1994), Europe (Rozanski, 1985; Stute and Deak 1989), North America (Claassen, 1986; Phillips et al., 1986; Stute et al., 1992; Plummer, 1993; Dutton, 1995) and Australia (Davidson and Airey, 1982; Jacobson et al., 1989). Elliot et al. (1999) discussed the hydrochemical trends, paleorecharge and groundwater ages in the fissured chalk aquifer of the London and Berkshire basins, UK. Isotope stratigraphy of the major ice caps has been examined by various groups to reconstruct the Holocene and late Pleistocene climate (Koerner, 1989). In Asia, Sukhija et al. (1998), obtained the isotopic signature of paleoclimate for the past 30 ka in ground waters of southern India. An attempt has been made in this article to identify the climate and nature of the groundwater during the time of recharge.

1.1 Study Area

India has a long coastal line of about 7560 km with East and West coasts. The East coast has rich biodiversity than the Western coast due to the presence of estuaries, mangroves and coral reefs. The study area falls within the latitudes 10°44′ to 12°00′ N and longitudes 79°42′- 80°38′ E covering an area of 1114 sq. km, along the north eastern coastal region of Tamilnadu (Fig. 1). The Vellar and Coleroon, the major rivers flowing in the study area, form an estuary with a marshy mangrove environment at Pichavaram. The northern part of the study area has an industrial estate (SIPCOT) near Cuddalore. The Cuddalore formation is one of the major aquifer apart from coastal alluvium in the northern part of the study area. The southern part of the study area (south of river Coleroon) is chiefly of alluvium (Chidambaram et al., 2008).

The study area enjoys humid and tropical climate with hot summers, significant to mild winters and moderate to heavy rainfall. The temperature in the study area is maximum during the month of May whereas minimum during the month of January. The study area receives rainfall almost throughout the year and especially during southwest and northeast monsoons. The normal annual rainfall of the study area is 1230 mm. The study area forms part of Cauvery river basin and is drained by a network of rivers like the Cauvery, the Kollidam, the Arasalar, the Vettar and the Pamniyar. The river Kollidam flows along the northern border of the study area.

The northern and north central parts of the study area is drained by the Cauvery and its distributaries viz. Kollidam, Palavar, Virasolanar and Kirttamani rivers. The rivers flow from west to east and enters into the Bay of Bengal. All the rivers are almost ephemeral in nature. At the estuaries, all the rivers are filled with the back-waters.

Table 1: Stratigraphic succession of geological formations in study area

ERA	*Period*	*Formation*	*Lithology*
Quaternary	Recent to pliestocene	Alluvium	Soils, coastal and river sand
Tertiary	Pliocene	Cuddalore formation	Sand and clays

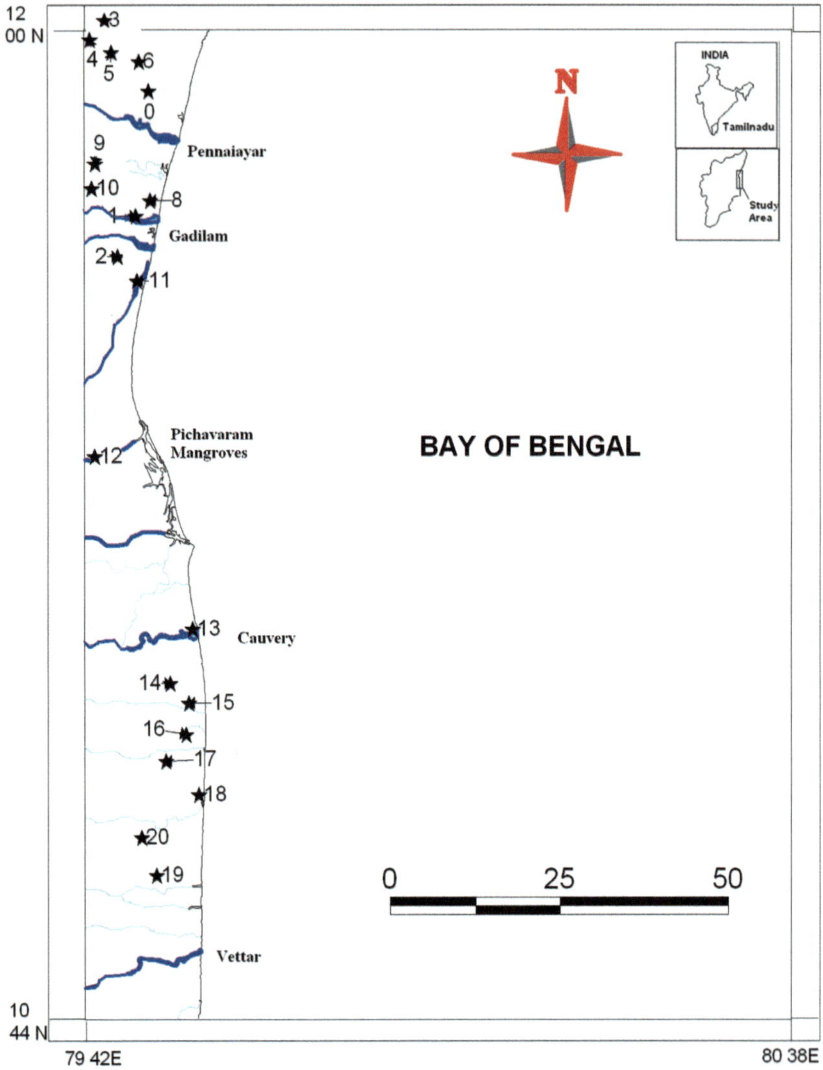

Figure 1: Location map of study Area.

1.2 Quaternary Formations

The formations of this group occupy almost entire study area. They are fluvial and semi-marine in origin. The fluvial deposits consisting of sands intercalated with brown and black clays are seen on the either side of the present Cauvery river course. The thickness of these deposits increases in the easterly direction and is about 40 m near the coast. The sandy silts of 3 to 6 m thickness on the top partly confine the coarser deposits beneath (Central Ground Water Board, 1997).

1.3 Depth to Water Table

The depth to water table was arrived, based on the water level data obtained from the network hydrograph stations of the Central Ground Water Board and State Ground Water Department. A study of this map revealed that the depth to water table in the study area ranged around 1.56 m. In major part of the study area, the depth to water table lies within 4.0 mbgl. The reduced water level shows that the major portion of the study area falls in the negative gradients, except along the northern boundary of the study area.

2. METHODOLOGY

One litre of water sample was collected in polyethylene bottle, sealed, brought to laboratory for analysis and stored properly (4°C) before analysis. 20 water samples were collected during June 2008, from the aquifer and were analysed for major and minor cation and anions using standard procedures (APHA, 1995; Ramanathan, 1992; Ramesh and Anbu, 1996). Water samples for stable isotope analyses were collected in glass bottles and analyzed at the same laboratory using mass spectrometry and related to the SMOW or PDB standards. The precision for $\delta^{13}C$ and $\delta^{18}O$ are 0.2 and 0.1, respectively. Stable isotope analysis ($\delta^{18}O$) was carried out for selected all groundwater samples. Tritium was determined on electrolytically enriched water samples by low-level proportional counting and the results are reported as ^3H units with a typical error of ^3H 1 TU. The $\delta^{13}C$ and Tritium was analysed for only selected samples. The analytical results are given in Table 2.

3. HYDROGEOCHEMISTRY

pH of water is an indicator of its quality and geochemical equilibrium for solubility calculation (Hem, 1985). pH indicates the state of equilibrium reaction in which the water precipitates. The pH ranges from 5.5 to 7.51 with an average of 6.86 and is higher in southern part of study area. The Electrical Conductivity (EC) ranges from 262 to 4790 µs/cm with an average of 1285 µs/cm. EC of pure water is .05 µs/cm (Hem, 1991). Highest EC was also observed in southern part of the study area. The order of dominance of the cation is $Na^+ > Ca^{2+} > Mg^{2+} > K^+$ and that of anion is $HCO_3^- > Cl^- > SO_4^{2-} > NO_3^-$.

3.1 Tritium

The strong seasonal and annual variation seen in precipitation represents a complicated input function and makes it difficult to determine the initial ^3H at the recharge. In the three decades since the last major tests, thermonuclear bomb tritium has been greatly attenuated by the oceans, and levels are now approaching that of natural atmospheric production. This evolution of the input

Table 2: Analytical results

S.No	*Location*	*pH*	*EC*	*Temp*	*Cl*	HCO_3	NO_3	SO_4	*Na*	*K*	*Mg*	*Ca*	$\delta^{18}O$	$\delta^{13}C$	^3H
1	Periyaganganakuppam	6.63	294	32.3	25	122	bdl	8.6	18	2.23	9.4	28.3	-6.9	-15.67	0.85±0.16
2	Caper quary	5.5	262	32.1	28	97.6	bdl	8.8	16	2.6	8.7	27.9	-6.7	-15.58	0.79±0.19
3	Vanoor	7	759	31.5	37.5	378	0.9	10.2	15	4.5	53.4	72.7	-5.5	-9.33	2.34±0.18
4	Katterikuppam	7	722	31.1	23.1	476	2.4	21.8	42	5.8	41.7	119	-5.6	-9.33	-
5	Karasoor	7	1080	32.5	37.5	439	11.6	187	78	1.9	35.9	150	-5.6	-11.14	2.35±0.18
6	Poothurai	7.4	638	30.9	42.4	342	8.6	39.2	55	1.6	26.8	56.4	-6.5	-13.8	-
7	Muthiraipalayam	6.25	756	30.7	100	195	2.9	101	79	4.2	12.7	57	-5.1	-14.01	3.88±0.26
8	Manapattu	6.57	364	32.1	24.2	134	bdl	9.3	24	3.3	12	33.7	-6.9	-15.45	0.61±0.19
9	Kaduvanoor	6.98	618	30.8	47.3	293	bdl	32.4	80	1.5	15.5	35.3	-4	-5.304	0.87±0.15
10	Sornavoor	7.3	753	29.7	70	378	14.7	27	72	1.8	24.7	59.2	-5.3	-8.983	-
11	Thaikathonithurai	6.37	674	31.3	197	97.6	3.5	30.5	104	6.8	10.3	10.4	-6.2	-13.96	0.58±0.12
12	Mutlur	6.55	1128	32.1	168	159	19.1	77.4	95	31	39.7	113	-6.2	-	-
13	Thirumullaivasal	7.51	2511	31.5	728	366	bdl	183	394	34.8	35.8	74.8	-5.8	-17.29	3.35±0.24
14	Thiruvengadu	7.22	496	30.9	55.2	146	29	71	29	47.7	13.8	22.1	-6.2	-16.37	3.04±0.25
15	Tharmakulam	6.97	1291	29.8	266	342	26.1	42.5	209	31.5	22.5	38.5	-6.5	-15.55	3.03±0.24
16	Karuvi	6.91	1874	29.4	280	464	0.8	380	323	6.2	33	55.4	-3.5	-	3.53±0.25
17	Thirukadaioor	6.6	2305	29.7	434	512	96	150	257	176	61.1	46.3	-7.3	-	3.06±0.23
18	Tarangambadi	6.98	4790	31.1	1549	464	121	281	672	96.8	91.5	150	-4.4	-	2.63±0.19
19	Thirunallaru	7.25	1941	30.6	452	512	1.9	77	410	19.1	12.2	14.6	-3.4	-	-
20	Nedungadu	7.29	2444	29.2	375	732	0.16	81	366	21.6	28.9	84.2	-3.7	-13.85	1.97±0.17

All values in mg/l, except pH and Temp (°C), stable isotopes (permil), and ^3H in pmc.

functions along with the mixing in some aquifers with pre-bomb groundwater ages. A quantitative interpretation of groundwater mean residence times may not be possible and only qualitative interpretations can be made (Clarke and Fritz, 1997).

Recharge is sporadic, and shallow groundwater can have a wide range of ages for coastal and low latitude regions

< 0.8 TU	Sub-modern, recharged prior to 1952
0.8 to ~ 2 TU	Mixture between sub-modern and recent recharge
2 to 8	Modern (<5 to 10 yrs)
10 to 20	Residual "bomb" ^3H present
>20 TU	Considerable component of recharge from 1960s or 1970s

The tritium in shallow alluvial formation in Oman shows that groundwater is recharged primarily by infiltration of flash flood events, whereas inter channel areas receive direct infiltration and sub-surface flow (Clarke, 1987).

A tritium result of the study area (Fig. 2) reveals groundwater of three different types.

Type A: Old groundwater with low Electrical Conductivity

Type B: Sub-modern recharged groundwater with moderate Electrical Conductivity

Type C: Modern groundwater with high Electrical Conductivity

Type A: The groundwater of this group has tritium units <0.8 and they are sub-modern waters recharged prior to 1932. These samples fall in the northern part of the study area representing the Cuddalore sandstone formation. The

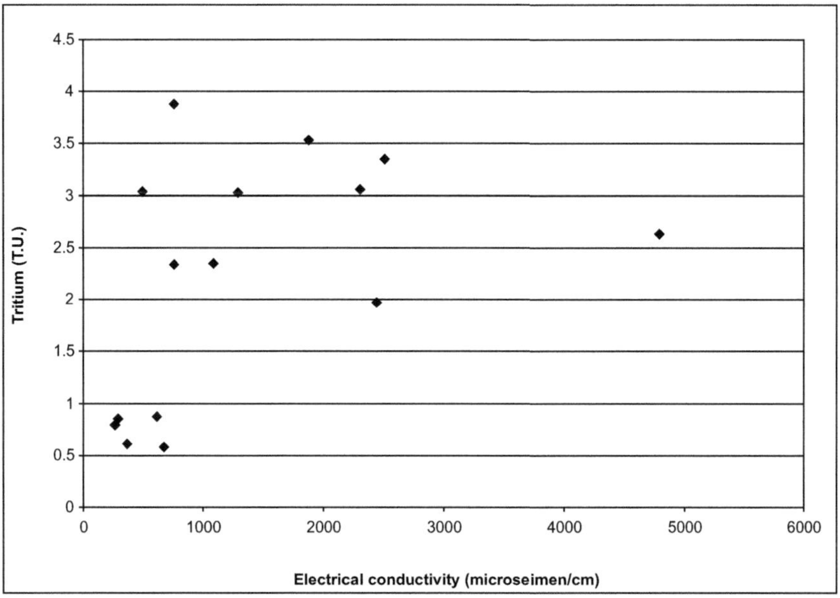

Figure 2: ^3H vs Electrical Conductivity.

Electrical Conductivity of these samples is less than 700 μs/cm. This means that the locally recharged groundwater existed with long residence time. It also reveals that the recharge process of this area is less active and lesser interaction between the groundwater and the aquifer matrix.

Type B: The samples of this type have tritium units 0.8 to 2. They are mixture between sub-modern and recent recharge with Electrical Conductance <1200 μs/cm. It is inferred that the groundwater is comparatively younger than type A and reacts with the medium to have higher ionic concentration.

Type C: The groundwater of this type is comparatively younger with >2 TU and designated as modern waters (Clark and Frits, 1997). Majority of these samples are found in the alluvial formations representing the southern part of the study area. The Electrical Conductivity of the samples range from 496 to 4790 μs/cm. Thus the residence time of this groundwater is estimated to be relatively short after infiltration. Groundwater flow system was thought to be active as well. The EC ranges from low to high, indicating the fresh and the saline nature of coastal aquifer.

3.2 Geochemical Nature

Piper plots depict not only the quality of groundwater but also pathways of freshwater and saline water movement and abstracted geochemical process. The percentages of cations (Na^+, Ca^{2+} and Mg^{2+}) and anions (SO_4^{2-}, Cl^- and $HCO_3^- + CO_3^{2-}$) are plotted in the trilinear diagram and then to the diamond field of piper. Figure 3 shows a conservative mixing line for sea water and fresh water (Appelo and Geirnaert, 1991).

The piper plot shows basic types of water.

1. Fresh water recharge, $CaHCO_3$ type
2. Exchange, Ca-Na-HCO_3-Cl; Ca-Mg-SO_4 –Cl-HCO_3
3. Mild saline water intrusion with exchange, Na-HCO_3-Cl type
4. Estuary intrusion, Na-SO_4-Cl type

Zone 1: The samples falling in zone (3, 4, 1, 8, 6, 10, 2 and 5) have water type; they indicate freshwater recharge. Most of the samples are recharged by the direct infiltration without mixing and are least saline. Their EC values are less than 800 μs and indicating low ionic strength. All the samples fall in the Cuddalore sandstone formation. The tritium values of these samples are >2.35 TU indicating older nature of these groundwater. This shows the older groundwaters are fresh in nature.

Zone 2: Samples (7, 14 and 9) fall in the exchange zone. The waters are in the process of constant exchange from $Ca^{2+} + Na^+$ and have mild salinity. The samples 7 and 14 are younger with modern tritium values and sample 9 with sub-modern tritium indicating that older EC values of these samples range between those of samples falling in Zones 1, 3 and 4.

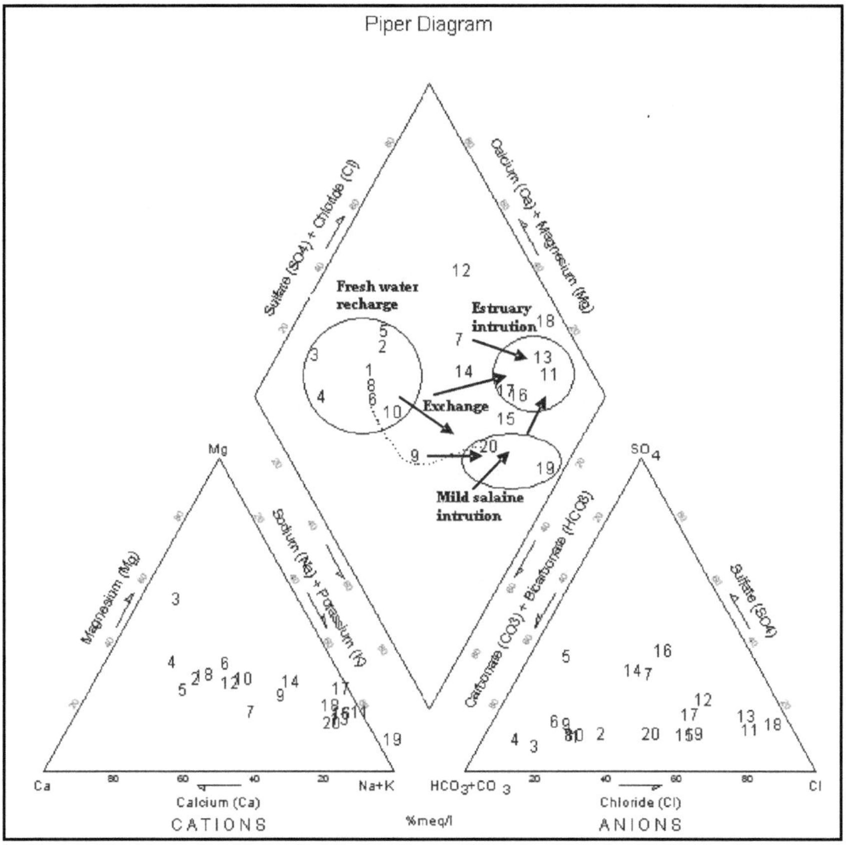

Figure 3: Piper Plot exhibiting the geochemical nature of the groundwater.

Zone 3 has dominantly Na-HCO_3-Cl type (20, 19, 15). The samples falling in this region indicate mild saline water intrusion. The sample 20 is older with 1.97±0.17 TU. This sample falls in the southern part of the study area, indicating for modern salinity in groundwater. Tritium values of the samples 19 and 15 show modern salinity. Higher EC values were noted in all these samples.

Zone 4 has dominantly Na-Cl-HCO_3 type (17, 16, 11, 3, 18). The samples falling in this zone also have higher EC indicating estuarine intrusion. All these samples falling in this zone are modern except sample 11, indicating sub-modern salinity.

It's inferred that the older waters falling in the northern part are dominated by fresh water and they are approaching mild salinity via exchange zone. The sub-modern waters are from the Cuddalore sandstone and the modern are of the alluvium, falling in the southern part of the study area.

3.3 $\delta^{18}O$

Depletion in groundwater reflects a bias in seasonality of recharge and the nature of summer precipitation. The warm pacific air masses rise over the western cordillera where orographic rainout over the Coast Mountains produces a steep $\delta^{18}O$ gradient (Yonge et al., 1989) and a regional Alpine effect. The study area is a flat terrain with not much topographical variation and a coastal groundwater. All the samples were collected from the alluvial formations. The recharge in arid regions shows high evaporation which is highly fractionating process (Clarke and Fritz, 1997). Evaporative enrichment in the falling droplets beneath the cloud base was effective during warm and dry months when rain amounts are small. This partially evaporated rain is characterised by relatively higher $^{18}\delta$ values. Dincer et al. (1974) showed that soil moisture between 0.9 and 6 m depth in dense sands from Saudi Arabia was strongly evaporated.

Three different groups were identified based on the isotopic composition (Fig. 4). Among the three groups of water identified the first group with $\delta^{18}O$ ranging from -3.6 % to -7.3% shows recharge in comparatively warmer period and the colder period. It also reflects that there is wide variation in the climate during this period of recharge. A major difference is observed between summer rain and winter rain. The strong evaporation observed in the summer recharge is due to the run-off during flow over a hot landscape (Clarke, 1987). These variations are observed in the younger waters in which the TU is >2, modern waters. The dependence of the isotope variations on the local temperature or the closely related parameter of the precipitable water content (Sonntag et al., 1983) appears as the overriding parameter (Yurtsever, 1975; Fricke and O'Neil,

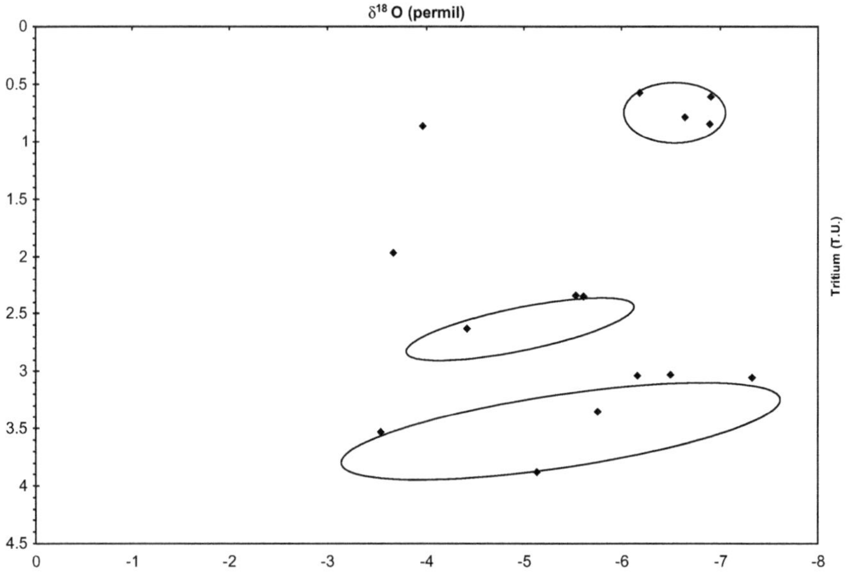

Figure 4: ^{3}H vs $\delta^{18}O$.

1999). Generally the temperature dependence of both δ values is smaller than shown by the latitude effect, Yearly average $\delta^{18}O$ values vary from year to year. In temperate climates the values generally do not vary by more than 1‰, and a large part of the spread is caused by variations in the average annual temperature. In semi-arid climates, with a less regular rain distribution in time, larger variations occur.

The second group have $\delta^{18}O$ ranging from –4.42% to –5.61%. There is a very short variation in the isotopic concentration and the samples of this representation are mixtures of sub-modern recharge and recent recharges. This reflects that the climatic variations are less. The third set of samples have a very narrow variation of $\delta^{18}O$ ranging from –6.18% to –6.91%. The low values of $\delta^{18}O$ indicate cool climates during the time of recharge and these samples show comparatively older age with tritium units <0.8. Meteoric waters originating from precipitation in the distant past, named palaeo-waters (Fontes, 1981), are encountered as deep groundwater in those areas where the present-day replenishment of groundwater is slow, predominantly in arid ones. In most cases their isotopic composition differs appreciably from that of the modern precipitation in the recharge zone.

It is clear that the older groundwaters were recharged only in cooler compared to the recent-day recharge. The recent-day recharge shows a wide variation in the climatic conditions.

3.4 $\delta^{13}C$

The concentration of HCO_3 is controlled initially by variations in dissolved CO_2 present in the soil zone and then taken up by the reaction within the aquifers. The different input CO_2 has different $\delta^{13}C$ values. Thus, it is necessary to determine the source of input CO_2. Assuming that the $\delta^{13}C$ of the CO_2 is in equilibrium with water source DIC which has $\delta^{13}C$ and pH values at the temperature of samples in the recharge area indicating the likely sources of CO_2 input.

The $\delta^{13}C$ values of samples have a range of –5.3% to –17.29%. The possible sources of $\delta^{13}C$ are dissolved CO_2 and the dissolution of calcite. The change of $\delta^{13}C$ values with increasing pH showed the contribution of different sources and the dilution of primary DIC from dissolved calcite. The extent of calcite dissolution depends on the openness of the system.

The combination of log PCO_2 and $\delta^{13}C$ (Fig. 5) can provide an indication of recharge conditions (Clark and Fritz, 1997). The calculated log PCO_2 is more than that in soil (PCO_2 about $10^{-1.8}$) (Zhang et al., 1987). Thus, calcite dissolution has been taking place under open system conditions and the initial carbonic acid from soil CO_2 has not been consumed by calcite dissolution. The closed soil layer was also demonstrated by Chen and Ni (1987), as an alternate process. The decreases of SIC (SI<0) indicate the dissolution of calcite. Ion exchange drives further calcite dissolution. All water samples except for those in the recharge area are unsaturated with respect to calcite. The SIC of the

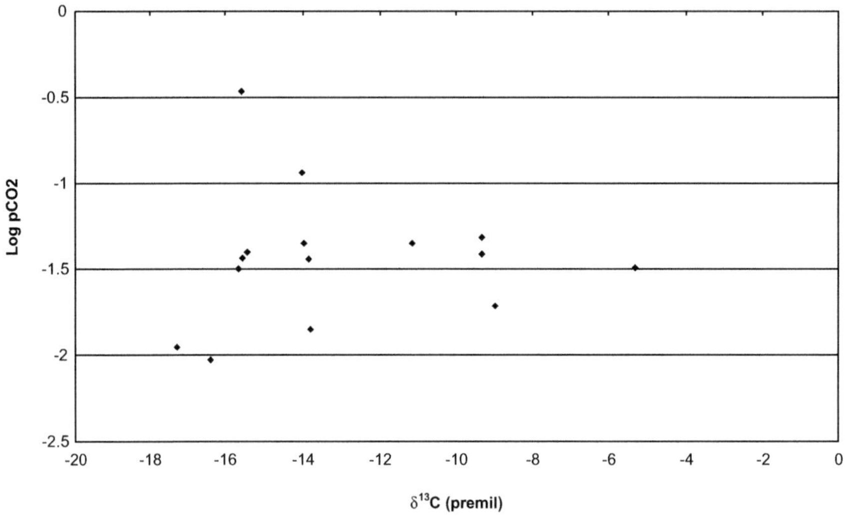

Figure 5: Log PCO_2 vs $\delta^{13}C$.

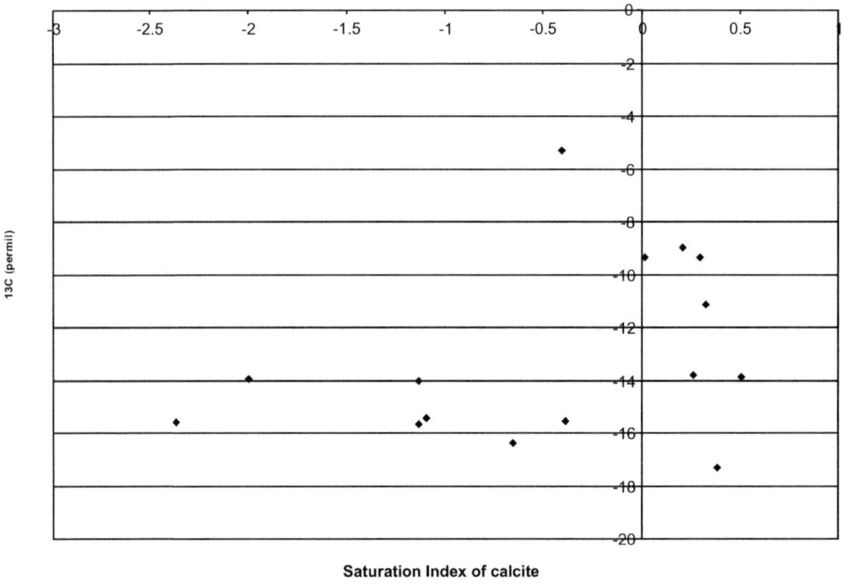

Figure 6: $\delta^{13}C$ vs Saturation index of Calcite (SIC).

samples (Fig. 6) show that most of the older (sub-modern) samples are saturated when compared to the modern groundwaters. These samples also fall in the charge zone of the Piper diagram. In general there are little differences in HCO_3 concentration and $\delta^{13}C$ values between saline water and fresh water. The influence of DIC by mixing is neglected. After the groundwater reached calcite saturation, cation exchange reduces Ca^{2+} concentration and induces calcite dissolution.

4. CONCLUSION

The groundwater of the region show three different characteristics in tritium ages as sub-modern, mixture of sub-modern and modern recharge and modern waters. The sub-modern waters are generally fresh and the modern waters are saline in nature. The mixing nature of these waters are also well established in the Piper diagram. The sub-modern waters are recharged in comparatively cooler climate than the modern waters. A wide variation in climate is noted in the modern recharge. The calcite dissolution has taken place in the open system and the sub-modern waters recharged in comparatively cooler environments are saturated with calcite, due to long residence time. The saturated water later by ion exchange converts to Na-HCO_3 type and then to saline groundwaters. This is clearly witnessed in certain sub-modern groundwaters where the ion exchange is predominant.

ACKNOWLEDGEMENTS

The authors are thankful to BRNS for financial support and Isotope Hydrology Application Division, BARC, Mumbai for helping to analyze the samples.

REFERENCES

APHA (1995). Standard methods for the examination of water and waste water, 19thedition, APHA, Washington DC, US.

Appelo, C.A.J. and Geirnaert, W. (1991) Processes accompanying the intrusion of salt water. *In:* Hydrogeology of salt water intrusion; A Selection of SWIM Papers, Vol. 11. IAH, Hanover, pp. 291-304.

Andrews, J.N., Fontes, J.F., Dodo, A., Edmunds, W.M., Joseph, A. and Travi, Y. (1994). The evolution of alkaline groundwater in the continental intercalaire aquifer of the Irhazer Plain, Niger. *Water Resour. Res.*, **30**, 45-61.

Chidambaram, S. (2008). Evaluation of the hydrogeochemistry of groundwater using factor analysis in the Cuddalore coastal region, Tamilnadu, India. *Indian Journal of Marine Sciences*, **37**, 181-185.

Chen, W.H. and Ni, M.Y. (1987). Quaternary Geology in Hebei. Geological Publishing House, Beijing.

CGWB (1997). Groundwater Resources and Development prospects in Nagapattinam District, Tamilnadu. Report.

Claassen, H.C. (1986). Late-Wisconsinan Paleohydrology of the west central Amargosa Desert, Nevada, U.S.A. *Chem. Geol.*, **58**, 311-323.

Clark, I.D. (1987). Groundwater resources in the Sultanate of Oman: Origin, circulation times, recharge processes and paleoclimatology. Isotopic and geochemical approaches; Unpublished doctoral thesis, Universite de Paris-sud, Orsay, France, 264 p.

Clark, I.D. and Fritz, P. (1997). Environmental Isotopes in Hydrogeology. Lewis Publishers, New York, pp. 112-244.

Davidson, M.R. and Airey, P.L. (1982). The effect of dispersion on the establishment of a paleoclimatic record for groundwater. *J. Hydrol.*, **58**, 131-147.

Dansgarrd, W. (1964). Stable isotopes in precipitation. *Tellus*, **16**, 436-468.

Dincer, T., Al-Mugrin, A. and Zimmermann, U. (1974). Study of the infiltration and recharge through the sand dunes in arid zones with special reference to the stable isotopes and thermonuclear tritium. *Journal of Hydrology*, **23**, 79-109.

Dutton, A.R. (1995). Groundwater isotopic evidence for paleorecharge in U.S. high plain aquifers. *Quatern. Res.*, **43**, 221-231.

Dray, M., Gonfiantini, R. and Zuppi, G.M. (1983). Isotopic composition of groundwater in the southern Sahara. *In:* Palaeoclimates and Palaeowaters; A Collection of Environmental Isotope Studies. Panel Proc. Series. IAEA, Vienna, pp. 187-199.

Elliot, T., Andrews, J.N. and Edmunds, W.M. (1999). Hydrochemical trends, palaeorecharge and groundwater ages in the fissured Chalk aquifer of the London and Berkshire Basins, UK. *Appl. Geochem.*, **14**, 333-363.

Fontes, J. Ch. (1981). Paleowaters. *In:* Stable Isotope Hydrology. Gat and Gonfiantini (eds.). IAEA Techn. Report no. 210, pp. 273-302.

Fontes, J. Ch., Stute, M., Schiosser, P. and Broecker, W.S. (1993). Aquifers as archives of paleoclimate. *EOS Trans. AGU.*, **74**, 21-22.

Fricke, H.C. and O'Neil, J.R.O. (1999). The correlation between $^{18}O/^{16}O$ ratios of meteoric water and surface temperature; its use in investigating terrestrial climate change over geologic time. *Earth planet. Sc. Lett.*, **170**, 181-196.

Hem, J.D. (1985). Study and interpretation of the chemical characters of natural water. USGS Water Supply 1473, 363 p.

Hem, J.D. (1991). Study and interpretation of the chemical characters of natural water. 3rd Edn. U.S. Geol. Survey; water supply paper 2254, Scientific pub, Jodhpur.

Jacobson, G., Calf, G.E., Jankowski, J. and Mc Donald, P.S. (1989). Groundwater chemistry and Palaeocharge in the Amadeus Basin, Central Australia. *J. Hydrol.*, **109**, 237-266.

Koerner, R.M. (1989). Ice core evidence for extensive melting of Greenland ice sheet in the last interglacial. *Science*, **224**, 964-968.

Phillips, F.M., Peeters, L.A., Tansey, M.K. and Davis, S.N. (1986). Paleoclimatic inferences from an isotopic investigation of groundwater in the Central San Juan Basin, New Mexico. *Quatern. Res.*, **26**, 179-193.

Plummer, L.N. (1993). Stable isotope enrichment in paleowaters of the Southeast Atlantic coastal plain, United States. *Science*, **262**, 2016-2020.

Ramanathan, AL. (1992). Geochemical studies in the Cauvery river basin. Ph.D thesis, Punjab University, Chandigarh.

Ramesh and Anbu (1996). Chemical methods for environmental analysis—water and sediments, 161 p.

Rozanski, K. (1985). Deuterium and oxygen-18 in European groundwaters: Inks to atmospheric circulation in the past. *Chem. Geol.*, **52**, 349-363.

Rozanski, K., Arans-Aragnas, L. and Gonfiantini, R. (1992). Relation between long-term trends of oxygen-18 isotope composition of precipitation and climates. *Science*, **258**, 981-984.

Sonntag, Ch., Munnich, K.O., Jacob, H. and Rozanski, K. (1983). Variation of Deuterium and Oxygen-18 in continental precipitation and ground water, and their causes. *In:* Variation in the Global water Budget. Street-Perrot et al. (eds.). Reidel Publ., pp. 107-124.

Stute, M., Schlosser, P., Clark, J.F. and Broecker, W.S. (1992). Paleotemperatures in southwestern United States derived from noble gases in groundwater. *Science*, **256**, 1000-1003.

Stute, M. and Deak, J. (1989). Environmental isotope study (^{14}C, ^{13}C, ^{18}O, D, noble gases) on deep groundwater circulation systems in Hungary with reference to paleoclimate. *Radiocarbon.*, **31**, 902-918.

Sukhija, B.S., Reddy, D.V. and Nagabhushanam, P. (1998). Isotopic fingerprints of paleoclimates during the last 30,000 years in deep confined groundwaters of Southern India. *Quatern. Res.*, **50**, 252-260.

Yonge, J.C., Goldenberg, L. and Krouse, H.R. (1989). An isotope study of water bodies along a traverse of southwestern Canada. *Journal of Hydrology*, **106**, 245-255.

Yurtsever, Y. (1975). Worldwide survey of isotopes in precipitation. IAEA report, Vienna.

Zhang, Z.G., Zhang, H.P., Sun, J.C., Yang, L.J., Cui, Y.L., Wu, L.H., Qi, J.X. and Xu, J.M. (1987). Environmental isotope study related to groundwater age, flow system and saline water origin in Quaternary aquifer of Hebei plain. *Hydrogeol Engin. Geol.*, **4**, 1-6.

10

Natural Arsenic in Coastal Groundwaters in the Bengal Delta Region in West Bengal, India

Prosun Bhattacharya[1], Gunnar Jacks[1], Bibhas Nath[2,3], Debashis Chatterjee[2], Ashis Biswas[1,2], Dipti Halder[1,2], Santanu Majumder[2], Subhamoy Bhowmick[2] and AL. Ramanathan[4]

[1]KTH-International Groundwater Arsenic Research Group, Department of Land and Water Resources Engineering, Royal Institute of Technology (KTH), SE-100 44 Stockholm, Sweden

[2]Department of Chemistry, University of Kalyani, Kalyani-741 235, India

[3]School of Environmental Systems Engineering, University of Western Australia, Crawley 6009, Western Australia

[4]School of Environmental Sciences, Jawaharlal Nehru University, New Delhi 110067, India

1. INTRODUCTION

Bengal Delta region is currently confronted with largest groundwater arsenic calamity in history of human kind (BGS-DPHE, 2001; Mukherjee and Bhattacharya, 2001; Bhattacharya et al., 2002a; McArthur et al., 2001; Smedley and Kinniburgh, 2002; Mukherjee et al., 2006; Nath et al., 2005, 2007, 2008). Concentrations of arsenic in drinking water wells in the region often exceed the WHO drinking water guideline value (10 $\mu g\ L^{-1}$) and the national safe limit of both India and Bangladesh for arsenic in drinking water (Smedley and Kinniburgh, 2002; RGNDWM, 2002; CGWB, 1999; Bhattacharya et al., 2002a). About one third (35 million) population inhabiting in this region (West Bengal and Bangladesh), currently at risk of long-term arsenic exposure

(Bhattacharya et al., 2001; RGNDWM, 2002; Chakraborti et al., 2004; Kapaj et al., 2006), are being diagnosed with a wide spectrum of adverse health impacts including skin disorders such as hyper/hypo-pigmentation, keratosis and melanosis and are also in hot-spot areas of BDP which is reflected in a rise in the number of cancer cases (Guha Mazumdar et al., 1988). The distribution pattern of arsenic occurrence in BDP is patchy and there are numerous hotspots of arsenic contamination in the semi-confined shallow Holocene aquifer (Bhattacharya et al., 1997; Smedley and Kinniburgh, 2002). The scale of the problem is serious both in terms of hotspots and geographic area coverage (173×10^3 km^2, eastern part of Hoogly-Bhagirathi/Western part of Ganga-Padma-lower Meghna flood plains).

In West Bengal, groundwater arsenic occurrences are reported from younger Gangetic flood plains (Bhagirathi-Hoogly and Ganga-Padma interfluve) covering eight districts (Malda, Murshidabad, Nadia, North and South 24 Parganas, Howrah, Hoogly and Bardwan) and includes 75 blocks affected with arsenic contamination (Bhattacharya et al., 1997). The arsenic affected areas of West Bengal extends (23,000 km^2) in a NNW-SSE direction on the eastern part of the Bhagirathi-Hoogly channel (CGWB, 1999). BDP consists of upward fining sequences of fluvio-deltaic sediments. Mineralogy of the sediment and the governing geochemical processes are mainly responsible for release of arsenic from sediment to groundwater.

In general, the extent of geographical area coverage of the natural contamination of arsenic in groundwater and related public health issues in Bengal Delta has been studied in considerable details (Mukherjee and Bhattacharya, 2001; Bhattacharya and Mukherjee, 2002; Ahmed et al., 2004; Chakraborti et al., 2004). The recent research focussed on understanding the underpinning mechanisms that control groundwater chemistry in the sedimentary aquifers of the Bengal Delta (Bhattacharya et al., 1997; Bhattacharya et al., 2002a; Bhattacharya et al., 2003; Chatterjee et al., 2003, 2004, 2005; Charlet et al., 2007; Hasan et al., 2007; Jana, 2004; Mukherjee-Goswami et al., 2008; Mukherjee et al., 2008; Nath, 2006; Nath et al., 2005, 2007, 2008). However, there is a considerable need of proper and reliable data bank of the complete spectrum of groundwater chemistry to understand the exact sequences of biogeochemical process that is responsible for release of redox-sensitive species in groundwater in the aquifers especially in the coastal aquifers where the groundwater quality is severely constrained due to saline water ingression. In this chapter we present a comparative study of the salient hydrochemical characteristics of groundwater in selected wells in two contrasting physiographic settings in the Bengal Delta region located in the districts of Nadia (Chakdaha block) and South 24 Pargana (Baruipur block) in West Bengal with an aim to understand the sequence of redox processes that have an important control on the translocation of As from sediment to groundwater. This study also advocates the role of groundwater chemistry in terms of natural As release in coastal and non-coastal areas.

2. MATERIALS AND METHODS

2.1 Study Area

The study area is an integral part of Ganges-Brahmaputra-Meghna (GBM) river system and it is located on the eastern bank of Bhagirathi-Hooghly River at Chakdaha in Nadia district and in the coastal belt of the GBM river system at Baruipur in South 24-Pargana district (Fig. 1). The sediments in the study area are mainly dominated by sand and silts sized fractions with occasional clay lenses with aquifer of semi-confined nature. BDP lithology indicates that the sedimentary successions include sand (channel facies) and silt as well as clay (overbank facies) and generally show a typical fining upward sequence (Bhattacharya et al., 1997; Ahmed et al., 2004). The sediments generally overlie the older deposits; and comprise a sequence of poorly oxidized to unoxidized successive layers laid on the dissected as well as partially eroded older alluvium (Chatterjee et al., 2005). The mineralogy of the sediments is dominated by quartz, feldspar, calcite, muscovite, biotite and with clay minerals, mainly chlorite and kaolinite.

Figure 1: Map of the study area marked with red circle.

2.2 Sampling and Analytical Techniques

Field investigations were carried out to collect groundwater samples from Chakdaha block of Nadia district and Baruipur block of South 24-Pargana district in West Bengal between May 1999 and January 2000. The samples represent domestic and irrigation wells ranging at depth intervals between 5-126 m and 14-305 m in Nadia and South 24-Pargana districts respectively (Table 1). At each sampling site, pH, redox potential (Eh), and electrical conductivity (EC) were also measured at the wellhead. Water sampling was carried out following the protocol (Bhattacharya, 2002b) which includes filtered samples for major anion analyses, filtered and acidified with suprapure HNO_3 for cations and other trace element analyses including As. The major anions are analyzed by a Dionex 120 Ion Chromatograph and the major and trace metals were analyzed on a Perkin Elmer ELAN 6000 ICP-MS at the University of Linköping, Sweden. Dissolved organic carbon (DOC) in groundwater was determined on a Schimatzu 5000 TOC analyzer.

3. RESULTS AND DISCUSSION

3.1 pH, Redox and Major Ion Characteristics

Chemical analyses of groundwater from selected wells at Chakdaha and Baruipur blocks of West Bengal, India, show a marked distinction in the groundwater chemistry (Tables 1 and 2). The well waters are circum-neutral to slightly alkaline with pH values in the range 6.0-8.42. Eh values for groundwater measured for few wells in Chakdaha region suggest moderate to strong reducing conditions (Eh = –0.26 to –0.34 V). On the other hand Baruipur wells indicate slightly elevated redox status (Eh = –0.07 to –0.14 V) and are thus mildly reducing environment. EC values of groundwater in Chakdaha and Baruipur wells (up to a depth of 74 m) varied between 336 and 1810 $\mu S\ cm^{-1}$, while in some of the deeper wells of Baruipur, EC values are significantly higher (1960-3440 $\mu S\ cm^{-1}$), may be due to higher chloride load (409-877 $mg\ L^{-1}$) of the groundwater. Groundwaters in both regions are characterized by high alkalinity (HCO_3 ~ 262-740 $mg\ L^{-1}$; Table 1) which contributed to the EC values. Chakdaha and shallow Baruipur groundwaters are in general Ca-HCO_3^- or Ca-Mg-HCO_3^- type, while wells at depth >74 m in Baruipur, yield Na-Ca-Cl or Na-Ca-Mg-Cl-HCO_3^- type water with elevated levels of chloride (409-878 $mg\ L^{-1}$) and sulfate (26.5-106 $mg\ L^{-1}$). Among other anions, chloride (6-104.50 $mg\ L^{-1}$), nitrate (0.1-2.33 $mg\ L^{-1}$) and sulfate (up to 14.25 $mg\ L^{-1}$) are low in the Chakdaha/Baruipur shallow wells (up to 40 m) (Table 1). Phosphate levels are high in most wells of Chakdaha (0.29-8.01 $mg\ L^{-1}$) as compared to Baruipur. Some of the shallow wells in Baruipur revealed high NH_4^+concentrations (up to 10.12 $mg\ L^{-1}$).

Table 1: Chemical composition of well waters from Chakdaha (Nadia)

S. No.	*Depth (m)*	*TDS mg/L*	*DOC mg/L*	*HCO_3 mg/L*	*Cl mg/L*	*NO_3 mg/L*	*SO_4 mg/L*	*PO_4 mg/L*	*Na mg/L*	*K mg/L*	*Mg mg/L*	*Ca mg/L*	*As mg/L*
1	19.1	462	4.2	268	27.9	0.18	0.31	3.61	12.3	4.8	19.0	88	0.185
2	20.1	641	3.8	476	50.7	2.33	2.95	6.35	21.6	4.6	27.2	103	0.183
3	20.1	627	3.3	473	42.6	1.08	3.18	5.64	14.0	6.3	25.0	101	0.175
4	20.9	513	5.5	323	32.2	3.05	0.06	5.17	21.1	3.5	24.9	104	0.182
5	21.3	504	1.2	396	15.4	1.07	2.73	3.58	16.5	3.3	17.5	85	0.131
6	22.3	412	3.3	263	7.5	1.63	0.03	2.76	17.6	2.7	16.0	88	0.136
7	21.1	588	2.0	448	32.5	1.04	8.90	2.95	18.9	2.7	23.4	94	0.015
8	22.9	616	3.5	218	14.6	3.28	1.57	3.78	22.4	4.7	20.1	94	0.085
9	117.8	430	1.9	262	6.0	2.7	0	0.29	21.0	3.6	27.0	105	0.121
10	26.2	678	3.5	500	26.9	0.2	0.21	0.33	21.5	3.9	25.2	100	0.007
11	20.1	626	1.8	458	25.6	0.3	0	6.07	14.0	5.3	22.3	95	0.154
12	33.8	939	4.7	700	12.7	0.2	0.13	4.55	56.3	4.1	36.7	125	0.267
13	121.9	530	0.8	409	6.5	0.1	0.03	0.19	11.8	2.9	21.7	78	0.086
14	4.6	602	3.3	471	25.5	0.58	2.50	8.01	23.3	4.4	24.0	90	0.236
15	19.2	400	1.3	218	36.5	0.16	0.08	3.81	19.4	1.8	20.9	98	0.033
16	28.0	714	7.0	584	19.1	0.18	2.92	6.50	22.1	4.2	27.4	100	0.180
17	21.0	846	5.8	307	131.7	0.58	9.57	2.57	37.1	2.5	38.0	152	0.026
18	23.2	592	2.3	431	60.9	0.16	2.36	7.99	19.0	7.0	27.4	89	0.149
19	23.0	295	2.3	160	15.8	0.70	1.52	3.12	6.9	2.1	11.9	87	0.036
20	125.6	520	1.6	414	6.3	0.30	0	1.45	18.0	2.3	18.7	98	0.048
21	46.3	439	3.2	324	26.8	0.22	3.00	4.16	12.1	3.5	14.5	79	0.138
22	45.7	853	5.6	654	50.8	0.19	2.67	3.71	35.0	4.1	36.0	140	0.096
23	25.6	765	5.2	607	24.7	0.17	2.74	2.82	32.5	3.5	32.1	127	0.119
24	46.3	563	3.4	439	23.5	0.11	1.78	0.35	12.9	2.5	20.7	97	0.001

3.2 Distribution of Arsenic, Iron, Manganese and DOC

Groundwater Fe_{tot} (1.09-20.25 mg L^{-1}), As_{tot} (1.0-1059 µg L^{-1}) and Mn (50-4740 µg L^{-1}) are varying with depth. In most of the wells (both Chakdah and Baruipur) As_{tot} concentrations are above the WHO drinking water guideline and it is noteworthy to mention that As_{tot} concentration of deeper wells of Baruipur is well within safe limit of both international as well as national standard. DOC levels are relatively high (up to 12.9 mg L^{-1}) in the wells of Baruipur and Chakdaha region. However, the levels of DOC in deeper wells are relatively low at both the sites. Since As(V)/As(III) ratios are lacking in the data sets, a plot of Eh-pH relationship reveals that nearly 80% of As_{tot} in the analyzed wells is expected to be present as As (III) species.

Ba μg/L	*Bi* μg/L	*Cd* μg/L	*Co* μg/L	*Cr* μg/L	*Cu* μg/L	*Fe* mg/L	*Mn* mg/L	*Ni* μg/L	*Pb* μg/L	*Sr* μg/L	*U* μg/L	*V* μg/L	*Zn* μg/L
282	0.04	0.85	0.5	5.9	10.9	6.51	0.18	5.3	13.61	343.8	0.02	0.51	248.1
250	0.02	0.04	0.3	48.9	4.8	8.69	0.17	9.0	3.23	383.2	0.02	0.29	60.8
274	0.06	0.03	0.3	140.4	3.7	6.70	0.16	2.9	2.11	329.9	0.01	0.35	31.6
239	0.09	0.20	0.5	5.6	7.9	8.10	0.16	7.4	17.57	375.0	0.02	0.47	294.7
221	0.06	0.03	0.2	1.8	4.9	3.93	0.35	2.9	1.86	238.2	0.01	0.11	52.4
253	0.05	0.08	0.4	4.0	5.6	3.85	0.37	7.2	12.28	255.7	0.01	0.35	213.2
207	0.06	0.05	0.2	2.6	4.5	15.50	0.20	5.1	3.06	258.3	0.01	0.18	41.8
237	0.02	0.10	0.5	26.1	7.9	5.61	0.28	9.8	64.13	269.4	0.03	0.76	219.8
191	0.02	0.26	0.4	5.2	7.2	1.61	0.12	7.4	14.25	591.6	0.03	0.49	1395.5
125	0.01	0.10	0.2	2.2	1.6	1.23	0.16	7.4	0.01	337.6	0	0.05	22.2
242	0.00	0.03	0.2	1.8	6.5	6.29	0.15	2.6	1.08	291.5	0.01	0.28	24.2
514	0.03	0.01	0.3	1.8	1.9	4.58	0.07	2.3	7.68	486.9	0.01	0.31	67.9
92	0.01	0.11	0.1	1.5	4.5	1.09	0.12	3.2	4.42	396.7	0.02	0.60	549.4
330	0.06	0.01	0.4	1.2	4.0	7.75	0.11	2.5	2.08	399.6	0.01	0.26	37.6
187	0.01	0.05	0.4	3.9	5.2	4.06	0.11	24.9	5.72	289.5	0.02	0.53	137.1
439	0.03	0.02	0.5	5.2	4.3	5.88	0.13	3.1	5.61	462.8	0.02	0.41	82.4
288	0.01	0.13	1.0	5.4	9.1	8.58	0.30	9.0	13.53	610.2	0.09	1.28	304.8
209	0.07	0.04	0.3	1.3	3.5	4.43	0.08	2.9	2.91	375.7	0.01	0.19	130.4
164	0.04	0.20	0.3	4.9	9.3	3.65	0.10	62.8	13.78	243.2	0.02	0.59	165.5
104	0.07	0.10	0.2	2.3	4.6	1.66	0.11	3.0	3.96	303.7	0.01	0.15	1565.3
169	0.06	0.11	0.4	3.0	6.4	3.47	0.30	5.0	9.04	237.3	0.01	0.21	210.6
464	0.06	0.01	0.5	5.0	3.9	7.13	0.16	7.9	2.24	551.7	0.01	0.34	53.2
383	0.06	0.25	0.6	24.1	8.3	5.12	0.23	7.3	12.02	472.4	0.07	0.66	87.6
71	0.06	0.04	1.4	1.2	3.5	1.20	4.74	2.7	3.18	361.3	1.52	1.12	86.9

3.3 Relationship between the Key Water Quality Parameters with Redox Sensitive Elements

The correlation of HCO_3^- with As_{tot} and Fe_{tot} in the two data sets is shown in Figs 2a and 2b respectively. Phosphate also has good correlation with As_{tot} in Chakdaha area (r^2-0.46), showing negative correlation in Baruipur area (Fig. 3a) whereas with Fe_{tot} has good correlation in Baruipur (r^2-0.37) and no correlation in Chakdaha area (Fig. 3b). DOC concentrations show negative correlation with both As_{tot} and Fe_{tot} in the wells of Baruipur and Chakdaha region (Figs 4 a,b). In general, a weak positive correlation is observed between As_{tot} and Fe_{tot} in both regions but in general the values of As_{tot} is much higher in Baruipur area (upto 1059 μg L^{-1}) as compared to Chakdaha area.

Table 2: Chemical composition of well waters from Baruipur (South 24 Parganas)

S. No.	*Depth (m)*	*HCO_3 mg/L*	*Cl^- mg/L*	*NO_3 mg/L*	*SO_4 mg/L*	*PO_4 mg/L*	*Na mg/L*	*K mg/L*	*Mg mg/L*	*Ca mg/L*	*DOC mg/L*	*$NH4^+$ mg/L*	*As mg/L*	*Ba μg/L*
1	32.3	496	29.53	0.59	0.96	1.08	33.98	7.38	25.70	92	4.59	3.84	0.693	400
3	18.3	597	59.99	0.57	0.12	5.60	46.52	5.94	28.24	120	6.61	7.36	0.969	248
4	24.4	619	37.22	0.57	0.11	6.19	44.00	5.90	29.86	128	5.23	7.45	0.961	254
5	36.6	578	16.80	0.77	0.31	2.62	38.16	5.91	31.73	126	3.45	3.86	0.878	198
6	243.8	503	119.90	0.85	0.03	0.17	114.61	5.58	32.37	99	1.70	0.20	0.003	226
7	15.2	444	31.82	0.80	14.25	0.41	14.29	11.33	28.92	145	2.34	0	0.002	198
8	36.6	518	57.54	0.65	0.08	6.81	39.40	7.44	36.57	108	4.68	8.64	0.470	235
9	42.7	740	67.17	0.95	0.14	3.03	62.73	8.86	46.66	111	12.90	4.16	0.416	164
10	30.5	617	22.52	0.41	0	2.92	41.31	8.22	37.57	101	3.16	3.65	0.434	106
11	43.3	547	51.88	0.35	0	4.38	98.47	7.35	38.21	96	5.85	2.80	0.498	48
12	106.7	270	538.10	0.40	6.42	0.15	127.81	5.59	46.87	214	0.38	0	0.001	301
13	73.8	179	528.69	1.76	7.23	0.14	118.93	5.32	43.84	207	0.50	0	0.002	239
14	18.3	814	87.49	0.29	0.13	26.16	315.63	5.59	9.33	23	11.18	4.82	0.176	42
15	39.6	653	104.40	0.24	0.69	1.20	188.55	5.82	21.99	68	1.12	1.02	0.078	123
16	24.4	641	16.88	1.20	0.06	7.67	55.03	6.26	27.51	111	7.00	10.12	0.683	268
17	15.2	413	26.31	1.34	5.54	3.02	9.82	3.23	20.56	100	4.56	0	0.012	126
18	14.0	307	15.70	1.65	4.66	0.30	15.41	2.76	13.58	80	2.03	0.10	0.110	81
19	243.8	431	182.89	0.31	8.64	0.13	133.08	6.05	41.49	91	0.61	0.66	0.001	97
20	30.5	401	14.68	1.04	0.49	2.20	16.11	4.72	17.66	84	3.90	2.15	0.006	210
21	54.9	291	12.48	0.62	0.72	4.34	18.20	4.20	21.00	101	2.28	2.25	0.366	245
22	15.2	548	60.55	0.34	0.09	3.02	38.62	5.34	27.43	120	2.19	1.90	0.242	387
23	27.4	410	40.53	0.82	8.73	0.75	25.66	2.89	18.68	100	1.11	0.49	0.096	264
24	714.0	408	17.60	2.14	0.01	1.46	12.81	3.84	17.08	94	2.86	1.25	0.178	140
25	34.1	503	29.28	0.12	1.53	0.54	25.67	5.36	28.76	111	4.31	3.05	0.133	80
26	30.5	399	9.82	0.73	0.61	0.10	11.93	1.34	29.02	79	0.76	0.05	0.001	31
27	30.5	665	110.79	0.96	1.97	10.00	194.91	9.20	39.83	54	4.05	3.18	0.030	267
28	304.8	421	409.93	0.89	7.85	0.08	199.78	7.40	54.44	111	0.40	1.13	0.001	157
29	36.6	429	55.59	0.16	3.24	0.19	29.16	3.23	26.41	100	1.54	0.25	0.017	85
30	259.1	458	877.91	0.77	5.60	0.11	467.11	7.64	69.39	140	0.92	1.28	0.001	253
31	36.6	655	122.09	0.41	0.41	1.82	118.26	7.77	37.45	99	3.89	5.90	1.059	529

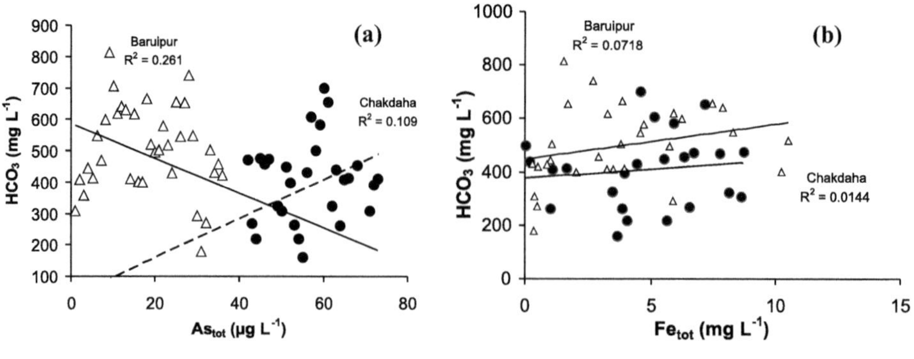

Figure 2: Relation of HCO_3^- with (a) As_{tot} and (b) Fe_{to}.

Bi µg/L	*Cd* µg/L	*Co* µg/L	*Cr* µg/L	*Cu* µg/L	*Fe* mg/L	*Mn* mg/L	*Ni* µg/L	*Pb* µg/L	*Rb* µg/L	*Sr* µg/L	*U* µg/L	*Zn* µg/L	*V* µg/L
0.26	0.08	0.50	12.60	5.43	5.72	0.11	15.92	8.30	1.87	405	0.19	115	0.36
0.10	0.33	1.16	8.42	7.93	6.23	0.68	15.12	16.77	2.42	495	0.02	245	0.43
0.09	0.10	0.69	9.14	6.40	5.88	0.41	15.37	8.50	1.88	502	0.02	94	0.27
0.04	0.14	0.60	4.94	7.04	4.70	1.08	11.89	5.79	1.67	484	0.07	121	0.32
0.03	0.23	0.78	148.80	9.23	1.70	0.15	43.46	12.67	2.34	509	0.12	1330	0
0.04	0.15	0.52	8.64	6.43	1.72	0.91	13.95	18.25	0.98	269	1.42	70	0.33
0.01	0.04	0.75	0	3.59	10.52	0.26	6.75	11.33	1.44	441	0.02	164	0.45
0.07	0.16	0.66	11.17	14.55	2.68	0.24	21.64	22.14	1.69	508	0.04	8168	2.85
0.02	0.06	0.33	1.32	3.39	3.26	0.09	6.70	8.44	1.93	442	0.01	24	0.39
0.04	0.04	0.51	3.16	4.37	4.57	0.61	9.97	7.96	1.10	457	0.02	94	0.39
0.01	0.15	0.33	2.55	55.23	1.50	1.34	3.96	11.18	1.05	933	3.16	44	0.16
0.02	0.05	0.39	3.18	2.12	1.36	1.26	4.40	9.73	1.07	901	4.11	36	0
0.05	0.06	0.42	9.29	2.41	1.55	0.05	9.30	7.74	0.82	78	0.02	15	2.53
0.02	0.14	0.34	4.90	27.09	1.70	0.21	7.98	12.96	1.43	327	0.34	54	0.28
0.05	0.10	0.95	11.30	3.67	7.83	0.39	15.79	9.17	1.67	479	0.02	107	0.41
0.01	0.05	0.20	1.35	5.26	3.95	0.21	7.47	8.60	1.00	167	0.02	3123	1.08
0.02	0.08	0.40	6.46	2.74	1.39	0.54	9.03	9.92	0.82	132	0.30	21	0.33
0.01	0.05	0.18	4.33	2.56	1.31	0.05	5.11	7.49	2.75	643	0.07	1153	0.09
0.03	0.05	0.75	3.02	5.96	20.25	0.52	11.38	7.68	1.94	294	0.02	21	0.23
0.01	0.06	0.23	1.21	7.03	5.85	0.88	7.41	8.57	1.32	343	0.02	281	0.33
0.02	0.06	0.39	1.76	7.41	8.25	1.39	11.85	21.19	2.18	416	0.02	245	0.41
0.01	0.04	0.23	1.47	3.16	3.24	0.89	5.55	6.94	0.92	165	0.07	28	0.27
0.03	0.08	0.43	11.72	3.05	3.50	0.83	10.92	10.24	1.57	244	0.10	148	0.12
0.03	0.19	0.76	12.12	6.62	3.81	0.37	12.52	17.88	1.21	415	0.26	1842	0.47
0	0.07	0.33	3.96	1.78	2.02	0.38	4.87	6.63	0.29	358	0.87	172	1.35
0	0.04	0.43	1.83	2.69	3.85	0.18	4.75	4.84	1.59	330	0.01	12	0.44
0	0.07	0.16	0.72	2.19	1.52	0.06	2.51	5.40	2.48	919	0.06	1132	0
0.01	0.10	0.34	4.76	4.88	1.87	0.40	6.47	10.62	0.49	420	5.96	94	0.97
0.01	0.11	0.28	4.58	2.61	2.91	0.08	5.92	9.29	4.34	1015	0.03	2072	0
0.02	0.07	1.11	2.82	3.99	7.44	0.22	10.44	3.12	1.47	526	0.07	99	0.73

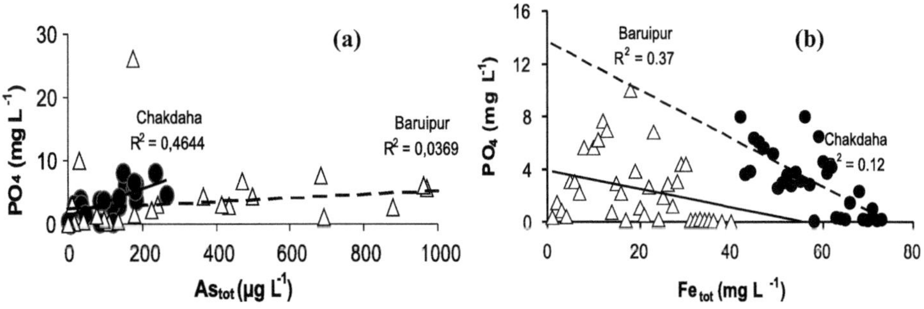

Figure 3: Relation of PO_4 with (a) As_{tot} and (b) Fe_{tot}.

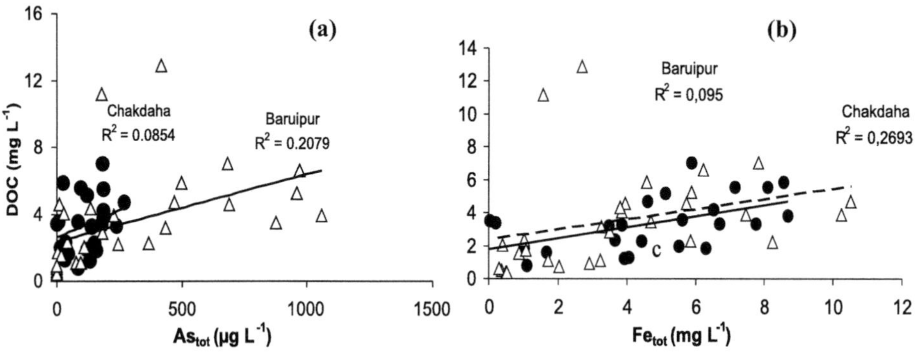

Figure 4: Relation of DOC with (a) As_{tot} and (b) Fe_{tot}.

3.4 Statistical Evaluation of the Data

In order to identify the different elemental association in the study area, statistical method is adopted to find the probable reason behind the evolution of arsenic with its relation to other elemental association in groundwater. The principles of this analysis are based on the application in geosciences as referred by different authors (Davis, 1986; Rock, 1988). The factor analysis has been carried out using the software STATISTICA (Stat Soft, Europe) using varimax rotation of the factor axis. The data is first being normalized to mean value of zero and standard deviation of one before used in the analysis. The analysis gives three dominant factors and expressed by the data matrix of 58% and 55% for Chakadaha and Baruipur area respectively. The result of the factor loading is presented in Tables 3a, b and Figs 5-7.

3.4.1 Factor 1

The total variance is 27% and 24% for Chakadaha and Baruipur area respectively. For Chakdaha the high loadings are expressed by the components like: TDS, DOC, HCO_3, PO_4, Cl, SO_4, Na, Mg, K, Ca, Ba, Fe and Sr (Fig. 5a). In case of Baruipur the components dominating are: TDS, DOC, K, NH_4, As, Ba, Bi, Cd, Co, Cr, Fe and Sr (Fig. 5b). This means that in both the areas Factor 1 is dominated by the main dissolved load of water and this may be due to interaction between aquifer material and the local reducing environment which leads to dissolution of arsenic and other redox sensitive species into groundwater. But in Baruipur region Factor 1 is also dominated by few trace elements As, Cd, Bi and all of this are adsorbed on Fe-oxyhydroxides and can be released when local reducing condition predominates.

3.4.2 Factor 2

The total variance is 18% for Chakdaha and Baruipur area respectively and the factor loading is dominated mainly by the elements PO_4, K, As, Ba, Fe and Cr

Table 3: Factor loading of the groundwater chemical parameters in (a) Chakadaha and (b) Baruipur

	Factor 1	*Factor 2*	*Factor 3*
(a) Chakdaha			
TDS (mg/L)	0.91	0.04	–0.22
DOC (mg/L)	0.76	0.06	0.20
HCO_3 (mg/L)	0.63	0.10	–0.69
Cl (mg/L)	0.61	–0.06	0.36
NO_3 (mg/L)	0.24	–0.08	0.76
SO_4 (mg/L)	0.46	–0.10	0.24
PO_4 (mg/L)	0.25	0.71	–0.06
Na (mg/L)	0.82	0.07	–0.10
K (mg/L)	0.23	0.58	–0.17
Mg (mg/L)	0.93	0.02	–0.11
Ca (mg/L)	0.87	–0.15	0.15
As (mg/L)	0.25	0.70	–0.20
Ba (μg/L)	0.74	0.46	–0.05
Bi (μg/L)	–0.06	0.04	–0.15
Cd (μg/L)	–0.19	0.07	0.50
Co (μg/L)	0.44	–0.71	0.28
Cr (μg/L)	0.12	0.28	0.00
Cu (μg/L)	–0.13	0.07	0.88
Fe (mg/L)	0.53	0.56	0.36
Mn (mg/L)	0.00	–0.82	–0.14
Ni (μg/L)	–0.34	–0.04	0.43
Pb (μg/L)	–0.03	0.04	0.70
Sr (μg/L)	0.76	–0.14	0.04
U (μg/L)	0.04	–0.83	–0.14
V (μg/L)	0.32	–0.62	0.61
Zn (μg/L)	–0.30	–0.19	0.07
Contribution to variance (%)	26.90	17.67	13.13
(b) Baruipur			
TDS (mg/L)	0.47	0.02	0.75
Cl (mg/L)	–0.23	–0.88	–0.23
NO_3 (mg/L)	–0.14	–0.01	–0.16
SO_4 (mg/L)	–0.36	–0.87	–0.07
PO_4 (mg/L)	0.08	0.18	0.72
Na (mg/L)	–0.21	–0.76	0.38
K (mg/L)	0.42	–0.48	0.22
Mg (mg/L)	0.04	–0.91	–0.11
Ca (mg/L)	0.16	–0.40	–0.77
DOC (mg/L)	0.50	0.19	0.67
NH_4 (mg/L)	0.70	0.06	0.43
As (mg/L)	0.76	0.15	0.16
Ba (μg/L)	0.54	–0.20	–0.27
Bi (μg/L)	0.54	0.10	0.09
Cd (μg/L)	0.57	–0.11	–0.24
Co (μg/L)	0.86	0.13	0.06
Cr (μg/L)	0.29	–0.04	–0.11
Cu (μg/L)	0.09	–0.14	–0.42
Fe (mg/L)	0.56	0.24	0.14
Mn (mg/L)	0.04	0.23	–0.71
Ni (μg/L)	0.64	0.09	0.00
Pb (μg/L)	0.39	0.04	–0.21
Rb (μg/L)	0.28	–0.76	0.10
Sr (μg/L)	0.07	–0.85	–0.36
U (μg/L)	–0.30	0.00	–0.53
V (μg/L)	0.03	0.23	0.60
Zn (μg/L)	0.17	–0.27	0.25
Contribution to variance (%)	23.85	16.49	12.41

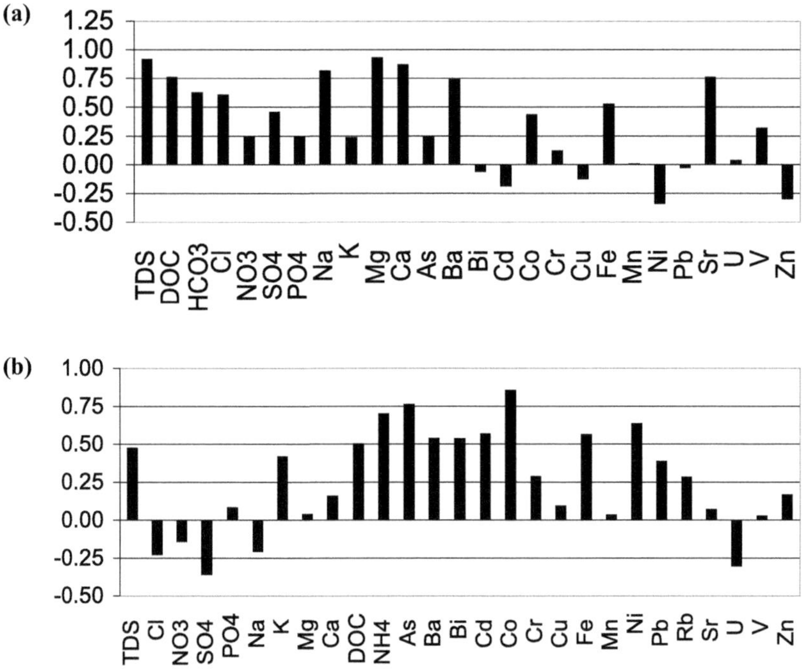

Figure 5: Scores of factor 1 (a) Chakdaha and (b) Baruipur.

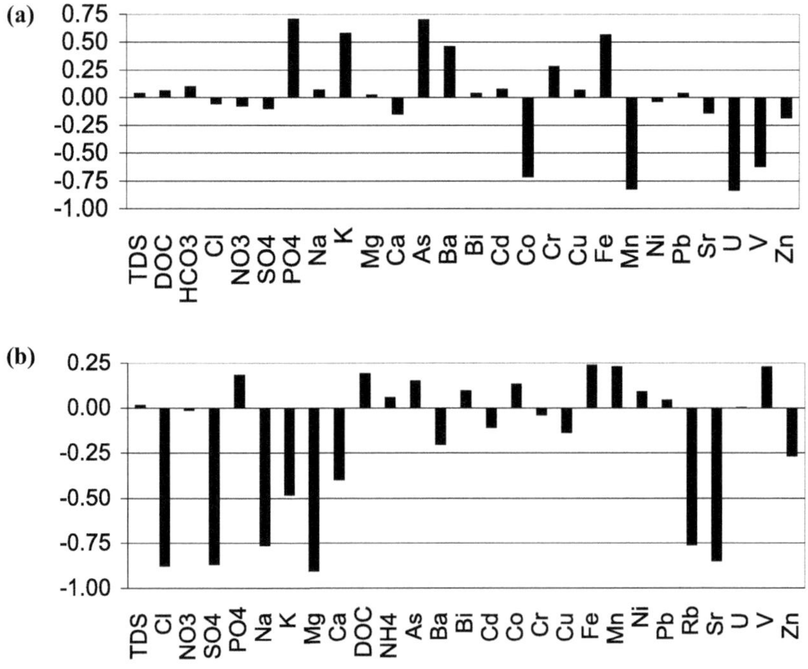

Figure 6: Scores of factor 2 (a) Chakdaha and (b) Baruipur.

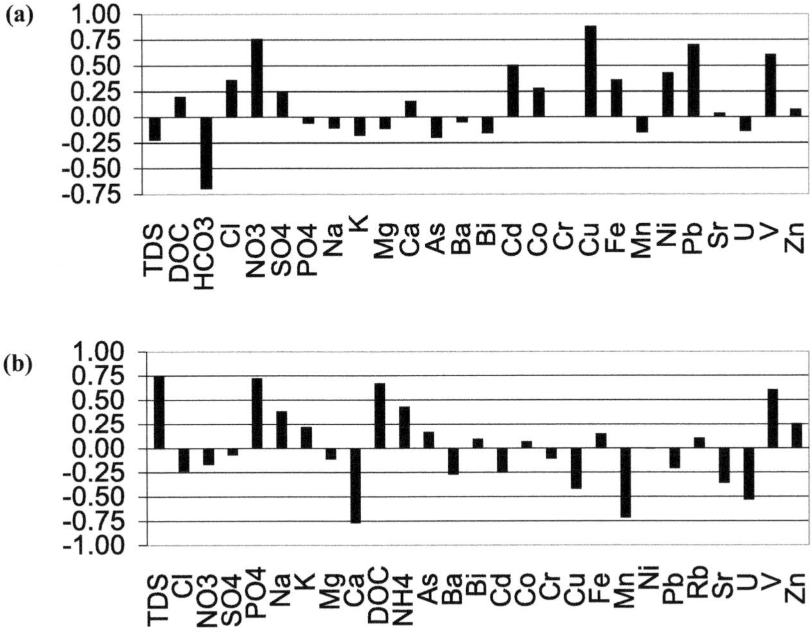

Figure 7: Scores of factor 3 (a) Chakdaha and (b) Baruipur.

in Chakdaha and PO_4, DOC, As, Fe and Mn in Baruipur groundwater (Figs 6a, b).

3.4.3 Factor 3

The total variance is 13% for both areas. The factor loading is dominated by trace elements Fe, Cd, Cu, Ni, V and Pb along with high loading of SO_4 and NO_3, in groundwater in Chakdaha area, but in case of Baruipur TDS. PO_4, Na, K, DOC and NH_4 dominate (Figs 7a, b).

From the statistical analysis of the results of groundwater samples it is clear that in both the study area the main dissolved load of groundwater plays a crucial role in mobilizing arsenic into groundwater and showing good correlation with respect to arsenic concentration in groundwater (Figs 2a, b). The results also reciprocate in the high loadings of bicarbonate, NH_4^+, Cl in groundwaters (Table 1).

4. CONCLUSIONS

The source of As in the coastal aquifer sediments is geogenic, and thus it is important to understand the complex biogeochemical sediment-water interactions in the BD aquifers which control the microbial mediated and thermodynamically favoured redox reactions. Several terminal electron accepting processes (TEAP) are active in the BDP aquifers (Routh et al., 2000) which control the correlation between Fe_{tot}-HCO_3^-, Fe_{tot}-PO_4^{3-}, Fe_{tot}-As_{tot}.and HCO_3^--DOC in the groundwaters of Chakdaha and Baruipur regions. High bicarbonate levels indicate the weathering of carbonate as well as degradation of organic matter under local reducing condition (redox traps). High NH_4^+ concentrations in Baruipur groundwaters may result from dissimilatory nitrate reduction or from in-situ degradation of redox traps. Sulphate reduction is one of the process which causes low sulphate concentrations in groundwaters, but not sufficient enough to cause precipitation of sulfides. Microbial reduction of sedimentary iron (III) phase produces soluble Fe(II) and Fe(III) catalyses arsenic reduction [As(V)-As(III)] that has already been demonstrated when Fe(II) being adsorbed on Fe(III) phases (mostly as nano particles).

ACKNOWLEDGEMENTS

The authors would like to acknowledge the Swedish Research Council (VR), and the Swedish International Development Cooperation Agency (Sida), Strategic Environmental Research Foundation (MISTRA) and Rajiv Gandhi National Drinking Water Mission (RGNDWM) for supporting the research on high arsenic groundwater of West Bengal, India.

REFERENCES

Ahmed, K.M., Bhattacharya, P., Hasan, M.A., Akhter, S.H., Alam, S.M.M., Bhuyian, M.A.H., Imam, M.B., Khan, A.A. and Sracek, O. (2004). Arsenic contamination in groundwater of alluvial aquifers in Bangladesh: An overview. *Applied Geochemistry*, **19(2)**, 181-200.

BGS, DPHE (2001). Arsenic Contamination of Groundwater. *In:* Kinniburgh, D.G. and Smedley, P.L. (ed.) Brit. Geol. Surv. (Tech. Rep., WC/00/19.4 volumes). British Geological Survey, Keyworth.

Bhattacharya, P., Chatterjee, D. and Jacks, G. (1997). Occurrence of arsenic contaminated groundwater in alluvial aquifers from Delta Plains, Eastern India: Options for safe drinking water supply. *International Journal of Water Resources Management*, **13(1)**, 79-82.

Bhattacharya, P., Frisbie, S.H., Smith, E., Naidu, R., Jacks, G. and Sarkar, B. (2002a). Arsenic in the Environment: A Global Perspective. *In:* B. Sarkar (ed.) Handbook of Heavy Metals in the Environment. Marcell Dekker Inc., New York, pp. 147-215.

Bhattacharya, P., Jacks G., Ahmed, K.M., Khan, A.A. and Routh, J. (2002b). Arsenic in Groundwater of the Bengal Delta Plain Aquifers in Bangladesh. *Bulletin of Environmental Contamination and Toxicology*, **69(4)**, 538-545.

Bhattacharya, P. and Mukherjee, A.B. (2002). Management of arsenic contaminated groundwater in the Bengal Delta Plain. *In:* M. Chatterji, S. Arlosoroff and S.G. Guha (eds). Conflict Management of Water Resources. Ashgate Publishers, UK, pp. 308-348.

Bhattacharya, P., Welch, A.H., Ahmed, K.M., Jacks, G. and Naidu, R. (2004). Arsenic in Groundwater of Sedimentary Aquifers. *Applied Geochemistry*, **19(2)**, 163-167.

Bhattacharyya, R., Jana, J., Nath, B., Sahu, S.J., Chatterjee, D. and Jacks, G. (2003). Groundwater arsenic mobilization in the Bengal Delta Plain—The use of ferralite as a possible remedial measure: A Case Study. *Applied Geochemistry*, **18:** 1435-1451.

Bhattacharya, P., Jacks G., Jana J., Sracek A., Gustafsson J.P. and Chatterjee D. (2001). Geochemistry of the Holocene Alluvial sediments of Bengal Delta Plain from West Bengal, India: Implications on arsenic contamination in groundwater. *In:* Jacks, G., Bhattacharya, P. and Khan, A.A. (eds.) Groundwater Arsenic Contamination in the Bengal Delta Plain of Bangladesh, KTH Special Publication, TRITA-AMI Report 3084, pp. 21-40.

CGWB (1999). High Incidence of As in groundwater in West Bengal. Central Groundwater Board, India, Ministry of Water Resources, Government of India.

Chakraborti, D., Sengupta, M.K., Rahman, M.M., Ahamed, S., Chowdhury, U.K., Hossain, M.A., Mukherjee, S.C., Pati, S., Saha, K.C., Dutta, R.N. and Quamruzzaman, Q. (2004). Groundwater arsenic contamination and its health effects in the Ganga-Meghna Brahmaputra plain. *Journal of Environmental Monitoring*, **6**, 74-83N.

Charlet, L., Chakraborty, S., Appelo, C.A.J., Roman-Ross, G., Nath, B., Ansari, A.A. et al., 2007. Chemodynamics of an As "hotspot" in a West Bengal aquifer: A field and reactive transport modeling study. *Applied Geochemistry*, **22**, 1273-1292.

Chatterjee, D., Chakraborty, S., Nath, B., Jana, J., Mukherjee, P. and Sarkar, M. (2004). Geochemistry of arsenic in deltaic sediment. *Geochimica et Cosmochimica Acta*, **68**, A514.

Chatterjee, D., Roy, R.K. and Basu, B.B. (2005). Riddle of arsenic in groundwater of Bengal Delta Plain – Role of non-inland source and redox traps. *Environmental Geology*, **49**, 188-206.

Chatterjee, D., Chakraborty, S., Nath, B., Jana, J., Bhattacharya, R., Basu Mallik, S. et al. (2003). Mobilization of arsenic in sedimentary aquifer vis-à-vis subsurface iron reduction processes. *Journal de Physique IV*, **107**, 293-296.

Davis, J.C. (1986). Statistics and data analysis in Geology. Wiley & Sons, New York, London, Sydney, Toronto.

Guha Majumder, D.N., Chakraborty, A.K., Ghosh, A., Chakraborty, D.P., Dey, S.B. and Chattopadhyay, N. (1988). Chronic arsenic toxicity from drinking tubewell water in rural West Bengal. *Bulletin of the World Health Organization*, **66**, 499-506.

Hasan, M.A., Ahmed, K.M., Sracek, O., Bhattacharya, P., von Brömssen, M., Broms, S., Fogelström, J., Mazumder, M.L. and Jacks, G. (2007). Arsenic in shallow groundwater of Bangladesh: Investigations from three different physiographic settings. *Hydrogeology Journal*, **15**, 1507-1522.

Jana, J. (2004). Genesis of arseniferrous groundwater in the Bengal Delta Plain in West Bengal, Eastern India. Ph.D. thesis, University of Kalyani, India. (Unpublished).

Kapaj, S., Peterson, H., Liber, K. and Bhattacharya, P. (2006). Human health effects from chronic arsenic poisoning – A Review. *Journal of Environmental Science and Health*, Part A., **41(10)**, 2399-2428.

McArthur, J.M., Ravenscroft, P., Safiulla, S. and Thirlwall, M.F. (2001). Arsenic in groundwater: Testing pollution mechanisms for sedimentary aquifers in Bangladesh. *Water Resources Research*, **37**, 109-117.

Mukherjee, A.B. and Bhattacharya, P. (2001). Arsenic in groundwater in the Bengal Delta Plain: Slow Poisoning in Bangladesh. *Environmental Reviews*, **9(3)**, 189-220.

Mukherjee, A., von Bromssen, M., Scanlon, B.R., Bhattacharya, P., Fryar, A.E., Hasan, M.A., Ahmed, K.M., Chatterjee, D., Jacks, G. and Sracek, O. (2008). Hydrogeochemical comparison and effects of overlapping redox zones on groundwater arsenic near the western (Bhagirathi sub-basin, India) and eastern (Meghna sub-basin, Bangladesh) margins of the Bengal Basin. *Journal of Contaminant Hydrology*, **99**, 31-48.

Mukherjee, A.B., Bhattacharya, P., Jacks, G., Banerjee, D.M., Ramanathan, AL., Mahanta, C., Chandrashekharam, D., Chatterjee, D. and Naidu, R. (2006). Groundwater Arsenic Contamination in India: Extent and severity. *In:* Naidu, R., Smith, E., Owens, G., Bhattacharya, P. and Nadebaum. P. (eds.) Managing Arsenic in the Environment: From soil to human health. CSIRO Publishing, Melbourne, Australia, pp. 533-594. (ISBN: 0643068686)

Mukherjee-Goswami, A., Nath, B., Jana, J., Sahu, S.J., Sarkar, M.J., Jacks, G., Bhattacharya, P., Mukherjee, A., Polya, D.A., Jean, J.-S. and Chatterjee, D. (2008). Hydrogeochemical behavior of arsenic-enriched groundwater in the deltaic environment: Comparison between two study sites in West Bengal, India. *Journal of Contaminant Hydrology*, **99**, 22-31.

Nath, B. (2006). Aquifers bearing low and high arsenic groundwater – A comparative study on litho- and hydrochemistry in Chakdaha block, Nadia district, West Bengal. Ph.D. thesis, Jadavpur University, India (Unpublished).

Nath, B., Berner, Z., Basu Mallik, S., Chatterjee, D., Charlet, L. and Stüeben, D. (2005). Characterization of aquifers conducting groundwaters with low and high arsenic concentrations: A comparative case study from West Bengal, India. *Mineralogical Magazine*, **69**, 841-853.

Nath, B., Sahu, S.J., Jana, J., Mukherjee-Goswami, A., Roy, S., Sarkar, M.J. and Chatterjee, D. (2007). Hydrochemistry of arsenic-enriched aquifer from rural West Bengal, India: A study of the arsenic exposure and mitigation option. *Water, Air and Soil Pollution*, **190**, 95-113.

Nath, B., Stuben, D., Basu Mallik, S., Chatterjee, D. and Charlet, L. (2008). Mobility of arsenic in West Bengal aquifers conducting low and high groundwater arsenic. Part I: Comparative hydrochemical and hydrogeological characteristics. *Applied Geochemistry*, **23**, 977-995.

RGNDWM, Ministry of Rural Water and Development (2002). Final report executed by University of Kalyani, Nadia, West Bengal, India.

Rock, N.M.S. (1988). Lecture Notes in Earth Sciences 18. Numerical Geology, Springer-Verlag, Heidelberg.

Routh, J., Bhattacharya, P., Jacks, G., Ahmed, K.M., Khan, A.A. and Rahman, M.M. (2000). Arsenic geochemistry of Tala groundwater and sediments from Satkhira District, Bangladesh. *Eos Trans American Geophysical Union*, **81**, 550.

Smedley, P.L. and Kinniburg, D.G. (2002). A review of the source, behaviour and distribution of arsenic in natural waters. *Applied Geochemistry*, **17**, 517-568.

11

Chemical Composition and Origin of the Coastal Zone Thermal Springs in Far East Russia

O.V. Chudaev, V.A. Chudaeva and I.V. Bragin

Far East Geological Institute, FEB RAS. Vladivostok, Prospect 100let, 159 Russia 690022

Pacific Institute of Geography, FEB RAS. Vladivostok, Radio st. 7 Russia 690022

1. INTRODUCTION

In the Far East Russia, nitric thermal springs are wide spread along coastal zone (sea of Japan and Okhotsk sea) and were formed owing to the youngest tectonic movements of the Sikhote Alin ridge (Primorye and Khabarovsk provinces) (Fig.1). Areas are typically confined to deep brecciation zones, which facilitate penetration of infiltration ground waters in the high temperature horizons of the Earth. The problem of the formation of nitric thermal waters of this region is poorly studied (Kurukhin and Reznikov, 1962; Avdeeva, 1976). The first reliable trace elements data on the Primorye nitric thermal springs are reported in Chudaeva et al. (1999) and Chudaev (2003), and those for the Khabarovsk in Boldovski et al. (2004) and Bragin et al. (2007). However, these materials are fragmentary and could not reveal the regional geochemical tendencies in the formation of nitric thermal waters. This article reports new geochemical data on the nitric thermal waters of the Khabarovsk region (the Annenskii and Tumninskii groups) and Primorye (the Chistovodnoe and Amgu groups).

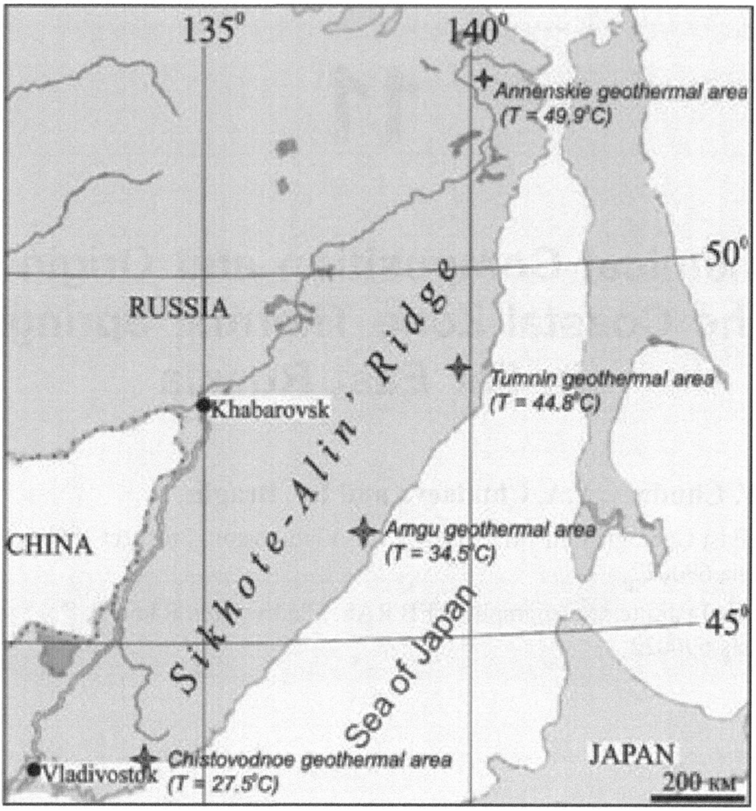

Figure1: Location of studied springs.

2. MATERIALS AND METHODS

Sampling of studied springs was done in summer seasons (July-August) of 2006, 2007, and 2008 years. Samples were filtered through 0.45 μ membrane filters. The water samples for the cation and silica determinations were acidified with HNO_3 to pH 2 to stabilize the solution and prevent the development of organics. To analyze the anions, the filtered sample was collected in a separate polyethylene bottle without oxidation. For the oxygen and hydrogen isotope composition, non-filtered water samples were loaded into a glass bottle and closed with a dense cap in order to avoid exchange with the atmospheric air. The unstable water parameters such as the pH, Eh, SEC (electro conductivity), DO (dissolved oxygen), and bicarbonate ion (HCO_3^-) were measured in situ. For this purpose, we used a portable field laboratory consisting of a microprocessor equipped with exchangeable pH, Eh, SEC, and DO electrodes. The HCO_3 was determined by titration using a "Hach" digital equipment. The main ions and trace elements were analyzed using ICP-AES and ICP-MS on

Thermoscientific iCAP 6000 and Agilent 7500c devices. The anion concentrations were measured using a LC-10Avp (Shimadzu) liquid ion chromatograph. The oxygen and hydrogen isotopes were determined using a Finnigan-MAT 252 mass spectrometer. All these analyses were carried out at the analytical centre of the Far East Geological Institute, Russian Academy of Sciences. The computer modelling and calculation of the equilibrium reactions were done using the SOLMINEQ software package.

3. STUDY AREA

The thermal waters of Sikhote Alin belong to the province of nitric thermal waters in the zones of young tectonic movements, which are typically distributed in the granite massifs and/or their contact zones. The tectonic factor primarily controls the localization of intrusions, which are penetrated by water along shears and fractures. Among the diverse nitric thermal waters of the Far East Russia Chistovodnoe and Amgu (including Saion) groups of springs (Primorye territory) and Tumninskii and Annenskii springs (Khabarovskii territory) were studied. Geographically these territories belong to Sikhote Alin ridge (Fig. 1). The studied springs of Sikhote Alin are formed in the contact zones between granite intrusions and volcanic rocks of the East Sikhote Alin volcanic belt. The surface water temperature at the springs gradually increases from southern Sikhote-Alin northward from 27°C to 50°C. This indicates the northward deepening of the water circulation. We also cannot exclude the influence of the geothermal gradient, which can be higher in the north owing to the emplacement of the younger Pliocene volcanism in this area (Martynov, 1999).

4. RESULTS AND DISCUSSION

4.1 Major Ions

The studied waters are characterized by pH close to 9 with Na as the predominant main cation and HCO_3 as the predominant anion. These are typical sodium waters with a sharp predominance of bicarbonate ion and sodium (Table 1). The sodium and calcium contents show opposite correlation.

The sodium increases with increasing water temperature. With increasing temperature (even of several degrees) in an aluminosilicate-water system, sodium is leached from Na-Ca-bearing minerals more efficiently than calcium. According to the data of Ryzhenko et al. (1996, 2000), HCO_3-Na waters are formed in the granite/water system at the initial stages if water predominates. The sodium concentration in the studied thermal waters varies within 61-19.6 mg/l (Table 1) and typically is two and more times lower than in the main European springs of this type and several times lower than in the similar waters of the Korean Republic (Michard, 1990; Yum, 1995). At the same time, these values are higher than those in the surrounding fresh shallow-circulation ground

Table 1: Selected data of average chemical elements contents for Chistovodnoe, Amgu-Saion, Tumninskii and Annenskii thermal waters

Group of springs	*Chistovodnoe* *n=21*	*Amgu-Saion* *n=12*	*Tumninskii* *n=12*	*Annenskii* *n=12*
T °C (mes./cal.)	27.5/61	34.5/81	44.8/86	49.9/99
		Mg/l		
TDS	157	183	195	300
HCO_3^-	63.4	70.7	78/1	112.8
SiO_2	32.1	53.71	60.78	80.9
SO_4^{2-}	5.7	13.6	7.1	25.4
Cl^-	2.4	3.6	1.4	4.0
Na^+	19.6	33.4	32	61
K^+	0.33	0.7	0.35	1.17
Ca_2^+	5.42	2.0	2.11	2.0
Mg_2^+	0.007	0.08	0.01	0.01
		µg/l		
Mo	15.6	21.78	13.89	6.23
Ge	0.72	0.65	0.65	2.65
F	4900	1000	300	2500
Cu	0.09	0.38	2.59	3.79
As	3.42	6.26	21.28	14.32
Al	11.7	22.9	11.3	14.5
Li	46.17	6.26	11.39	71.49
Rb	0.25	1.65	1.15	6.91
Sr	42.93	24.35	17.14	69.99

waters of the same area. According to the thermodynamic calculations, sodium mainly occurs in the ionic form, and the complex ions of $NaCO_3^-$ and $NaSO_4^{2-}$ is lower than the ionic form.

The potassium contents show no significant variations (1.17-0.33 mg/l) and practically do not differ from those in the water of shallow circulation (Table 1). Its concentration is several times lower than that in the European springs of the same type and 5-10 times lower than that in the similar waters of the Korean Republic (Michard, 1990; Yum, 1995). Based on the constant potassium content in the variably cooled thermal waters, Michard (1990) concluded that this element is weakly involved in the water-rock interaction. The thermal waters of Sikhote Alin also have stable calcium contents within 2.0-5.4 mg/l (Table 1), which is within the calcium variations in the thermal waters of the granite massifs of Europe and corresponds to the similar springs of the Jangwon area in the Korean republic (Michard, 1990; Yum, 1995). Calcium shows no correlation with the temperature or HCO_3^-. As was mentioned above, it negatively correlates with sodium.

Magnesium in the alkaline thermal waters occurs in small amounts (Table 1). The waters of Primorye and the studied springs of Khabarovskii region contain less than 0.08-0.007 mg/l Mg. Similar thermal waters of Europe and South Korea also have a low Mg content. Among the studied springs, the cold fresh ground waters of the same area are significantly higher in their Mg content than the thermal waters.

The silica (as SiO_2) content in the thermal waters of Primorye is 32.1-80.9 mg/l (Table 1). Silica migrates in the solutions as polymeric compounds, whose proportions vary depending on the pH. The silica content in the thermal waters is controlled by the equilibrium with quartz or chalcedony. The thermal waters have a three times higher bicarbonate ion content than the surrounding cold waters.

The loss and changes of the chlorine during its migration with the water through the soil and fractures in the rocks are less significant than those of other tracers. Such an immobile behaviour of chlorine is used to identify the chemical components of the ground waters. The sources of chlorine in waters are sea water intrusions, salt deposits, biotite of host rocks and others (Chudaev, 2003). In studied thermal waters marine influence is presumably absent (Chudaev, 2003). The thermal waters contain low and weakly variable Cl contents, which correspond to those in surface waters: 1.4-4 mg/l with the highest content in the northern group (Annenskii springs, 4 mg/l) and the lowest contents in the Tumninskii springs (1.4 mg/l). The Cl content in the studied thermal waters is lower than similar waters in the People's Republic of Korea.

Sulfate ion behaves as a mobile complex. No saturation in calcium sulfate (gypsum or anhydrite) is attained in the alkaline waters. Its lowest contents were found in the springs of the Chistovodnoe group (5.7 mg/l), and the highest contents, in the Annenskii waters (25.4 mg/l) (Table 1). Some researchers suggest that they are formed through oxidation of sulfides in the host sequences, as in Baikal rift zone (Zamana, 2000). In finalizing the description of the major ions in the nitric thermal waters, we note that the TDS (total dissolved solids) is mainly provided by sodium and, partly, by silica, whose content depends on the water temperatures. The latter increases from the south to the north of Sikhote Alin along with the proportional increase of TDS.

4.2 Trace Elements

According to the generally accepted geochemical classification of elements, trace elements are subdivided into three main groups: siderophile, chalcophile, and lithophile. Rare earth elements (REE) will be described separately owing to their particular geochemical features, though they belong to lithophile elements. Among the obtained trace element data (>50 elements), only the elements that emphasize the geochemical features of each of the selected water groups are considered. Representative data of some minor elements are shown in Table 1.

4.2.1 Siderophile Elements

Of most interest among the siderophile elements are iron, manganese, molybdenum, cobalt and nickel. The thermal waters have a low Mn content within 0.013-0.005 mg/l, which somewhat increases only in the surrounding cold ground waters reaching 0.04 mg/l. At the available Eh and pH, the main mode of iron occurrence in aqueous solutions is insoluble $Fe(OH)_3$. Though manganese does not belong to this group, it is considered together with iron in view of their high geochemical affinity. The manganese concentration in the thermal waters, as that of iron, is extremely low, mainly below 0.1 μg/l with its increase from 0.06 to 0.28 μg/l in the northern direction. All waters have high molybdenum contents, which are the highest in the Amgu springs (21.78 μg/l) and the lowest (6.23 μg/l) in the Annenskii springs (Table 1). As is known, the molybdenum content in an alkaline environment is significantly higher than that in an acid environment. In the fresh surface water of the closely spaced Amgu springs, the Mo content decrease down to 0.37 μg/l and lower. Cobalt varies within 0.005-0.009 μg/l. Its highest content was found in the Annenskii springs and the lowest in the Chistovodnoe springs. The nickel content increases northward from 0.033 μg/l in the Chistovodnoe springs to 0.43 μg/l in the Annenskii spring. The surrounding fresh waters contain 0.1-0.2 μg/l nickel.

4.2.2 Chalcophile Elements

Among the chalcophile elements, in this section, we will consider copper, arsenic and zinc. The copper content in the studied thermal waters is low, no more than 3.79 μg/l (Annenskii). The lowest contents are observed in the Chistovodnoe of springs (0.09 μg/l). The content of dissolved copper gradually increases from the south northward. The lowest arsenic content was also obtained in the Chistovodnoe springs (3.42 μg/l) and the highest content in the Annenskii springs (21.28 μg/l). The Amgu springs contain 6.26 μg/l As (Table 1).

4.2.3 Lithophile Group

Among the lithophile elements, we will consider Al, F, Li, Rb, Sr, Ge and Ba. The Al content in the studied waters varies within 6.26-46.17 μg/l (Table 1). The highest contents were found in the Chistovodnoe springs (46.17 μg/l) and the lowest contents, in the Tumniskii springs (11.3 μg/l). These values are insignificantly higher than those in the fresh waters. The F content in the studied thermal waters is within 0.3-4.9 mg/l with the lowest contents found in the Tumninskii springs (0.3 mg/l) and the highest contents, in the Chistovodnoe group (4.9 mg/l) (Table 1). In the Annenskii springs, the F concentration reaches 2.5 mg/l. In general, this is higher than that in the cold waters (0.1-0.4 mg/l). The F content in the studied thermal waters is within the values reported by Yum (1995) and Michard (1990) for the thermal waters of Korea and Europe. According to Krainov (1973), the highest F waters in the former USSR are nitric alkaline low-salinity thermal waters. The F content in them can reach 27 mg/l. The data reported by Krainov (1973) for Primorye (Chistovodnoe group) of 15 mg/l are significantly higher than our values. Taking into account that the thermal waters of Primorye were monitored by us for ten years, our data obtained by different analytical methods seem to be more plausible. The Li content varies within 6.26-71.49 μg/l. The highest content was obtained in the waters of Annenskii springs and the lowest content in the Amgu springs (Table 1). In general, the Li content in the thermal waters is significantly higher (by almost an order of magnitude) than that in the fresh cold ground waters. The thermal waters have a low Rb content (a few μg/l), which, however, is higher than that in the fresh ground waters and surface waters of the studied areas. The highest Rb content occurs in the Annenskii springs (6.91 μg/l) and the lowest content, in the Chitovodnenskii springs (0.25 μg/l). There is a trend of increasing Rb content from the south northward of the studied area. The Sr content in the thermal waters is close to that in the ground fresh waters, reaching the highest values (69.99 μg/l) in the Annenskii thermal waters and the lowest (17.14 μg/l) in the Tumninskii springs. Strontium typically migrates in the ionic form. The lowered Li, Rb and Sr concentrations in the studied thermal waters relative to some European thermal waters can be related, first, to the low temperatures of Sikhote Alin and, second, to some differences in the composition of the host rocks.

Researchers also have reported relatively low contents of trace alkaline elements in nitric thermal waters (Lomonosov, 1974; Chudaeva et al., 1999). The Ga and Ge contents in the thermal waters are several times higher than their contents in the ground waters. The greatest Ga content occurred in the Amgu springs (2.27 μg/l) and the lowest contents in the springs (1.32 μg/l) (Chudaeva, 2010). Ge content is also significantly higher than that in the fresh waters. The lowest contents (0.65 μg/l) were found in the Tumnin and Amgu springs and the highest contents in the Annenskii springs (2.65 μg/l). The fresh waters of the studied area contain <0.01 μg/l Ge. At the same time, Krainov (1973) reported up to 10 μg/l of Ge in the thermal waters of Primorye, which is not confirmed by our data. Boron shows a notable increase from the south northward. In particular, the Chistovodnoe springs contains no more than 6.2 μg/l, whereas the Annenskii waters contain up to 21 μg/l. The surrounding fresh waters have B content around 1.4 μg/l. Among the actinides, we found only uranium, whose maximal content is noted in the Chistovodnoe springs (2.94 μg/l). The minimal content of 0.04 μg/l was noted in the Amgu springs. The results of Yum (1995) showed an insignificant U content (0.02-0.09 μg/l) in the thermal waters of the Pochenga biotite granites (South Korea).

4.2.4 Rare-earth Elements (REE)

The REE level in the thermal waters is close to that in the atmospheric precipitation (Chudaev, 2003). All the waters have some HREE (heavy REE) enrichment and Eu anomaly. Eu is supplied in the thermal waters from plagioclase during its interaction with the water. As is known, plagioclases have a distinct positive Eu anomaly as compared to other REE-bearing minerals (Rollinson, 1993). According to Moller et al. (1999), the HCO_3-Na thermal waters of the Czech Republic, Germany and Turkey demonstrated a diverse Eu anomaly. In addition to plagioclase, the host rock (granites) of the Chistovodnoe waters contain up to 7.2 mg apatite, 10 mg monazite, and 19 mg titanite in a 5-kg granite sample (Chudaev, 2003). These minerals are the main REE carriers. At the same time, the REE supply in the water via dissolution of these minerals is limited by their low solubility at the given physicochemical conditions and relatively high water exchange rate. The thin-sections study showed that the plagioclases in the water-hosting rocks were strongly altered and transformed into mica, secondary albite and zeolite.

4.3 Gas Composition

No significant deep supply of volatile components along the faults is observed at the discharge sites of the thermal waters (Chudaeva et al., 1999; Chudaev 2003). The absence of the influence of juvenile gases is confirmed, for instance, by the low He isotopic ratio $^3He/^4He$ (0.1-0.24 $\times$ 10^{-6} for the thermal waters of the Chistovodnoe springs (Bogolybov et al., 1984). Our data show that the dissolved gases of studied thermal waters are dominated by nitrogen (up to 99%). There are also insignificant amounts of H_2S. The gas component presumably had an atmospheric origin and was transformed during deep circulation. As can be calculated, the Ar $\cdot$ $100/N_2$ ratio in the spontaneously emanating gas in the Primorye thermal waters is, respectively, 1.26, 1.27, and 1.15, which, given the measurement error, is very close to the atmospheric ratio (1.18). The insignificant northward increase of this ratio in the similar waters can be related to the addition of dissolved gas, where the Ar/N_2 ratio is higher owing to the higher Ar solubility.

4.4 Origin of Thermal Waters

The oxygen and hydrogen isotopic compositions signify the meteoric origin of the water component in the thermal waters and distinct latitudinal variations (Fig. 2). In particular, the southern Chistovodnoe springs are characterized by $\delta^{18}O$ of –10.8‰ and δ^2H of –70‰. Northward, the values of the oxygen and hydrogen isotopic composition become lower. The northern Annenskii group has $\delta^{18}O = -18.8$ and $\delta D = -136.1$‰. The obtained data confirm the view of a significant contribution of meteoric waters in the formation of the hydrothermal systems in different geological tectonic settings, including the zones of modern volcanism (Chudaev, 2003).

The $^{87}Sr/^{86}Sr$ ratios are about 0.70638 in the Chistovodnoe group of thermal waters and 0.71027 in the fresh ground waters of this area. This ratio in the thermal springs of the Amgu group is within 0.70458–0.70483. In our opinion

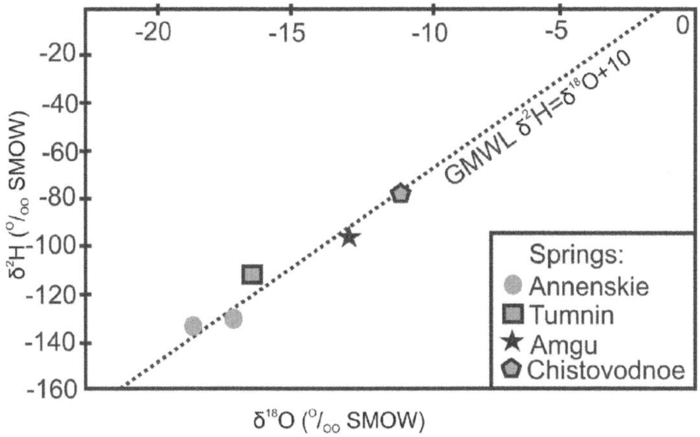

Figure 2: Variation of the oxygen and hydrogen isotopes in studied thermal waters.

this difference in the Sr isotopic ratios between the northern and southern groups of the thermal waters of Primorye is caused by the different water exchange rate and composition of the host rocks. Taking into account the difference in the TDS, its northward increase, and the drilling data, the rate of water exchange in the studied thermal waters of Primorskii region is higher than that of the Khabarovsk district (Bragin et al., 2007). Our data indicate that the cations in the atmospheric precipitation of the Far East are dominated by sodium, whereas the ground waters have a mixed cation composition (Chudaev, 2003). This can be explained by the fact that the HCO_3-Cl-Na rain water in the zone of the groundwater formation accumulates mainly calcium and carbonate ions owing to the decomposition of the soil organics, which leads to the formation of HCO_3-Ca-Na-Mg ground waters. Further subsidence and heating of the waters is accompanied by the predominant accumulation of sodium via mainly plagioclase dissolution.

The results of Ryzhenko et al. (1996) of the modelling of the granite-water system showed that, at the initial stage of the interaction (the water/rock ratio >> 1), weakly mineralized HCO_3-Na waters are formed. Note that, according to the calculations of Ryzhenko et al. (1996), the liquid phase in the water-granite system at a temperature >25°C and pressure up to 5 kbar has an alkaline reaction. The thermodynamic calculations showed that the studied thermal waters are oversaturated with respect to clay minerals (smectite, illite and kaolinite), the group of low-temperature zeolites and albite. The calculation of the forms of the main ions in the solution showed that up to 85% are represented by the ion form and 15% occurs as complexes with bicarbonate ions, whose content increases from the south northward. The waters of the Amgu and Annenskii groups contain insignificant amount of forms with sulfate ion. The temperature determined from the quartz was close to 61°C in the Chistovodnoe group, 81°C in the Amgu group, 86°C in the Tumninskii group, and 99°C in the Annenskii group; i.e., the temperature equilibrium with quartz increases from the south northward of Sikhote Alin. This indicates a northward increase of the temperature gradient, on the one hand, and deepening of the water circulation in the same direction, on the other. However, the studied thermal waters, judging from their chemical composition, belong to hydrothermal systems with sufficiently high water exchange rate, which follows from the TDS and REE data.

5. CONCLUSIONS

1. The considered nitric thermal waters are low-salinity waters of the HCO_3-Na type. The level of most of the elements is lower than in similar waters of Europe and Korea. The content and behaviour of the chemical elements in the studied waters is controlled by the residence time and the formation of secondary equilibrium minerals.
2. The first obtained REE data showed their low contents close to those in the atmospheric precipitation. The REE distribution patterns indicate that their main source in the water were plagioclases. The low salinity of the waters and the REE distribution pattern attest to the rapid water circulation in the rock sequence. This is also confirmed by the data on the TDS.
3. The oxygen and hydrogen isotopic data indicate that the main water component is meteoric water. No influence of sea water intrusions. Nitrogen, the main gas phase, is of atmospheric origin.
4. The obtained data show increasing water temperature from the Chistovodnoe (south Sikhote Alin) to the Amgu, Tumninskii, and Annenskii (north Sikhote Alin) springs, which presumably indicates the increasing depth of the water circulation to the north and an increase of the thermal gradient owing to the occurrence of young Pliocene volcanism there.

ACKNOWLEDGEMENTS

This work was supported by the Russian Foundation for Basic Research (project no. 07-05-00282) and by DVO FEB RAS project (09-2-CO-08-005) and 09-1-P16-03.

REFERENCES

Avdeeva, A.B. (1976). Main types of mineral waters of the south Far East Russia (Primorskii and Khabarovskii territory). *In:* Proceedings of Central Institute of Physiotherapy. Moscow, **31**, pp.19-30.

Bogolyubov, A.N., Korplyakov, O.P., Benkevich, L.G. and Yudenich, V.S. (1984). Helium Isotopes in the Groundwaters of Primorye. *Geokhimiya*, **22(8)**, 1241-1244 [in Russian].

Boldovski, N.V. and Kulakov, V.V. (2004). Continental Hydrothermal Systems in the South of the Far East. *In:* Proceedings of the International IAGOD Conference on Tectonics, Magmatism and Metallogeny. Dalnayka.Vladivostok, pp. 623-625.

Bragin, I.V., Chelnokov, G.A., Chudaev, O.V. and Chudaeva, V.A. (2007). Low-Temperature Geothermal Waters of Continental Margin of the Russian Far East. *In:* Proceedings of International Symposium on Water-Rock Interaction/Balkema. Kunming. **1**, 481-484.

Chudaev, O.V. (2003). Composition and Formation Conditions of the Modern Hydrothermal Systems of the Russian Far East. Dalnauka, Vladivostok, 216 p. [in Russian].

Chudaeva, V.A., Chudaev, O.V., Chelnokov, A.N., Edmunda, M. and Schand, P. (1999). Mineral waters of Primorye (Chemical Aspects). Dalnauka, Vladivostok, 213 p. [in Russian].

Chudaeva, V.A. (2010). REEs in coastal zone thermal springs far-east Russia (unpublised report) personal communication.

Kirukhin, D.A. and Reznikov, A.A. (1962). New data on the chemical composition of nitric thermal waters of the south Far East Russia. Hydrogeology of Siberia and Far East Russia. Irkutsk, pp. 71-83. [in Russian].

Krainov, S.R. and Shvets, V.M. (1992). Hydrogeochemistry. Nedra, Moscow, 312 p. [in Russian].

Krainov, S.R. (1973). Geochemistry of Rare Earth Elements in the Ground waters. Nedra, Moscow, 203 p. [in Russian].

Lomonosov, I.S. (1974). Geochemistry and Formation of Modern Hydrothermal Springs of the Baikal Rift Zone. Nauka, Novosibirsk, 204 p. [in Russian].

Martunov, Yu. A. (1999). Geochemistry of basalt of active continental margin and island arks. Dalnauka, Vladivostok, 189 p.

Michard, G. (1990). Behavior of Major Elements and Some Trace Elements (Li, Rb, Cs, Sr, Fe, Mn, W, F) in Deep Hot Waters from Granitic Areas. *Chemical Geology*, **89**, 117-134.

Moller, P., Dulski, P., Morteani, G. Gerstenberger and Fuganti, A. (1999). Rare Earth Elements, Yttrium and H, O, C, Sr, Nd and Pb Isotope Studies in Mineral Water and Corresponding Rocks from NW-Bohemia, Czech Republic. *Applied Geochemistry*, **13**, 975-994.

Rollinson, H. (1993). Using Geochemical Data: Evaluation Presentation, Interpretation. Longman, Singapore, 421 p.

Ryzhenko, B.N. and Krainov, S.R. (2000). On the Influence of Mass Proportion of Reacting Rock and Water on the Chemical Composition of Natural Aqueous Solutions in Systems Open to CO_2. *Geokhimiya,* **38(8)**, 803-815.

Ryzhenko, B.N., Barsukov, Vik. and Knyazeva, S.N. (1996). Chemical Characteristics (Composition, pH, and Eh) of a Rock-Water System. 1: Granitoids-Water System. *Geokhimiya*, **34(5)**, 436-454.

Yum, B.W. (1995). Movement and Hydrochemistry of Thermal Waters in Granite at Cosung, Republic of Korea. *In:* Proceedings of the 8th International Symposium on Water-Rock Interaction. (eds.) Y. Kharaka and O. Chudaev. Balkema, Vladivostok, pp. 324-327.

Zamana, L.V. (2000). Genesis of Sulfate Composition of Nitric Springs in the Baikal Rift Zone. *Doklady Akademii Nauk*, **372(3)**, 361-363.

Section 3

Biodiversity of Coastal Zones and Its Sustainability

12

Organic Matter and Mangrove Productivity

M. Bala Krishna Prasad, Thorsten Dittmar[1] and AL. Ramanathan[2]

Earth System Science Interdisciplinary Center, University of Maryland College Park, MD 20740–3823, USA

[1]Max Planck Research Group for Marine Geochemistry, Carl von Ossietzky University, Institute for Chemistry and Biology of the Marine Environment (ICBM), Carl-von-Ossietzky-Str. 9-11, D-26129 Oldenburg, Germany

[2]School of Environmental Sciences, Jawaharlal Nehru University, New Delhi 110067, India

1. INTRODUCTION

Mangrove forests function as a link between marine and terrestrial ecosystems; are important for the stability and maintenance of various adjoining ecosystems such as seagrass, coral reef, and other marine ecosystems; and provide habitats for juvenile crustaceans and fish. Mangroves also contribute significant quantities of organic matter (OM) via litter fall that can be transported offshore (Odum and Heald, 1972; Alongi, 1990). Depending upon the geomorphology of the basin and tidal amplitudes, mangrove stands receive nutrients and OM from terrestrial runoff, estuarine outflows (Boto and Bunt, 1981; Alongi, 1990), marine inputs and water column production by phyto-plankton and -benthos (Bouillon et al., 2004). As a result, mangrove ecosystems are considered sites of intensive biogeochemical processes with a potentially impacting global element cycles (Dittmar et al., 2006; Bouillon et al., 2008).

Litter fall is an important index of potential carbon export to coastal marine ecosystems (Twilley et al., 1997) and a valuable indicator of mangrove productivity (Clough, 1992). The amount of litter produced in a mangrove forest directly affects the outwelling of particulate organic matter (POM) between mangrove forests and adjacent marine habitats. Ambient temperature is a regulating factor for the dynamics and cycling of leaf litter in mangroves (Mfilinge et al., 2002). Given the wide temperature range, there is little quantitative information of a comparative nature available for the effect of temperature and other climatic factors on mangrove outwelling and its impact on sedimentary organic matter. Low temperatures result in slowing important ecological processes such as litter production (Tam et al., 1998) and turnover, detritus decomposition, and activity of leaf eating crabs.

Direct grazing by crabs substantially reduces export of leaf litter by tidal transport and accelerates its breakdown (Robertson, 1991). Reduced crab activity in winter slows down leaf breakdown (Mia et al., 2001). Therefore, at that time, most litter decomposes by microbial action or is transported by tides to adjacent habitats. On this basis, it was assumed that even though a smaller amount of litter might be produced in mangroves during winter, a larger fraction might be available for export. Throughout the year, not all OM is exported by tidal action to adjacent coastal waters. The retained OM accumulates in the sediment where it is degraded and chemically modified by both aerobic and anaerobic processes. A very minor portion of OM escapes degradation and is permanently buried in the sediments. The major pathways of OM cycling in mangroves are described in Fig. 1.

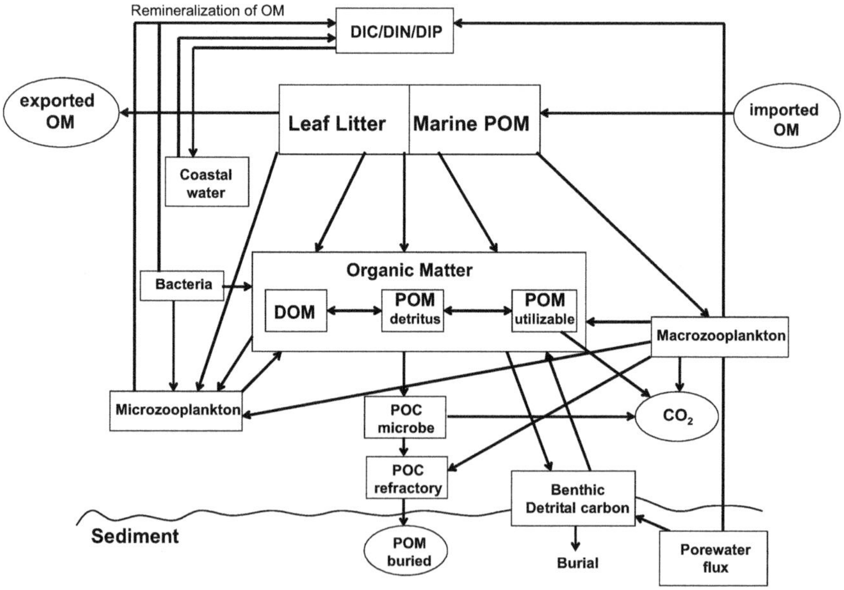

Figure 1: Major sources and pathways of organic matter in the mangrove ecosystem. Squares represent pools of organic matter. Ovals represent pools that are permanently displaced from mangrove by the biological transformation and arrows represent transfer of material between pools through key processes like transport, leaching, microbial degradation and faunal grazing (modified from Hansell and Carlson, 2002; Kristensen, 2008).

In this article, authors have reviewed and evaluated the current understanding of OM dynamics and its impact on mangrove productivity. The various sources of OM in mangroves are discussed and subsequently the molecular and chemical characteristics of OM are described. Finally, the fate of mangrove-derived OM in the sediments and water column is described.

2. ORGANIC MATTER SOURCES AND SINKS

The relative contribution of mangrove-derived OM to mangrove food webs varies widely with forest type and the amount of clear, open waters. In general, mangrove ecosystems have a high proportion of forest floor to open water, and waters are usually turbid (Blader et al., 1985; Robertson and Duke, 1987). In these systems, carbon fixed by mangrove plants is usually the dominant contributor of OM to the food web. In systems with a larger proportion of clear, open water, benthic and planktonic algae can be another significant primary producer. In addition to litter fall and phytobenthos, riverine supply of OM to mangroves can be significant and impact ecosystem productivity (Singh et al., 2005). Nevertheless, mangrove litter fall is often used as a standard proxy to estimate mangrove productivity. It is reported that the global litter fall rates are in the order of approx. 38 mol C m^{-2} yr^{-1} (Twilley et al., 1992; Jennerjahn and Ittekkot, 2002). However, this underestimates the total net CO_2 fixation by mangrove vegetation and benthic algae, because it does not incorporate wood and belowground biomass production (Middleton and McKee, 2001) nor dissolved organic matter (DOM) released from roots. Belowground biomass can comprise a substantial part (up to 55%) of the total mangrove biomass (Twilley et al., 1992; Alongi and Dixon, 2000). In several studies net primary production (e.g. Alongi et al., 2005) and wood and root production rates (e.g. Twilley et al., 1992) were estimated. According to these estimates litter fall constitutes probably less than 25% of net primary production.

The supply of OM from microphytobenthos can be significant, and values between 7 and 73 mol C m^{-2} yr^{-1} were reported (Gattuso et al., 1998; Alongi et al., 1998; Holmer et al., 2001; Koch and Madden, 2001; Kristensen and Alongi, 2006). In other systems the contribution of phytobenthos is low mainly because of light limitation or possibly inhibition by tannins (see Alongi, 1994). The diversity, distribution and productivity of microbenthos are highly variable and regulated by geomorphology, hydrodynamics and nutrient chemistry. It has been reported that microbenthos productivity is higher in mangrove-dominated lagoons (e.g. Ivory Coast, approx. 146 mol C m^{-2} yr^{-1}; Robertson and Blader, 1992) than in mangrove-fringed estuaries (e.g. Fly River delta, 0.7-21 mol C m^{-2} yr^{-1}; Robertson et al., 1992). In spite of a potentially high production of microbenthos in mangrove creeks, high turbidity, tannins and active microbial degradation regulate many creeks as net heterotrophic systems, even during day times (Kristensen and Surawadi, 2002).

During the course of the biogeochemical transformation in intertidal mangrove ecosystems, a significant amount of OM is microbially transformed, mineralized or buried in sediments. The initial degradation of leaf litter is often initiated by various invertebrates, e.g. amphipods, nematodes, turbellarians, isopods and polychaetes (Padma Dorothy et al., 2003; Bosire et al., 2005). The shredding and macerating activities of these animals apparently increase the degradability of leaf litter by increasing the effective surface area accessible to microbial attack (Robertson et al., 1992).

Twilley et al. (1997) estimated that within 10–14 days newly fallen leaves lose 20–40% of organic carbon by leaching when submerged in sea water. The leached compounds are mainly non-lignocellulose compounds (Neilson and Richards, 1989). Most of DOM leached from plant litter is labile and degraded efficiently under oxic and nutrient satiated conditions with conversion efficiencies into the microbial loop of up to 90% (Benner et al., 1986). In contrast to this, in nutrient-limited and anoxic/hypoxic mangrove soils, the microbial incorporation is not efficient and accounts roughly to about 35% (Boto et al., 1989). The remaining lignocellulose compounds in plant litter can be degraded via extracellular enzymatic hydrolysis by aerobic and anaerobic bacteria, marine fungi and protoctista (Newell, 1996). Fungi and protoctista have a higher degradative potential for cellulose compounds in leaf detritus than for lignin, thus degraded plant litter is relatively enriched in lignin (Marchand et al., 2005). These lignin compounds are more easily degraded under oxic than anoxic conditions, and it has been estimated that in anoxic sediments the half-life of lignin is more than 150 years (Dittmar and Lara, 2001b).

Benthic and pelagic respiration are important mechanism that removes OM from mangroves. In the intertidal and subtidal regions of Hinchinbrook Channel, Australia, benthic and pelagic respiration accounted for >50% of all OM sinks (Alongi et al., 1998). Also in the Fly Delta of Papua New Guinea, benthic and pelagic respiration in the delta was considered a major sink for mangrove-derived organic carbon (Robertson and Alongi, 1995). Burial of OM in sediments is also a potential sink for OM in mangrove environments. The net burial of OM in sediments is a function of both deposition rate to the sediment and mineralizaton rates within the sediment. The latter is a function of the reactivity of the OM to be decomposed (e.g. vegetation source) and the sedimentary environment (e.g. burial rates, grain size, redox state, microbial population, etc.). Gonneea et al. (2004) estimated that OM burial rates in the mangroves of Yucatan Peninsula were higher in fringe mangrove sediments (55–70 g C m^{-2} yr^{-1} or 4.6-5.8 mol C m^{-2} yr^{-1}) than in seagrass-dominated mid-lagoon sediments (40 g C m^{-2} yr^{-1} or 3.3 mol C m^{-2} yr^{-1}). They also observed vertical decrease in burial rates in sediments which may suggest either that climatic fluctuations or anthropogenic perturbations negatively impacted the mangrove ecosystem and reduced OM burial. Alongi et al. (1998) developed a mass-balance model for OM in mangroves of Hinchinbrook Channel, Australia,

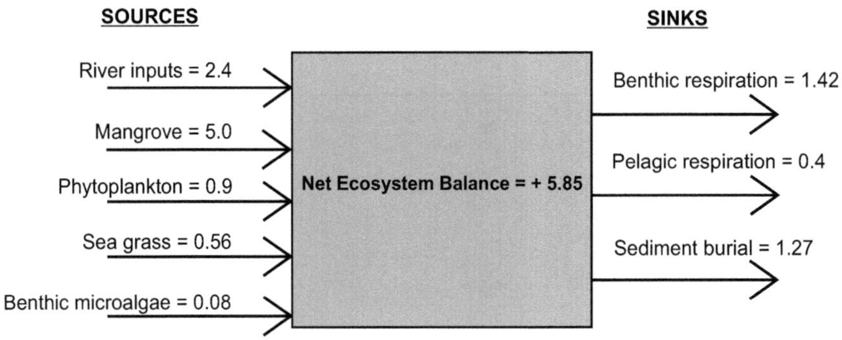

Figure 2: Mass balance of organic carbon in Hinchinbrook Channel. Units: mol organic carbon per year (modified from Alongi et al., 1998).

and reported that total sources of OM are 8.94×10^9 mol OC yr^{-1} and sinks are 3.09×10^9 mol OC yr^{-1}. The difference of 5.85×10^9 mol OC yr^{-1} is available for export to adjacent coastal water (Fig. 2).

3. CHEMICAL CHARACTERISTICS OF ORGANIC MATTER

3.1 Atomic C:N Ratio and Stable Carbon Isotopes

The abundance and ratios of important elements in biological cycles (e.g. C, H, N, O, S, Si and P) provide basic information on OM dynamics and cycling in mangrove ecosystems. Atomic C-to-N ratios can provide preliminary source information about OM (see Meyers, 1997). The broad range of C:N ratios across divergent sources of OM in the biosphere demonstrate how such ratio can provide an initial proxy for determining source information (Table 1). The basic reason for big difference in C:N ratio between vascular plants (>17) and micro-algae (5-7) is primarily due to carbohydrate-rich/protein-poor and protein-rich/carbohydrate-poor nature of each source, respectively. Several studies established quantitative mixing models using C:N ratios of 6 for the marine (Müller, 1977) and 13 for the terrestrial (Parrish et al., 1992) end members. However, C:N ratios can be misleading in determining OM sources in the absence of additional proxies. For instance, selective utilization of N by microbes can result in artificially high C:N ratios resulting in misidentification of OM sources. The source and fate of OM in mangroves can be traced more specifically with stable carbon isotopes ($\delta^{13}C$). Many consumers in mangrove habitats have carbon isotope ratios close to that of mangrove litter, which indicates the relative importance of leaf litter for various trophic levels.

Table 1: Approximate C:N ratios in some terrestrial and marine producers*

	C:N
Terrestrial	
Leaves	100
Wood	1000
Marine vascular plants	
Zostera marina	17-70
Spartina alterniflora	24-45
S. patens	37-41
Avicennia marina	37
A. germinas	53
Rhizophora apiculata	42
R. mangle	75
Laguncularia racemosa	73
Marine macroalgae	
Browns (Fucus, Laminaria)	16-68
Greens	10-60
Reds	20
Microalgae and microbes	
Diatoms	6.5
Greens	6
Blue-greens	6.3
Peridineans	11
Bacteria	5.7
Fungi	10

* Data compiled from Fenchel and Jørgensen (1977), Alexander (1977), Fenchel and Blackburn (1979), Valiela and Teal (1976), Valiela (1995), Dittmar and Lara (2001b), Prasad and Ramanathan (2009).

3.2 Molecular Biomarkers

Due to complexity of OM in estuaries and mangroves and the aforementioned problems associated with bulk measurements and C:N ratios, the application of chemical biomarkers has become a powerful tool in estuarine and mangrove research (see Bianchi and Canuel, 2001). Carbohydrates, amino acids, tannins, lignin-derived phenols, fatty acids and *n*-alkanes are generally used biomarkers to study the chemical nature of OM in estuaries and mangroves (Dodd et al., 1995; Opsahl and Benneer, 1999; Dittmar and Lara, 2001b; Hernes et al., 2001; Versteegh et al., 2004; Marchand et al., 2005; Meziane et al., 2007). Carbohydrates are present in abundant quantities in the cell walls of vascular plants. Total carbohydrate yields can be up to 65% of organic carbon in mangrove wood (Opsahl and Benner, 1999). Several studies show that even though mangrove litter have typical carbohydrate (i.e. glucose, arabinose, galactose, xylose, etc.) signatures, these tend to disappear rapidly in the detritus pool since carbohydrates are highly reactive compounds compared to bulk

OM (Opsahl and Benner, 1995; Marchand et al., 2005). Nevertheless, neutral carbohydrates show selective degradation in mangrove sediments, which may provide some source information, in spite of their low concentrations.

Amino acids can represent up to 9% of mangrove leaf biomass (Hernes et al., 2001), but very few studies have described their composition in mangrove tissues. Zieman et al. (1984) reported concentration of total amino acids in Rhizophora leaves of 833 mmol g^{-1} with glutamic acid, leucine, and glycine representing each more than 10%. Avicennia leaves, on the other hand, contain mostly glycine, glutamic acid and aspartic acid (Tremblay and Benner, 2006). The concentration of amino acids tends to increase during decomposition due to microbial accumulation (Hernes et al., 2001; Tremblay and Benner, 2006). Since amino acids of prokaryote origin then become prevalent, their use as mangrove tracers appears limited.

Tannins in vascular plants occur as two types, condensed and hydrolysable. They are more abundant in plant leaves than in woody tissues, and contribute to the colour and astringency of the bulk OM. Hernes et al. (2001) found that green leaves of *R. mangle* may contain more than 6% tannins, being third in abundance after polysaccharides (21%) and amino acids (9%). The tannin content of mangrove leaves is higher and more polymerized than found in most other dicotyledonous plant species (Hernes et al., 2001). Condensed tannin consists of 80% procyanidin and 20% prodelphinidin. The latter has a higher degree of hydroxylation and appears to be more labile than procyanidin tannin. Leaching, which induces an increase in polymerization of condensed tannin, is an important mechanism for tannin removal from leaves. Hernes et al. (2001) concluded that while the tannin composition of brown and partly decomposed leaves is recognizable as dicotyledonous in origin, it is difficult to attribute the composition specifically to mangrove leaves.

To date lignin is the only established molecular tracer for terrigenous OM in the coastal and ocean waters. It is an unambiguous tracer for vascular plants and can even distinguish vegetation types. The CuO oxidation of lignin yields a suite of compounds (Hedges and Ertel, 1982) that include vanillyl phenols (vanillin, acetovanillone and vanillic acid), syringyl phenols (syringaldehyde, acetosyringone and syringic acid) and cinnamyl phenols (*p*-coumaric acid and ferulic acid) and also *p*-hydroxy phenols (*p*-hydroxybenzaldehyde, *p*-hydroxyacetophenone and *p*-hydroxybenzoic acid). Alkaline CuO oxidation is the most established analytical method for molecular level analyses of lignin and it has been widely used in the last decades to trace the transport and fate of terrigenous OM in coastal ecosystems (Dittmar et al., 2001; Gordon and Goñi, 2003; Tremblay et al., 2007). Benner et al. (1990) reported that lignin-derived phenols were leached from mangrove leaves during early diagenesis. However, lignin concentration often decreased within just a few days in open waters due to photolytic degradation, making the quantification of terrigenous OM in the coastal systems difficult (Opsahl and Benner, 1998). Lignin-associated parameters, in particular the ratio of syringyl phenols to vanillyl phenols (S/V) and the ratio of cinnamyl phenols to vanillyl phenols (C/V), can be used to distinguish angiosperm from gymnosperm and woody from non-woody tissue (Fig. 3) (Goñi et al., 2000; Dittmar and Lara, 2001b; Marchand et al., 2005). The acid-to-aldehyde ratios of both vanillic and syringic phenols in mangrove sediment are indicative for oxic versus inoxic pathways of degradation (Dittmar and Lara, 2001b; Prasad and Ramanathan, 2009). Lignin is generally thought to be more refractory than other molecular compounds, and Marchand et al. (2005) found that lignin-derived phenols were lost at a lower rate during decomposition than total neutral sugars and bulk organic carbon. Decomposition pathways changing the monomer ratios are known to be dependent of the redox conditions (Dittmar and Lara, 2001b; Marchand et al., 2005).

Fatty acids are building blocks of lipids and are ubiquitous in living systems. Due to their biological specificity some act as biomarkers for prokaryotes, fungi, diatoms, dinoflagellates or vascular plants. They are therefore useful tracers of the origin and flow of mangrove-derived organic carbon trough estuarine food webs. The most common fatty acids found in nature are saturated

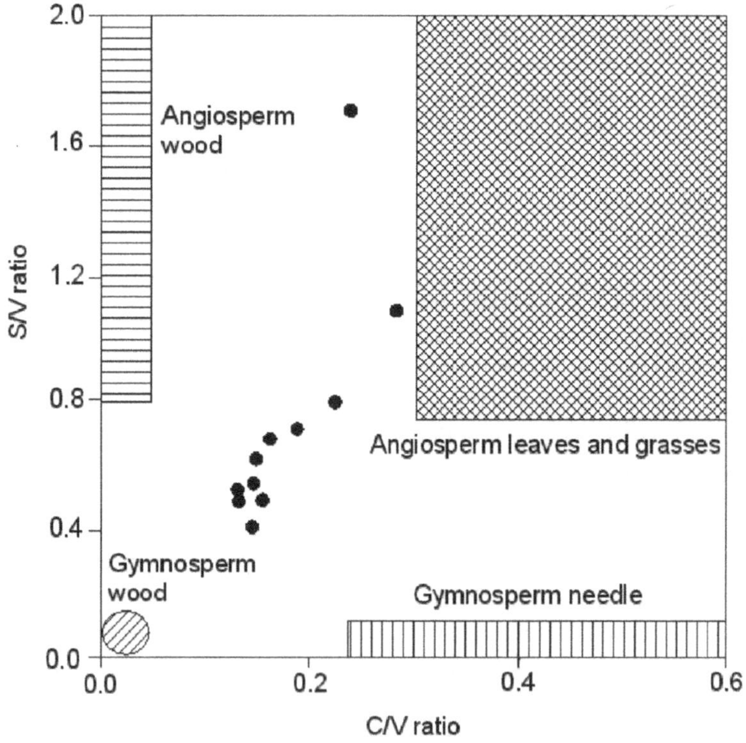

Figure 3: Lignin-derived phenols are indicative for different groups of plants and different kinds of plant tissue. Syringyl-to-vanillyl (S/V) and cinnamyl-to-vanillyl (C/V) ratios provide information about the relative importance of angiosperm versus gymnosperm sources and woody versus non-woody plant sources in the Beaufort Shelf (modified from Goñi et al., 2000).

and unsaturated compounds with chain lengths of C_{16} to C_{18} (Pulchan et al., 2003). Meziane et al. (2007) demonstrated that leaves of six mangrove species can be differentiated using their fatty acid profiles, and that geographically-separated populations of the same species can be identified. Saturated fatty acids (SAFA) dominate the fatty acid composition of mangrove leaves with palmitic acid as the most abundant (Mfilinge et al., 2005; Hall et al., 2006). Mfilinge et al. (2003) suggested that the amount of C_{16} fatty acid in mangrove leaves may be an indicator of degradation state since the concentration of SAFA in detritus declines constantly with age. Mangrove leaves also contain the long chain fatty acids (LCFA) C_{24}, C_{26} and C_{28}, which are typical vascular plant markers (Alfaro et al., 2006; Hall et al., 2006; Meziane et al., 2007). The concentrations of LCFA do not change with detritus age, suggesting that these vascular plant-markers can be useful biomarkers (Mfilinge et al., 2005). Also polyunsaturated fatty acids (PUFA) (with two double bonds 18:2ω6 and three double bonds 18:3ω3) have been identified as useful biomarkers of mangrove leaves in estuarine food chains (Hall et al., 2006; Meziane et al., 2007).

Long-chain *n*-alkanes (between C_{25} and C_{35}), that are characteristic components of epicuticular waxes of mangrove leaf surfaces, can also be used as tracers of higher plant remains (Dodd et al., 1998; Versteegh et al., 2004). Versteegh et al. (2004) and Mead et al. (2005) found that the most abundant lipid at the *Rhizophora* leaf surface, C_{29} n-alkane, accounts for 0.22% of the dry leaf material. However, it seems that the n-alkanes composition in mangrove plants is susceptible to biogeographic variations. Rafii et al. (1996) and Dodd et al. (1998) reported unusually high concentration of C_{28} *n*-alkane in Avicennia and Rhizophora from French Guiana, whereas C_{31} is also important in plants of these genera in West Africa. Dodd et al. (1999) suggested that the *n*-alkane composition of mangroves can be linked to environmental conditions and also used to establish evolution pattern.

4. OUTWELLING AND TRANSPORT OF MANGROVE ORGANIC MATTER

The export of detritus and faunal biomass from mangroves has long been considered as an important support for offshore secondary production and has been widely used as an argument for mangrove conservation. Because of the regular tidal flooding and draining in most mangrove areas, the material exchange between the forests and coastal waters can be very efficient (e.g. Dittmar and Lara, 2001a; Rezende et al., 2007). Many of the most productive mangrove forests in the world lose a significant fraction of their net primary production to coastal waters (Robertson et al., 1992; Jennerjahn and Ittekkot, 2002). Large differences occur between mangrove forests with respect to litter production and export rates, and some largely retain detritus within their sediments (Woodroffe, 1992), which is then mineralized or buried. The amount of OM available for export is influenced not only by the rate of production, but also by geomorphological, hydrological and hydrodynamic characteristics of each ecosystem. Globally, export of OM from mangrove forests ranges from 2 to 420 g C m^{-2} yr^{-1} (Alongi, 1998) and Jennerjahn and Ittekkot (2002) estimated the global export rate of mangrove litter to be 19 mol C m^{-2} yr^{-1} which is approximately half of the total litter production.

Large fluxes of mangrove OM to adjacent coastal waters can have a tremendous effect on aquatic food webs (Alongi, 1990; Dittmar and Lara, 2001c; Delgadillo-Hinojosa et al., 2008). However, some studies challenged the outwelling phenomena in some area (Taylor and Allanson, 1995). Hemminga et al. (1994) recorded a tight coupling between mangrove forests and seagrass meadows in Gazi Bay (Kenya) where strong outwelling of particles from the mangrove environment is evident. However, during flood tides overturned transport of organic particles to the mangrove forest from the seagrass zone can be observed. Respiratory CO_2 derived from mangrove detritus can be a major inorganic carbon source for the seagrass meadows as observed by Hemminga et al. (1994) in Gazi Bay, Kenya, and Lin et al. (1991) for a mangrove-seagrass system in Florida. In contrast to the tight coupling between mangroves and adjacent seagrasses, particle outwelling is often restricted to the reef line (Schwamborn et al., 1999) while nearby coral reefs can exist in relative isolation from mangrove influence (Hemminga et al., 1994). Thus, the role of mangrove litter on sediment processes and the tight coupling with adjacent ecosystems is mostly restricted to the direct vicinity of the forests. However, few kilometres offshore, mangrove detritus is not significant for burial and not available for consumption (Hemminga et al., 1994; Jennerjahn and Ittekkot, 2002).

During low tide periods, a significant amount of dissolved nutrients are released from sediments to the overlying water column by the action of tidal pumping and this can significantly increase the amount of OM exported from mangroves (Moran et al., 1991; Bava and Seralathan, 1999; Alongi et al., 2004).

Global estimates indicate that a significant fraction of net primary productivity is exported to coastal waters in the form of dissolved organic matter (DOM; Dittmar et al., 2006). Degradation and leaching products of leaf litter are likely sources of the exported mangrove-DOM (Dittmar et al., 2001). While the contribution of root exudates or decomposing belowground biomass may also be significant, their quantity is not known. The total export rate of organic carbon from mangrove forests may significantly exceed the estimates of litter export by Jennerjahn and Ittekkot (2002), 19 mol C m^{-2} $year^{-1}$, if the export of DOM is taken into account. Tidal DOM export from mangrove in Sepetiba Bay (Rezende et al., 2007) was estimated to be 1.17 kg C ha^{-1} day^{-1} (3.6 mol C m^{-2} yr^{-1}); while Florida mangrove (Twilley, 1985) exported 3.1-3.7 mol C m^{-2} yr^{-1} and mangrove in Australia around 1.8 mol C m^{-2} yr^{-1} (Ayukai et al., 1998). Dittmar et al (2006) quantified the export of DOM from Brazilian mangroves on continental shelf scale to be 12 mol C m^{-2} yr^{-1} and also recorded a lower export quantity in the same area on a small spatial scale of a single tidal creek to be 4 mol C m^{-2} yr^{-1} (Dittmar et al., 2001). The difference between these two estimates may be the gradual release of DOM from floating and suspended detritus in water column that was not considered in any previous estimates.

Considerable amounts of nutrients can be released from mangrove sediments to the water column, because of tidal pumping. During ebb nutrient-rich porewater enters the water column of tidal creeks following the gradual release of hydrostatic pressure towards low-water conditions. This causes tidal fluctuations of nutrient concentrations in the tidal creeks which in turn regulates water column productivity (Dittmar and Lara, 2001a). Diurnal dynamics of DOM in Brazilian and also Tanzanian mangroves were consistent with DOM-rich porewater seepage out of mangrove sediments during the ebb cycle (Figs 4 and 5) (Dittmar and Lara, 2001a; Bouillon et al., 2007b; Rezende et al., 2007). Furthermore, Bouillon et al. (2007a) demonstrated that there are two sources of DOM in the studied Tanzanian mangroves, based on $\delta^{13}C$ analysis over a diurnal cycle. During flood tide ^{13}C-enriched OM produced from seagrass was delivered to the mangrove and during ebb tide mangrove OM with a ^{13}C-depleted signature left the ecosystem. DOM dynamics in mangrove creek water are also driven by salinity. In Tanzanian mangroves (Bouillon et al., 2007b), high concentrations of DOM were recorded at high-saline conditions, this was probably because of high-saline porewater intrusion at low tide and sediment water exchange during tidal inundation, while in Brazilian mangrove, brackish water seems to be the major source of DOM (Dittmar and Lara, 2001a).

It has been proposed that mangroves supply >10% of the terrestrially derived DOC transported to the ocean despite the fact that only <0.1% of the continents' surface are covered by mangroves (Dittmar et al., 2006). Organic carbon export from mangroves to coastal waters is more than one order of magnitude higher in proportion to their net-primary production compared to any other terrestrial ecosystem which is trained through rivers (Fig. 6) (Kristensen et al., 2008). The rapid decline in mangrove cover over the recent decades (Alongi, 2008) may have significantly impacted the flux of terrigenous DOM to the ocean.

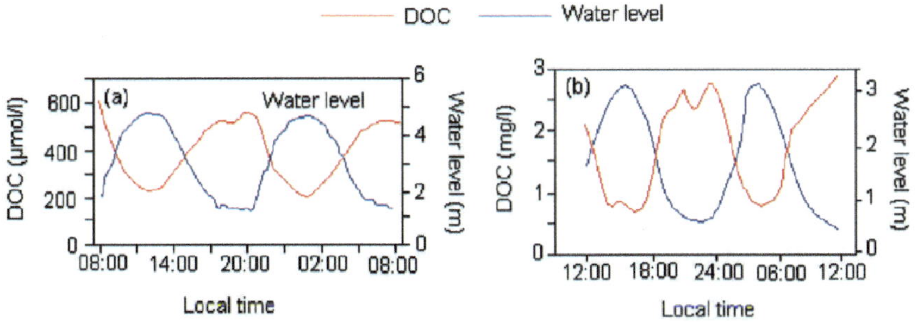

Figure 4: Diurnal variability of dissolved organic carbon (DOC) and water level in (a) a mangrove in Brazil, Bragança and (b) a mangrove in Tanzania, Ras Dege (redrawn from Dittmar and Lara, 2001a; Bouillon et al., 2007b).

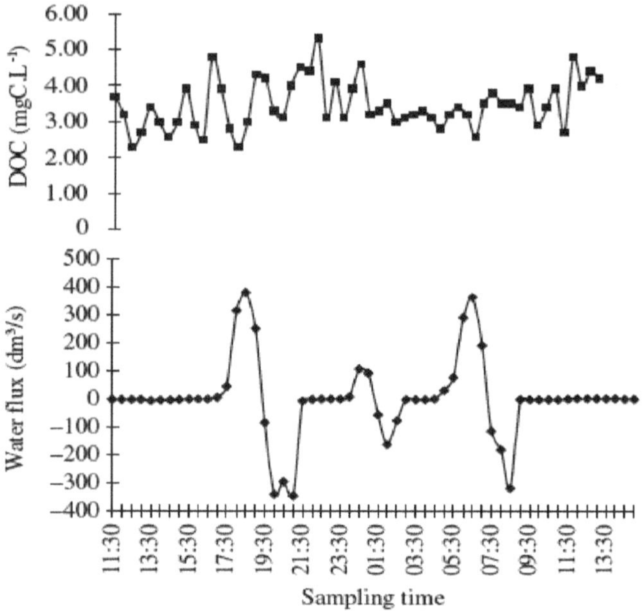

Figure 5: Diurnal variability of DOC and water flux in the Itacuruçá mangrove of the Sepetiba Bay, Brazil (redrawn from Rezende et al., 2007).

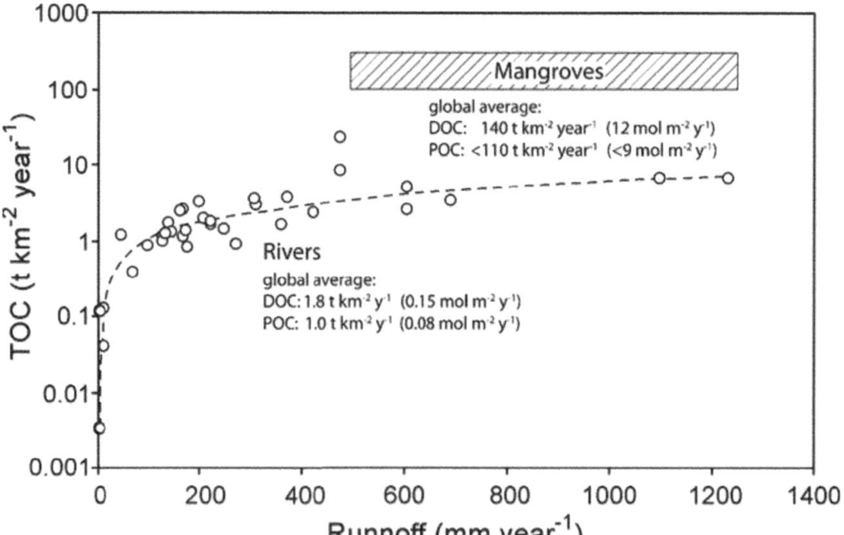

Figure 6: Comparison between the area-normalized organic carbon discharge from mangrove forests and from all climate regions draining a variety of continental ecosystems. Particulate organic carbon (POC) export rates from mangrove forests assume that >50% of the reported litter export is rapidly lost to the DOC pool and/or assimilated and not exported as suspended or floating matter off the inner-coastal zones (redrawn from Kristensen et al., 2008).

5. SUMMARY AND CONCLUSION

A large number of studies have been conducted in many mangroves around the world to understand and delineate the driving forces behind OM dynamics and the effects of external perturbations on the biogeochemistry of these ecologically vulnerable ecosystems. It has been proposed that some mangrove ecosystem retain nutrients while others supply considerable amounts of nutrients to the adjacent coastal waters, effecting biogeochemical processes in the adjacent ocean.

Globally, over the past decades mangroves have been destroyed at an alarming rate from various developmental activities. Given the potentially high impact of mangroves on element cycles in the coastal zone, effecting sediment and water column composition on large scales, the loss or severe degradation of mangrove functioning will go along with significant changes in coastal nutrient budgets and productivity. These changes in the ecosystem structure and function will also impact human society and economics. Mangrove ecosystems deliver a number of economic goods to rural populations, and degradation of mangroves will certainly have a profound negative economic impact on these population who depend upon mangroves for instance for fishery. The indirect economic effects of mangrove destruction via modified biogeochemical cycles is expected to be large, but unknown to date. Hence, a solid knowledge on biogeochemical processes and the factors driving carbon dynamics in mangrove systems is of prime importance for ecosystem management and restoration (McKee and Faulkner, 2000; Bosire et al., 2008). Further, a more fundamental understanding of nutrient cycling and factors influencing the nutrient biogeochemical processes will facilitate the policy makers to determine the carrying capacity of these ecosystems and the long-term response of mangroves to increased nutrient inputs.

REFERENCES

Alexander, M. (1977). Introduction to Soil Microbiology. John Wiley, New York.

Alfaro, A.C., Thomas, F., Sergent, L. and Duxbury, M. (2006). Identification of trophic interactions within an estuarine food web (northern New Zealand) using fatty acid biomarkers and stable isotopes. *Estuarine, Coastal and Shelf Science*, **70**, 271-286.

Alongi, D.M. and Dixon, P. (2000). Mangrove primary production and above and below-ground biomass in Sawi Bay, southern Thailand. Phuket Marine Biology Center Special Publication, **22**, 31-38.

Alongi, D.M., Sasekumar, A., Tirendi, F. and Dixon, P. (1998). The influence of stand age on benthic decomposition and recycling of organic matter in managed mangrove forests of Malaysia. *Journal of Experimental Marine Biology and Ecology* **225**, 197-218.

Alongi, D.M., Clough, B.F. and Robertson, A.I. (2005). Nutrient-use efficiency in arid-zone forests of the mangroves *Rhizophora stylosa* and *Avicennia marina*. *Aquatic Botany*, **82**, 121-131.

Alongi, D.M., Ayukai, T., Brunskill, G.J., Clough, B.F. and Wolanski, E. (1998). Sources, sinks and export of organic carbon through a tropical, semi-enclosed delta (Hinchinbrook Channel, Australia). *Mangroves and Salt Marshes*, **2**, 237-242.

Alongi, D.M., Sasekumar, A., Chong, V.C., Pfitzner, J., Trott, L.A., Tirendi, F., Dixon, P. and Brunskill, G.J. (2004). Sediment accumulation and organic material flux in a managed mangrove ecosystem: Estimates of Land-ocean-atmosphere exchange in peninsular Malaysia. *Marine Geology*, **208**, 383-402

Alongi, D.M. (1990). Effect of mangrove detrital outwelling on nutrient regeneration and oxygen fluxes in coastal sediments of the Central Great Barrier Reef Lagoon. *Estuarine Coastal and Marine Science*, **31**, 581-598.

Alongi, D.M. (1994). Zonation and seasonality of benthic primary production and community respiration in tropical mangrove forests. *Oecologia*, **98**, 320-327.

Alongi, D.M. (1998). Coastal Ecosystem Processes. CRC Press, USA.

Alongi, D.M. (2008). Mangrove forests: Resilience, protection from tsunamis, and responses to global climate change. *Estuarine Coastal and Shelf Science*, **76**, 1-13.

Ayukai, T., Miller, D., Wolanksi, E. and Spagnol, S. (1998). Fluxes of nutrients and dissolved and particulate organic matter in two mangrove creeks in northeastern Australia. *Mangroves and Salt Marshes*, **2**, 223-230.

Bava, K.A. and Seralathan, P. (1999). Interstitial water and hydrochemistry of a mangrove forest and adjoining water system, south west coast of India. *Environmental Geology*, **38(1)**, 47-52.

Benner, R., Weliky, K. and Hedges, J.I. (1990). Early diagenesis of mangrove leaves in a tropical estuary: Molecular-level analyses of neutral sugars and lignin derived phenols. *Geochimica et Cosmochimica Acta*, **54**, 1991-2001.

Benner, R., Peele, E.R. and Hodson, R.E. (1986). Microbial utilization of dissolved organic matter from leaves of the Red Mangrove, Rhizophora mangle, in the Fresh creek estuary, Bahamas. *Estuarine, Coastal and Shelf Science*, **23**, 607-619.

Bianchi, T.S. and Canuel, E.A. (2001). Organic geochemical tracers in estuaries. *Organic Geochemistry*, **32**, 451.

Blader, S.J.M., Young, S.W. and Dunning, M.C. (1985). Community structure and zoogeographic affinities of coastal fishes of the Dampier region of North-Western Australia. *Australian Journal of Marine and Freshwater Research*, **36**, 247-266.

Bosire, J.O., Dahdouh-Guebas, F., Kairo, J.G., Kazungu, J., Dehairs, F. and Koedam, N. (2005). Litter degradation and CN dynamics in reforested mangrove plantations at Gazi Bay, Kenya. *Biological Conservation*, **126**, 287-295.

Bosire, J.O., Dahdouh-Guebas, F., Walton, M., Crona, B.I., Lewis III, R.R., Field, C., Kairo, J.G. and Koedam, N. (2008). Functionality of restored mangroves: A review. *Aquatic Botany*, **89**, 251-259.

Boto, K.G. and Bunt, J.S. (1981). Tidal export of particulate organic matter from a Northern Australian mangrove system. *Estuarine Coastal and Shelf Science*, **13**, 247-255.

Boto, K.G., Alongi, D.M. and Nott, A.L.J. (1989). Dissolved organic carbon-bacteria interactions at sediment-water interface in a tropical mangrove system. *Marine Ecology Progress Series*, **51**, 243-251.

Bouillon, S., Moens, T., Overmeer, I., Koedam, N. and Dehairs, F. (2004). Resource utilization patterns of epifauna from mangrove forests with contrasting inputs from local versus imported organic matter. *Marine Ecology Progress Series*, **278**, 77-88.

Bouillon, S., Dehairs, F., Schiettecatte, L.-S. and Borges, A.V. (2007a). Biogeochemistry of the Tana estuary and delta (northern Kenya). *Limnology and Oceanography*, **52**, 46-59.

Bouillon, S., Middelburg, J.J., Dehairs, F., Borges, A.V., Abril, G., Flindt, M., Ulomi, S. and Kristensen, E. (2007b). Importance of intertidal sediment processes and porewater exchange on the water column biogeochemistry in a pristine mangrove creek (Ras Dege, Tanzania). Biogeosciences, **4**, 311-322.

Bouillon, S., Borges, A.V., Castañeda-Moya, E., Diele, K., Dittmar, T., Duke, N.C., Kristensen, E., Lee, S.Y., Marchand, C., Middelburg, J.J., Rivera-Monroy, V.H., Smith III, T.J. and Twilley, R.R. (2008). Mangrove production and carbon sinks: a revision of global budget estimates. *Global Biogeochemical Cycles* **22**, doi:10.1029/ 2007GB003052.

Clough, B.F. (1992). Primary production and growth of mangrove forests. *In:* Robertson, A.I. and Alongi, D.M. (eds.), Tropical Mangrove Ecosystems. Coastal and Estuarine, Series 41. American Geophysical Union, Washington, DC, pp. 225-249.

Delgadillo-Hinojosa, F., Zirino, A., Holm-Hansen, O., Hernández-Ayón, J.M., Boyd, T.J., Chadwick, B. and Rivera-Duarte, I. (2008). Dissolved nutrient balance and net ecosystem metabolism in a Mediterranean-climate coastal lagoon: San Diego Bay. *Estuarine Coastal and Shelf Sciences,* **76**, 594-607.

Dittmar, T. and Lara, R.J. (2001a). Driving forces behind nutrient and organic matter dynamics in a mangrove tidal creek in north Brazil. *Estuarine, Coastal and Shelf Science*, **52**, 249-259.

Dittmar, T. and Lara, R.J. (2001b). Molecular evidence for lignin degradation in sulfate reducing mangrove sediments (Amazônia, Brazil). *Geochimica et Cosmochimica Acta*, **65**, 1403-1414.

Dittmar, T. and Lara, R.J. (2001c). Do mangroves rather than rivers provide nutrients to coastal environments south of the Amazon River? Evidence from long term flux measurements. *Marine Ecology Progress Series*, **213**, 67-77.

Dittmar, T., Hertkorn, N., Kattner, G. and Lara, R.J. (2006). Mangroves, a major source of dissolved matter sources to the oceans. *Global Biogeochemical Cycles*, **20**, GB1012. doi:10.1029/2005GB002570.

Dittmar, T., Lara, R.J. and Kattner, G. (2001). River or mangrove? Tracing major organic matter sources in tropical Brazilian coastal waters. *Marine Chemistry*, **73**, 253-271.

Dodd, R.S., Blasco, F., Rafii, Z.A. and Torquebiau, E. (1999). Mangroves of the United Arab Emirates: ecotypic diversity in cuticular waxes at the bioclimatic extreme. *Aquatic Botany*, **63**, 291-304.

Dodd, R.S., Fromard, F., Rafii, Z.A. and Blasco, F. (1995). Biodiversity among west African Rhizophora: foliar wax chemistry. *Biochemistry and System Ecology*, **23**, 859-868.

Dodd, R.S., Rafii, Z.A., Fromard, F. and Blasco, F. (1998). Evolutionary diversity among Atlantic coast mangroves. *Acta Oecologia*, **19**, 323-330.

Fenchel, T.M. and Blackburn, T.H. (1979). Bacteria and Mineral Cycling. Academic Press, New York.

Fenchel, T.M. and Jørgensen, B.B. (1977). Detritus food chains of aquatic ecosystems: The role of bacteria. *Advances in Microbial Ecology*, **1**, 1-57.

Gattuso, J.P., Frankignoulle, M. and Wollast, R. (1998). Carbon and carbonate metabolism in coastal aquatic ecosystems. *Annual Review of Ecology and Systematics,* **29**, 405-434.

Goñi, M.A., Yunker, M.B., Macdonald, R.W. and Eglinton, T.I. (2000). Distribution and sources of organic biomarkers in arctic sediments from the Mackenzie River and Beaufort Shelf. *Marine Chemistry*, **71**, 23-51.

Gonneea, M.E., Paytan, A. and Herrera-Silviera, J.A. (2004). Tracing Organic matter sources and carbon burial in mangrove sediments over the past 160 years. *Estuarine, Coastal and Shelf Science*, **61**, 211-227.

Gordon, E.S. and Goñi, M.A. (2003). Sources and distribution of terrigenous organic matter delivered by the Atchafalaya River to sediments in the northern Gulf of Mexico. *Geochimica et Cosmochimica Acta*, **67**, 2359-2375.

Hall, D., Lee, S.Y. and Meziane, T. (2006). Fatty acids as trophic tracers in an experimental estuarine food chain: Tracer transfer. *Journal of Experimental Marine Biology and Ecology*, **336**, 42-53.

Hansell, D.A. and Carlson, C.A. (2002). Biogeochemistry of Marine Dissolved Organic Matter. Academic Press, New York.

Hedges, J.I. and Ertel, J.R. (1982). Characterization of lignin by gas capillary chromatography of cupric oxide oxidation products. *Analytical Chemistry*, **54**, 174-178.

Hemminga, M.A., Slim, F.J., Kazungu, J., Ganssen, G.M., Nieuwenhuize, J. and Kruyt, N.M. (1994). Carbon outwelling from a mangrove forest with adjacent seagrass beds and coral reefs (Gazi Bay, Kenya). *Marine Ecology Progress Series*, **106**, 291-301.

Hernes, P.J., Benner, R., Cowie, G.L., Goni, M.A., Bergamaschi, B.A. and Hedges, J.I. (2001). Tannin diagenesis in mangrove leaves from a tropical estuary: A novel molecular approach. *Geochimica et Cosmochimica Acta*, **65**, 3109-3122.

Jennerjahn, T.C. and Ittekkot, V. (2002). Relevance of mangroves for the production and deposition of organic matter along tropical continental margins. *Naturwissenschaften*, **89**, 23-30.

Koch, M.S. and Madden, C.J. (2001). Patterns of primary production and nutrient availability in a Bahamas lagoon with fringing mangroves. *Marine Ecology Progress Series*, **219**, 109-119.

Kristensen, E. and Alongi, D.M. (2006). Control by fiddler crabs (*Uca vocans*) and plant roots (*Avicennia marina*) on carbon, iron and sulfur biogeochemistry in mangrove sediment. *Limnology and Oceanography*, **51**, 1557-1571.

Kristensen, E. and Suraswadi, P. (2002). Carbon, nitrogen and phosphorus dynamics in creek water of a Southeast Asian mangrove forest. *Hydrobiologia*, **474**, 197-211.

Kristensen, E., Bouillon, S., Dittmar, T. and Marchand, C. (2008). Organic carbon dynamics in mangrove ecosystems: a review. *Aquatic Botany*, **89**, 201-219.

Kristensen, E. (2008). Mangrove crabs as ecosystem engineers; with emphasis on sediment processes. *Journal of Sea Research*, **59(1-2)**, 30-43.

Lin, G.H., Banks, T. and Sternberg, L.D.L.O. (1991). Variation in $\delta^{13}C$ values for the seagrass *Thalassia testudinum* and its relations to mangrove carbon. *Aquatic Botany*, **40**, 333-341.

Marchand, C., Disnar, J.-R., Lallier-Verges, E. and Lottier, N. (2005). Early diagenesis of carbohydrates and lignin in mangrove sediments subject to variable redox conditions (French Guiana). *Geochimica et Cosmochimica Acta*, **69**, 131-142.

McKee, K.L. and Faulkner, P.L. (2000). Restoration of biogeochemical function in mangrove forests. *Restoration Ecology*, **8**, 247-259.

Mead, R., Xu, Y., Chong, J. and Jaffè, R. (2005). Sediment and soil organic matter source assessment as revealed by the molecular distribution and carbon composition of *n*-alkanes. *Organic Geochemistry*, **36**, 363-370.

Meyers, P.A. (1977). Organic geochemical proxies of paleooceanographic, paleolimnologic and paleoclimatic processes. *Organic Geochemistry*, **27**, 213-250.

Meziane, T., Lee, S.Y., Mfilinge, P.L., Shin, P.K.S., Lam, M.H.W. and Tsuchiya, M. (2007). Inter-specific and geographical variations in the fatty acid composition of mangrove leaves: Implications for using fatty acids as a taxonomic tool and tracers of organic matter. *Marine Biology*, **150**, 1103-1113.

Mfilinge, P.L., Meziane, T., Bachok, Z. and Tsuchiya, M. (2003). Fatty acids in decomposing mangrove leaves: Microbial activity, decay and nutritional quality. *Marine Ecology Progress Series*, **265**, 97-105.

Mfilinge, P.L., Atta, N. and Tsuchiya, M. (2002). Nutrient dynamics and leaf litter decomposition in a subtropical mangrove forest at Oura Bay, Okinawa, Japan. *Trees Structure and Function*, **16**, 172-180.

Mfilinge, P.L., Meziane, T., Bachok, Z. and Tsuchiya, M. (2005). Total lipid and fatty acid classes in decomposing mangrove leaves of *Bruguiera gymnorrhiza* and *Kandelia candel*: Significance with respect to lipid input. *Journal of Oceanography*, **61**, 613-622.

Mia, Y., Shokita, S. and Watanabe, S. (2001). Stomach content of two grapsid crabs, *Helice formosensis* and *Helice leachi*. *Fisheries Science*, **67**, 173-175.

Middleton, B.A. and McKee, K.L. (2001). Degradation of mangrove tissues and implications for peat formation in Belizean island forests. *Journal of Ecology*, **89**, 818-828.

Moran, M.A., Wicks, R.J. and Hodson, R.E. (1991). Export of dissolved organic matter from a mangrove swamp ecosystem: Evidence from natural fluorescence, dissolved lignin phenols, and bacterial secondary production. *Marine Ecology Progress Series*, **76**, 175-184.

Müller, P.J. (1977). C/N ratios in Pacific deep-sea sediments: Effects of inorganic ammonium and organic nitrogen compounds sorbed to clays. *Geochemica et Cosmochemica Acta*, **41**, 765-776.

Neilson, M.J. and Richards, G.N. (1989). Chemical composition of degrading mangrove leaf litter and changes produced after consumption by mangrove crab *Neosarmatium smithi* (Crustacea: Decapoda: Sesarmidae). *Journal of Chemical Ecology*, **15**, 1267-1283.

Newell, S.Y. (1996). Established and potential impacts of eukaryotic mycelial decomposers in marine/terrestrial ecotones. *Journal of Experimental Marine Biology and Ecology*, **200**, 187-206.

Odum, W.E. and Heald, E.J. (1972). Trophic analyses of an estuarine mangrove community. *Bulletin of Marine Science*, **22**, 671-738.

Opsahl, S. and Benner, R. (1999). Characterization of carbohydrates during early diagenesis of five vascular plant tissues. *Organic Geochemistry*, **30**, 83-94.

Opsahl, S. and Benner, R. (1995). Early diagenesis of vascular plant tissues: Lignin and cutin decomposition and biogeochemical implications. *Geochimica et Cosmochimica Acta*, **59**, 4889-4904.

Opsahl, S. and Benner, R. (1998). Photochemical reactivity of dissolved lignin in river and ocean waters. *Limnology and Oceanography*, **43**, 1297-1304.

Padma Dorothy, K., Satyanarayana, B., Kalavathi, C., Raman, A.V. and Dehairs, F. (2003). Protozoa associated with leaf litter degradation in Coringa mangrove forest, Kakinada Bay, east coast of India. *Indian Journal of Marine Sciences*, **32(1)**, 45-51.

Parrish, C.C., Eadie, B.J., Gardner, W.S. and Cavaletto, J.F. (1992). Lipid class and alkane distribution in settling particles of the upper Laurentian Great Lakes. *Organic Geochemistry*, **18**, 33-40.

Prasad, M.B.K. and Ramanathan, AL. (2009). Organic matter characterization in a tropical estuarine-mangrove ecosystem of India: Preliminary assessment by using stable isotopes and lignin phenols. *Estuarine, Coastal and Shelf Science*, doi: 10.1016/j.ecss.2009.07.029.

Pulchan, K.J., Helleur, R. and Abrajano, T.A. (2003). TMAH thermochemolysis characterization of marine sedimentary organic matter in a Newfoundland fjord. *Organic Geochemistry*, **34**, 305-317.

Rafii, Z.A., Dodd, R.S. and Fromard, F. (1996). Biogeographic variation in foliar waxes of mangrove species. *Biochem. System Ecology*, **24**, 341-345.

Rezende, C.E., Lacerda, L.D., Ovalle, A.R.C. and Silva, L.F.F. (2007). Dial organic carbon fluctuations in a mangrove tidal creek in Sepetiba bay, Southeast Brazil. *Brazilian Journal of Biology*, **67(4)**, 673-680.

Robertson, A.I. and Alongi, D.M. (1995). Role of riverine mangrove forests in organic carbon export to the tropical coastal ocean: A preliminary mass balance for the Fly Delta (Papua New Guinea). *Geo-Marine Letters*, **15**, 134-139.

Robertson, A.I. and Blaber, S.J.M. (1992). Plankton, epibenthos and fish communities. *In:* Robertson, A.I. and Alongi, D.M. (eds.), Tropical Mangrove Ecosystems. American Geophysical Union, Washington, DC, pp. 173-224.

Robertson, A.I. and Duke, N.C. (1987). Insect herbivory on mangrove leaves in north Queensland. *Australian Journal of Ecology*, **12**, 1-7.

Robertson, A.I., Alongi, D.M. and Boto, K.G. (1992). Food chains and carbon fluxes. *In:* Robertson, A.I. and Alongi, D.M. (eds.), Tropical Mangrove Ecosystems. Coastal and Estuarine Series 41. American Geophysical Union, Washington DC, USA, pp. 293-326.

Robertson, A.I. (1991). Plant-animal interaction and the structure and function of tropical mangrove forest ecosystems. *Australian Journal of Ecology*, **16**, 433-443.

Schwamborn, R., Voss, M., Ekau, W. and Saint-Paul, U. (1999). Stable isotope composition of particulate organic matter and zooplankton in North-East Brazilian shelf waters. *Archives of Fisheries and Marine Research*, **47**, 201-210.

Singh, G., Ramanathan, AL. and Prasad, M.B.K. (2005). Nutrient cycling in the mangrove environment: A brief review. *International Journal of Ecology and Environmental Sciences*, **31(3)**, 231-244.

Tam, N.F.Y., Wong, Y.S., Lan, C.Y. and Wang, L.N. (1998). Litter production and decomposition in a subtropical mangrove swamp receiving wastewater. *Journal of Experimental Marine Biology and Ecology*, **226**, 1-18.

Taylor, D.I. and Allanson, B.R. (1995). Organic carbon fluxes between a high marsh and estuary and the inapplicability of the outwelling hypothesis. *Marine Ecology Progress Series*, **120**, 263-270.

Tremblay, L. and Benner, R. (2006). Microbial contributions to N-immobilization and organic matter preservation in decaying plant detritus. *Geochimica et Cosmochimica Acta*, **70**, 133-146.

Tremblay, L.B., Dittmar, T., Marshall, A.G., Cooper, W.J. and Cooper, W.T. (2007). Molecular characterization of dissolved organic matter in a North Brazilian mangrove porewater and mangrove-fringed estuaries by ultrahigh resolution Fourier Transform-Ion Cyclotron Resonance mass spectrometry and excitation/emission spectroscopy. *Marine Chemistry*, **105**, 15-29.

Twilley, R.R., Pozo, M. and Garcia, V.H. (1997). Litter dynamics in riverine mangrove forests in the Guayas River estuary, Ecuador. *Oecologia*, **111**, 109-122.

Twilley, R.R. (1985). The exchange of organic carbon in basin mangrove forests in a southwestern Florida estuary. *Estuarine, Coastal and Shelf Science*, **20**, 543-557.

Valiela, I. (1995). Marine Ecological Processes. 2nd edition. Springer, New York.

Valiela, I. and Teal, J.M. (1976). Production and dynamics of experimentally enriched salt marsh vegetation: Below ground biomass. *Limnology and Oceanography*, **21**, 245-252.

Versteegh, G.J.M., Schefuß, E., Dupont, L., Marret, F., Sinninghe-Damstè, J.S. and Jansen, J.H.F. (2004). Taraxerol and Rhizophora pollen as proxies for tracking past mangrove ecosystem. *Geochimica et Cosmochimica Acta*, **68**, 411-422.

Woodroffe, C.D. (1992). Mangrove sediments and geomorphology. *In:* Robertson, A.I. and Alongi, D.M. (eds.), Tropical mangrove ecosystems. (Coastal and Estuarine Studies 41) AGU, Washington, pp. 7-41.

Zieman, J.C., Macko, S.A. and Mills, A.L. (1984). Role of seagrasses and mangroves in estuarine foodwebs: Temporal and spatial changes in stable isotope composition and amino acid content during decomposition. *Bulletin of Marine Science*, **35**, 380-392.

13

Influence of Terrestrial Inputs on Mangrove and Coral Reef: Primary Productivity of the Andaman Islands

R. Ramesh*, B. Senthilkumar, Jennifer Immaculate Divia and R. Purvaja
Institute for Ocean Management, Anna University
Chennai 600 025, India

1. BACKGROUND

The marine environment of the Andaman Islands is characterized by warm tropical conditions varying at the surface temperature between 25°C and 31°C, stable salinity regimes and moderately high nutrient levels from terrestrial runoff. The semi-diurnal tidal regimes vary from 1.5 to 4 m amplitude from neap to spring tides, creating extensive intertidal platform and rocky-shore communities exposed twice daily during low tides. Fringing reef and barrier reefs dominate the Andaman Islands (960 km^2) forming a natural barrier to the wave energy from the Bay of Bengal. Coral reefs form the dominant ecosystem creating habitats for sea grasses and mangroves in the lagoons and creeks protected by the reef. The mangroves in the Andaman and Nicobar Islands are estimated to occupy about 115,000 ha.

This marine environment faces a number of threats from the island's growing population, extraction of fish and other resources from the coral reef and mangrove ecosystems. The carbon and nitrogen cycles are critical for humans and ecosystems and have strong links to climate. These cycles have been perturbed by human activity both in terrestrial and coastal aquatic

*Corresponding Author

ecosystems. The rate at which water moves through terrestrial and aquatic coastal ecosystems often have major impacts, such as rates at which dissolved and particulate materials are leached from the land to water courses.

Land use affects the hydrological response of a system and thus nutrient fluxes. Better understanding on the interactions and feedback mechanisms among C and N cycles and their linkages with ecosystems is essential to simulate variability and change in these linked processes. Changes in C and N contents are connected to changes in the land use pattern, agricultural practices, vegetation etc. Changes in C-N cycles in turn induce further changes in the water cycle itself, which can have adverse impacts on terrestrial and coastal ecosystems.

Several reviews on the biogeochemistry of the individual ecosystems viz., mangroves, coral reef etc. pertaining to the coastal zones have been published. However, a holistic approach to all these ecosystems (terrestrial, mangrove and reefs) is clearly lacking. This study focuses on the ecosystem approach with special emphasis on primary production, nutrient exchange and their transfer from one ecosystem to another in the Andaman Islands.

2. STUDY AREA

The Andaman and Nicobar Islands are a group of islands situated off the eastern coast of India; known for its tropical rain forest, they lie amidst the waters of the Bay of Bengal and Indian Ocean on one side and the Andaman Sea on the other. They are part of a submerged mountain range related to the Arakan Yoma Range of Myanmar. These pockets of islands, though diminutive in area in comparison with the sub-continent, are biologically diverse in terms of flora, fauna and ecosystems and are politically divided into South Andaman, Middle Andaman and North Andaman. The Andaman and Nicobar Islands are a group of about 572 islands, islets rocks and keys situated off the eastern coast of India in a junction with the Bay of Bengal and Indian Ocean on one side and the Andaman Sea on the other. Covered with lush green tropical rain forests, about 1200 km away from the mainland of the Indian subcontinent, these islands are also called "the Bay Islands".

The islands are mountain peaks that have emerged from the sea. They are part of a submerged mountain range related to the Arakan Yoma range of Myanmar. Andaman Islands, where inhabitation is found, are comprised of South Andaman, Middle Andaman and North Andaman (Fig. 1). The Andaman Islands are situated between 10°30′ and 13°42′ North latitude and 92°14′ and 94°16′ East longitude in the southeast Bay of Bengal. They are located between Chennai city of Indian sub-continent and the Mergui Archipelago lying off the town of the same name in the Tenasserim peninsula in lower Myanmar. These islands are the summits of a submarine range of hills 1120 km long that connects the Arakan Yoma of Myanmar with Achin head in Sumatra. A deep oceanic trench more than 1000 fathoms deep runs between Nicobar and Sumatra and

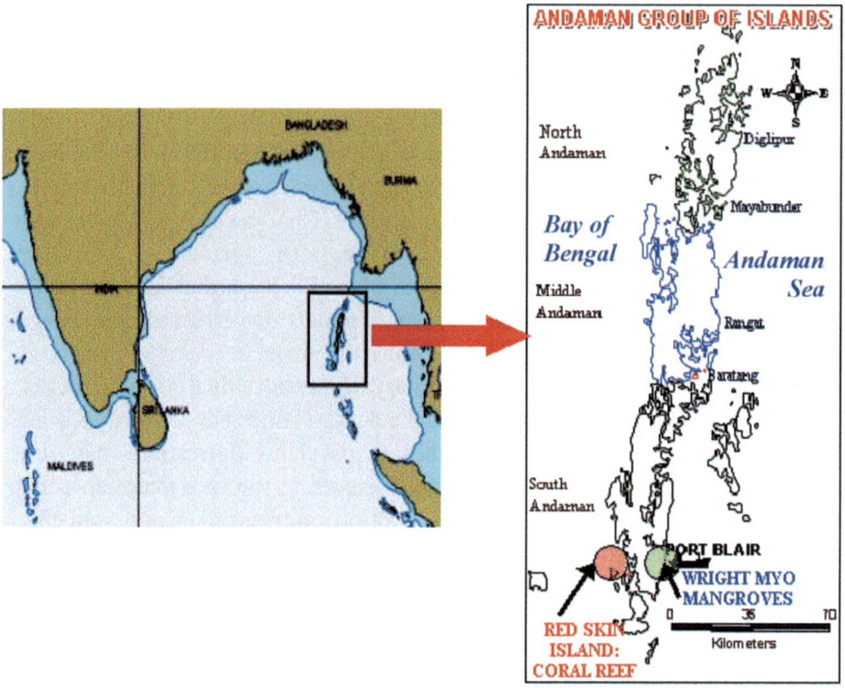

Figure 1: Study area in the Andaman (South Andaman) Islands
Green dot depicts the mangrove site sampled (Mangrove areas of Wright Myo) and Red dot indicates the coral reef area sampled.
Note: Both the sampling locations are connected to the terrestrial evergreen forest area.

extends up to Narcondam. The Ten Degree Channel of the sea divides the islands into two distinct islands: Andaman to the north and Nicobar to the south separated by about 160 km. The length between the extremities is about 355 km, while the maximum width is about 60 km.

This group of islands is made up of a large number of islands amounting to as much as 572 of which 26 are inhabited by human beings. Out of a total geographical area of 8239 km^2, 7464 km^2 are covered under reserve and protected forests. 36% of the reserve forests have been earmarked for primitive aborigines, 13% of the total area all along the coast has to be protected against soil erosion and for mangroves and about 12% of the area have been occupied under revenue settlements. The islands are categorized under the XV agro-climatic zone of the country called "island zone".

Mangrove Forests of the Andaman Islands

The mangroves in the Andaman and Nicobar Islands are estimated to occupy 115,000 ha of which 50,000 ha are in the Andaman group (Saldanha, 1989).

The coastline of the Andaman Islands is irregular and deeply indented thereby giving rise to a number of tidal creeks. The mangrove plant community densely populates these creeks. Singh et al. (1986) collected a total of 40 plant species from mangroves belonging to 28 genera distributed over 20 families. Among the plant species recorded, 25 were true mangroves and 15 were other common associated species. Community of these islands is greatly dominated by the family *Rhizophoraceae* with nine species and the seaward line is always occupied by *Rhizophora* spp. Based on the structure and composition of mangrove forests of different Andaman Islands sites, the six types of mangrove communities have been recognized by Mall et al. (1985).

- *Rhizophora mucronata—Rhizophora apiculata* community
- *Bruguiera gymnorhiza—Ceriops tagal* community
- *Rhizophora mucronata—Bruguiera gymnorhiza* community
- *Lumnitzera littorea—Avicennia officinalis* community
- *Heritiera littoralia—Pongamia pinnata* community
- *Acanthus ilicifolius—Acrostichum aureum* community

As dense formations well adapted to salinity, tidal fluctuations and marshy soil, mangrove forests interface between land and marine ecosystems. The mangroves at Wright Myo (Fig. 1) are distributed in a small tidal creek (8 km long; 50 metres wide) located on the South Andaman Islands, India. It is flanked with dense mangrove vegetation dominated by *Rhizophora* and *Avicennia sp.* This mangrove area is subjected to semi-diurnal tides with tidal amplitude of over 2.5 m. The terrestrial evergreen forest present in this location is directly connected to the mangrove forest, which in turn extends out into the Wright Myo creek and to the Bay of Bengal. The mangroves of Wright Myo have mixed vegetation being dominated by *Rhizophora* and to a lesser extent by *Bruguira*, *Ceriops* and *Avicennia* in the order of dominance.

Coral Reef Ecosystems of the Andaman Islands

In the Andaman Islands, coral reefs are of the fringing type, often several hundred metres wide and separated from shore by a lagoon of similar width. The commonest corals are *Acropora, Porites, Pocillopora, Montipora, Heliopora, Tubipora* and *Favia.* The Red Skin Islands located on the Andaman Sea will be our sampling location for the terrestrial-mangrove-coral reef transition. The tidal range is ~3 m and is semi-diurnal.

Significance of the Terrestrial and Coastal Ecosystems

The coastal zone exhibits a wide diversity of geomorphological types and ecosystems, each displaying great variability in terms of physical and biogeochemical forcing. Despite its relatively modest surface area, the coastal zone plays a considerable role in the biogeochemical cycles because it receives massive inputs of terrestrial organic matter and nutrients, is among the most geochemically and biologically active areas of the biosphere, and exchanges large amounts of matter and energy with the open ocean. The primary production, respiration, carbon burial and exchange with the adjacent systems will be the focus of this study.

Mineralization and Immobilization

One of the most interesting questions regarding decomposition is: What controls how nutrients are released when microbes consume organic matter? (1) The first aspect of litter quality concerns stoichiometry, specifically the C:N ratio. Because N is limiting, microbes immobilize all available N from the litter, and we usually say that the litter has poor quality because it is so low in N. As they decompose the litter, CO_2 is lost in respiration, and nutrients are retained in the growing microbial biomass. Eventually, the C:N ratio of the litter approaches that of the microbial biomass. At this stage, microbes are decomposing litter of the same quality as themselves. If the C:N ratio falls below 12-30, microbes become C limited and they cast off extra N in the process called N mineralization. (2) Second, plants use defense compounds like lignin to lower the palatability of their tissues. Lignin is extremely resistant to microbial attack, and it's thought that only a group of fungi are capable of breaking it down. Decomposition is one of the ways in which certain nutrients become available to plants to utilize for their survival and reproduction. Carbon is used for photosynthesis. Other nutrients are important too, including nitrogen (N) and phosphorus (P). Nitrogen is a major component of proteins, nucleic acids and chlorophyll while phosphorus is an important part of energy biochemical reactions. These nutrients become available when organic matter is decomposed. The leaves of trees decompose at different rates. Some of the factors that contribute to these varying rates are:

- Temperature and moisture
- Concentration of phosphorus and nitrogen
- Ratio of carbon to nitrogen (C:N)
- Content of lignin/ tannin

There is a positive correlation between temperature and moisture and higher rates of decomposition. This indicates that forests in different climates would be expected to have varying rates of decomposition. Studies have shown that tropical forests lose three times the mass that temperate forests do, showing a higher rate of decomposition. This has more to do with the chemical composition of the leaves.

Mangroves (to a great extent the seagrass ecosystems) and coral reefs often coincide in tropical coastal zones. These ecosystems are important for coastal stabilization and they have a high economic value, as they are sources of forestry and fishery products. In many areas, however, these systems are threatened by increasing resource consumption by the human populations. There is a growing awareness that many interrelations between mangroves, and coral reefs exist and that these should have implications for management strategies aimed at the sustainable use of coastal resources.

During the last few decades, human activities have almost doubled the annual amount of nutrients available in a form usable by living organisms. In coastal waters, human sources of N and P now rival or exceed natural inputs. By increasing supplies of nutrients usable by plants, animals, human activities have resulted in ecological impacts with significant and far-reaching, economic, social and cultural consequences.

This study addresses a few key issues of global significance on a regional scale, emphasizing the interactions between land-based activities, coastal water and sediment quality and coastal ecosystem productivity. The primary objectives are detailed below:

- To estimate the amount of carbon, and nitrogen input from the terrestrial ecosystem to the mangrove and coral reef ecosystems.
- To assess the composition of leaf litter from the terrestrial and mangrove vegetation.

3. MATERIALS AND METHODS

Sampling of Surface Water

Surface water samples were collected and analyzed from the coastal mangrove and coral reef ecosystem in the Andaman Islands. Samplings were done both in the South and North Andaman isles. In South Andaman surface water samples were collected from two different spatial transects. From each spatial transect 15 separate samples were collected with varying salinity gradient, one is from Wandoor isle to Chidiatapu towards the Bay of Bengal and the other from Wright-Myo mangrove creek towards the Andaman sea. Also diurnal study was done in a small tidal creek (11° 47′ 27.7″ N, 92° 42′ 24.3″ E) of 8 km long and 50 m wide. This mangrove creek experiences a tidal range of nearly 80 cm and is flanked by dense mangrove vegetation dominated by *Rhizophora* sp., surrounded by hilly and terrestrial vegetation coverage.

Similarly, in North Andaman sampling was made in two different spatial transects, (i) extending from Mayabunder to Kalighat creek towards the Andaman Sea and (ii) from Mayabunder to the North reef in the Bay of Bengal. Sampling was made for a full tidal cycle (24 hours) in the Kalighat creek (13° 07′ 29.2″ N, 92° 56′ 48.1″ E), which is 50 m wide and almost 10 km long, having a tidal influence of nearly 1 m. Like Wright Myo, the Kalighat creek is also quite thickly vegetated mangrove environment and is very sparsely populated.

Samples for the analysis of dissolved nutrients were filtered through Whatman GF/C 47 mm filter paper to remove the suspended sediments and collected in 250 ml Nalgene bottles that were pre-rinsed in 0.1N HCl followed by Milli-RO® (Millipore Corp; conductivity ~10 µS, bacterial rejection >99%) water. Filtrates were stored on dry ice during transport and subsequently frozen at -40°C for a maximum of one week prior to analysis. These samples were analysed for dissolved inorganic nitrogen namely NH_4^+, NO_3^-, NO_2^- and PO_4^+, using the following methods: IndoPhenol Blue method for ammonia (Sasaki and Sawada,1980). Cadmium reduction followed by treatment of colour reagent (Sulphanilamide and NED) which is then read in the spectrophotometer for nitrate and without the cadmium reduction for nitrite and Ortho Phosphate method is used for phosphate according to the APHA standards methodology. All analyses are carried out with Millipore distilled water.

For the analysis of chlorophyll, water samples were filtered in the field with the help of GF/C Whatman filter paper and stored in ice until analysis. Chlorophyll-a (Chl a), Chl-b and Chl-c were extracted in 90% acetone and measured spectrophotometrically by Parsons et al. (1984). NO_3^- and NH_4^+ were determined according to Brewer and Riley (1965) and Mantoura and Woodward (1983) respectively; analytical precisions (100ó/*x*) were ±1%. Salinity, temperature, pH and dissolved oxygen were determined simultaneously using a pre-calibrated Horiba water quality-monitoring probe W–22.23. At frequent intervals samples for Dissolved Oxygen were collected and analysed using Winkler's method

Sediment Denitrification

Sediment denitrification rates were measured using acetylene block (Law and Owens, 1990) in the present study, for this method has high reproducibility, even with brief periods of incubation time (Law et al., 1992) and also because logistical constraints dictated the use of a simple technique in the field. Sediment was sampled using 5 cm × 3 cm perspex tubes equipped with a series of 10 mm diameter silicon rubber ports for acetylene injection. At each site, six sediment cores were randomly collected and sealed using airtight bungs. The cores were divided into subsets of two for treatment prior to incubation. Two cores were left untreated to act as controls and to assess in situ rates of N_2O production by the sediment. The second subset was amended with seawater saturated with acetylene, to give a final pore water and overlying water saturation of 10% v/v, for calculation of denitrification rates. The remaining cores were treated with the nitrification inhibitor allythiourea (ATU), which prevents the microbial oxidation of NH_4^+ to NO_2^-. ATU was added to give a final concentration of 10 mg l^{-1} in the pore and overlying waters. This allows an indirect method for calculation of the nitrification contribution to the N_2O flux as the difference between N_2O flux in the control and ATU amended cores. Cores were incubated for four hours in the dark in an insulated environment at in situ temperature with no gaseous headspace.

Samples were collected at hourly intervals. 50 ml samples of water overlying the core were carefully removed and simultaneously replenished through two 3-way stopcock fitted in the rubber bung and immediately dispensed into a calibrated and graduated 100 ml glass sample syringe fitted with a similar stopcock. The sample syringe was then connected to a supply of ultra high purity (UHP) nitrogen to create a headspace of known volume. The gaseous and aqueous phases were then equilibrated by vigorously shaking for five minutes at the room temperature. The equilibrated N_2O in the headspace was collected in vaccutainers and measured using gas chromatograph (14B Shimadzu GC and HP GC 5890) with electron capture detector (ECD).

A sediment core was taken in Wright Myo, which was dominated by Rhizophora mangroves. Sediment core was taken using a cylindrical PVC pipe with a diameter of 6 cm only up to a depth of 36 cm, due to sediment and root compaction beyond this point. The core was taken by pushing the pipe through the sediment, and then taking it out without disturbing the sediment trapped within the pipe. The core was then cut at 2 cm intervals and packed into airtight polythene covers and brought to the laboratory in dry ice for further analysis. The sub samples were dried at 40°C for three days, to remove all the water content, grounded and sieved using a sieve of 100 μm mesh size, and then placed in a desiccator to prevent moisture contamination. Total carbon and nitrogen were then analyzed in a NC thermo electron soil analyzer.

Primary Productivity

Primary productivity was estimated by the oxygen technique and measured in situ using the light and dark bottle method (Gattuso, 1988). Gross primary production (GPP) was measured as the difference between light bottle (LB) and dark bottle (DB) and community respiration (R) was measured from the difference between initial bottle (IB) and dark bottle (DB) oxygen. NPP was calculated simply as the difference between GPP and CR (NPP = GPP–CR).

Leaf Litter C:N Estimation

To estimate the C:N composition of the mangroves and surrounding terrestrial ecosystems, the green leaves (senescent leaf) and litter were collected from the Wright Myo mangrove creek and Red Skin Island. After collection the green leaves and litter were gently washed with distilled water to remove the attached sediments. They were then oven dried at 70° C for 24 hours, packed in polythene cover and stored in the dessicator until further analysis. The samples were then ground into fine powder for analysis. The organic carbon and nitrogen composition were measured on aliquots of each 15-20 mg dry weight sample, using Thermo Finnigan FlashEA 1112Series NC Soil analyzer. Standardization of the analysis was done using the chemical standard L-Aspartic Acid (C = 36.09%, N = 10.52%) and the recovery and precision of the analysis was checked using soil reference material (SRM) (C = 1.998%, N = 0.196%).

4. RESULTS AND DISCUSSION

4.1 Spatial Variations

Table 1 represents the data from the transect Wandoor to Chidiyatappu (near Bay of Bengal) in South Andaman. The salinity values range from 32.38 to 34.89 during the intermonsoon season of April 2005, the monsoon (August 05), salinity ranges from 26.90 to 33.62 and the following intermonsoon of March 06 records 22.5 to 33.4, whereas during the monsoon (July 06), salinity ranges from 26.92 to 29.2. The pH during April varies from 8.08 to 8.21 while in August 05 it varies from 7.56 to 8.11 and it is 8.09 to 9.22 and 6.7 to 8.5 in March and July 06 respectively. Dissolved oxygen varies from 6.65 to 8.89 mg/l in the dry season (April 05), and during wet season (August 05) it ranges from 5.12 to 8.00 mg/l and the following March and July '06 it records 7.42 to 9.6 mg/l and 8.6 to 10.8 mg/l.

In the spatial sampling from Wright Myo towards Andaman Sea, the salinity values range from 20.00 to 31.00 during the dry season of April '05 and the same range from 4.83 to 31.70 during the wet season of August 05 and the following March and July 06 records 23.3 to 30.8 and 14.2 to 25.4 respectively. The pH values range from 7.41 to 7.98 during the summer of April 05, whereas it varies from 6.72 to 7.15 and 6.5 to 7.5 during August 05 (wet season) and March '06 summer respectively, while it records 7.0 to 7.5 in July 06. The

Table 1: Concentration range (mean ± SD) of physico-chemical parameters and nutrients at South Andaman Wandoor to Chidiyatappu (Bay of Bengal)

Parameters	*Dry period (April, 2005)*	*Wet period (August, 2005)*	*Dry period (March, 2006)*	*Wet period (July, 2006)*
Salinity	32.38-34.89	24.99-33.62	22.5-33.4	26.2-29.2
	33.49±1.03	30.36±2.58	30.36±0.79	28.8±0.9
pH	8.08-8.21	7.56-8.12	8.09-9.22	6.7-8.5
	8.15±0.05	7.99±0.18	8.21±0.07	7.4±0.473
Water	28.35-30.50	25.50-27.50	29.4-31.6	28.6-29.6
Temperature (°C)	28.90±0.64	26.40±0.66	30.34±0.18	29.0±0.224
Air	34.5-38.00	27.00-28.00	31.2-32.5	28.9-30.6
Temperature (°C)	36.97±1.31	27.67±0.45	31.94±0.11	30.0±0.519
Dissolved	6.65-8.89	5.12-8.00	7.42-9.6	8.6-10.8
Oxygen (mg/l)	7.73±0.80	7.08±0.80	8.77±0.15	10.0±0.561
NO_3 (μM)	0.20-5.05	0.53-15.11	5.05-14.2	12.4-21.0
	1.70±1.51	6.39±5.63	10.28±0.73	15.00±1.904
NO_2 (μM)	0.16-0.45	0.06-0.34	0.04-0.31	-0.3
	0.26±0.09	0.14±0.08	0.15±0.02	0.2±0.64
NH_4 (μM)	250-475	0.99-17.53	16.2-46.42	7.8-20.7
	350±59	6.04±4.89	31.23±2.2	16.3±3.268
PO_4 (μM)	0.04-6.87	0.29-0.80	0.262-1.84	1.7-5.0
	0.95±1.79	0.53±0.17	0.67±0.12	2.6±0.754

Table 2: Concentration range (mean ± SD) of physico-chemical parameters and nutrients at South Andaman (Wright Myo mangrove creek, Andaman Sea)

Parameters	*Dry period (April, 2005)*	*Wet period (August, 2005)*	*Dry period (March, 2006)*	*Wet period (July, 2006)*
Salinity	20.00-31.00	4.83-31.70	23.3-30.8	14.2-25.4
	26.62±3.11	19.91±7.39	27.35±0.64	22.0±2.755
pH	7.41-7.55	6.72-7.15	6.503-7.473	7.0-7.5
	7.51±0.04	7.02±0.12	7.19±0.07	7.2±0.153
Water	27.90-29.30	25.00-27.00	30-32.8	28.2-29.4
temperature (°C)	28.81±0.41	26.07±0.62	31.11±0.20	29.1±0.292
Air	31.00-33.00	27.00-28.50	30.2-37	29.8-30.4
temperature (°C)	32.07±0.70	27.67±0.49	34.31±0.59	30.0±0.154
Dissolved	7.89-9.91	1.60-7.04	4.43-6.01	3.3-5.6
oxygen (mg/l)	8.75±0.67	4.03±1.84	5.27±0.09	4.2±0.686
NO_3 (μM)	1.24-10.58	0.81-18.75	2.41-11.67	8.7-60.5
	3.50±2.31	12.16±4.87	6.70±0.70	14.8±2.731
NO_2 (μM)	0.18-0.58	0.12-1.91	0.05-0.18	0.1-0.7
	0.36±0.13	0.69±0.51	0.1±0.01	0.4±0.191
NH_4 (μM)	325-500	0.17-7.10	17-121	6.0-16.0
	381±52	6.04±4.98	33±6.64	8.1±2.826
PO_4 (μM)	0.01-4.00	0.54-2.15	0.38-1.82	1.6-3.5
	0.88±0.88	0.53±0.17	0.92±0.13	2.3±0.559

dissolved oxygen varies from 7.89 to 9.11 mg/l during dry season of 2005 and during wet season it varies from 1.60 to 7.04 mg/l, in March 06 it records 4.43 to 6.01 mg/l and in July 06 it shows 3.3 to 5.6 mg/l (Table 2). The comparatively low value of physico-chemical characteristics during wet season in both transects at South Andaman may be due to the influx of rain water to the system.

Seasonal variations of nutrients and atmospheric trace gases are concomitant with the seasonal variation of physico-chemical characteristics of the mangrove ecosystem (Purvaja and Ramesh, 2000). The nutrient concentration shows significant variation with respect to seasonal dynamics. The nitrate concentration (NO_3) during intermonsoon (April 2005) in South Andaman transect from Wandoor to Chidiyatappu varies from 0.20 to 5.05 μM, whereas the concentration varies from 0.53 to 15.11 μM and 5.05 to 14.2 μM in wet season August 05 and intermonsoon March 2006 respectively, while in July it ranges from 12.4 to 21.0 μM. The nitrite concentration (NO_2) varies from 0.17-0.45 μm during dry season (April 05) and it ranges from 0.06 to 0.34 μM in wet season (August 05) and the system records 0.04 to 0.31 μM and 0.1 to 0.3 μM in seasons of March and July 06 respectively. The phosphate concentration (PO_4) during the dry season (April 05) varies from 0.04 to 6.87 μM, and it records 0.33-1.00 μM in August 05 while it ranges from 0.26 to 1.84 μM in March 1.7 to 5.0 in July 06.

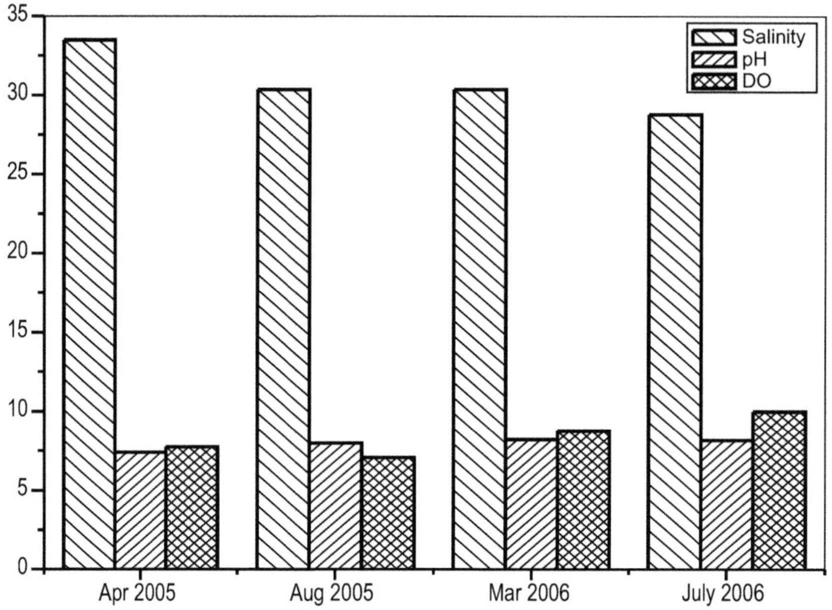

Figure 2: Seasonal variation of salinity, pH and dissolved oxygen in the transect Wandoor to Chidiyatappu (DO in mg/l).

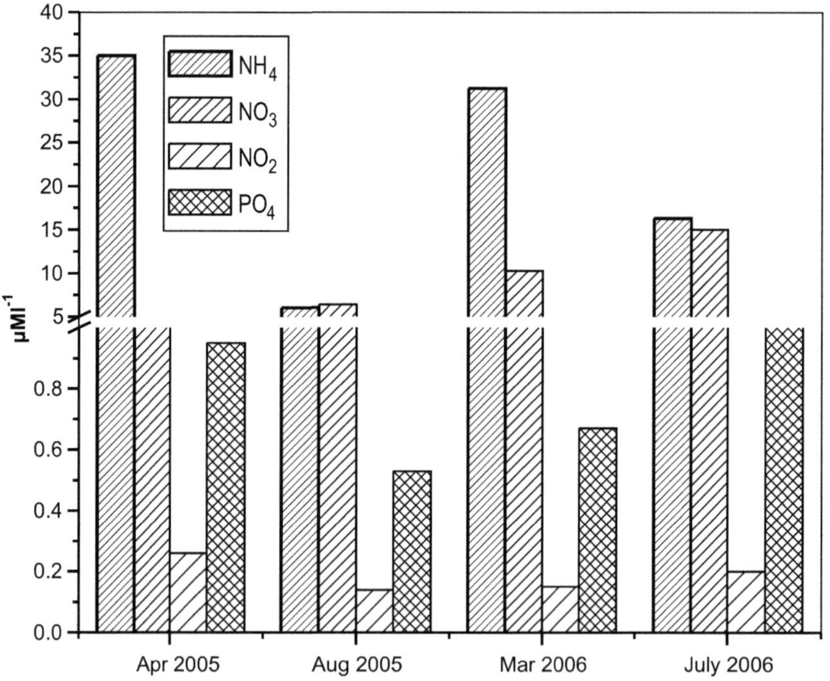

Figure 3: Seasonal variation of nutrients in the transect Wandoor to Chidiyatappu.

In the transect from Wright Myo mangrove creek towards Andaman Sea, the concentration of nitrate (NO_3) varies from 1.18 to 10.58 μM during the dry season (April 05) and the same shows 0.81-18.75 μM in the wet season (August 06) and 2.41 to 11.67 μM and 8.7 to 60.5 μM in the following intermonsoon (March 06) and monsoon (July 06) respectively. The nitrite (NO_2) concentration varies from 0.18 to 0.58 μM in dry season (April 05) and it ranges from 0.12-1.91 μM during the wet season (August 05) and 0.05 to 0.18 μM and 0.1 to 0.7 μM in the following intermonsoon (March 06) and monsoon (July 06) respectively. The concentration of phosphate (PO_4) ranges from 0.01 to 4.00 μM in dry season (April 05) and it varies from 0.29 to 0.80 μM during the wet season (August 05) whereas it records 0.38 to 1.82 μM in the following March 2006 and 1.6 to 3.5 μM in July 2006.

Generally, it found that in South Andaman the spatial transect at Wright Myo mangrove creek towards Andaman Sea shows maximum seasonal variability of physico-chemical characteristics, especially the salinity gradient that this mangrove environment experiences may be due to the tidal influx. Whereas, the transect from Wandoor to Chidiyatappu does not show such variation seasonally and also there is not much variability in salinity. This may be due to the topography of the region, which is covered by many small isles that may deflect the tidal influx and mixing in the system. Dissolved oxygen showed a reverse trend with higher oxygen saturations observed in T2 and T3 and lower concentrations in T1 and T4. The higher organic matter input into the mangrove surrounding waters make them heterotrophic leading to O_2 consumption and CO_2 production (Alongi et al., 1992).

4.2 Denitrification

The denitrification rate for Wright Myo (South Andaman) was 9.0 μmol m^{-2} d^{-1} for the monsoon (August 2005), 11.2 μmol m^{-2} d^{-1} and 10.2 μmol m^{-2} d^{-1} for the intermonsoon season of March 2006 and monsoon of July 06 respectively. In Kalighat Creek (North Andaman) and Red Skin Island (Coral Island) it was 10.4 μmol m^{-2} d^{-1} and 7.4 μmol m^{-2} d^{-1} for the same seasons respectively. The high dissolved oxygen concentration (6.12-7.40 ml/l) in the overlying waters is supposed to have caused the highly oxidized conditions of the sediment and suppressed the denitrification processes in August 2005; in contrast highest denitrification rate observed in March, may be explained by high temperature (30°C-32.8°C) compared to other seasons. The denitrification rates at these sites are lower in comparison with their counterparts in the temperate systems (Barnes and Owens, 1998), consistently with low concentrations of nitrate in the mangrove surrounding waters and the process of denitrification is generally considered to be an anaerobic process and requires organic C as substrate (Seitzinger, 1993), in addition to nitrite or nitrate. For this reason, denitrification is not thought to extend into deeper sediments.

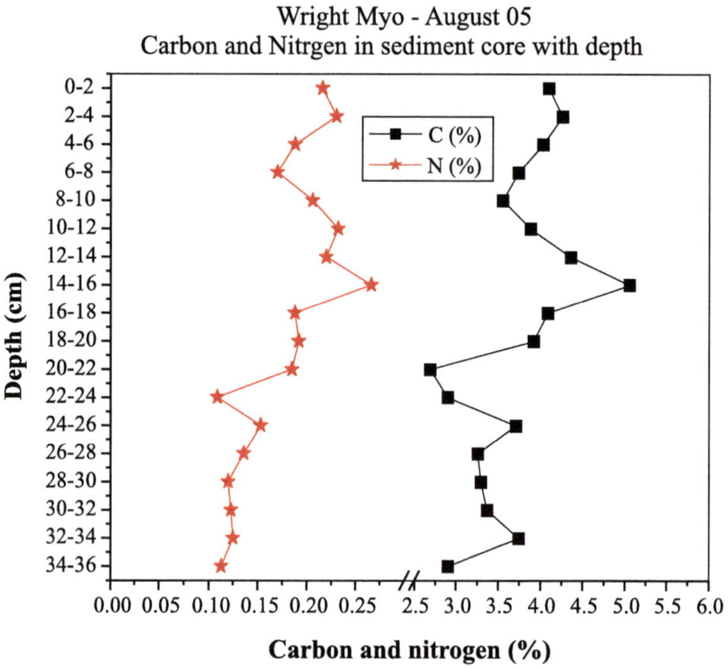

Figure 4: Carbon and nitrogen variation in the core against depth.

Direct comparison between sites is slightly off the mark because of differences in geomorphology and hydrodynamics. For example Kalighat Creek in North Andaman is a Creek much far off the main coastal waters and is highly impacted by human settlement and data are limited to a single season due to logistical constraints. While Redskin Islands are part of the Mahatma Gandhi Marine National Park free from anthropogenic input but the mangrove patch in the coral island are in direct contact with the seawater during the high tides. In the present study the flux pertaining to nitrous oxide (measured through other methods) is higher than that accounted for, through denitrification and so the Wright Myo mangroves are thought to be a potential system for nitrification rather than denitrification.

4.3 SEDIMENT CARBON AND NITROGEN

The total carbon values showed an average concentration of 3.71% with values ranging from 2.68 to 5.05%. Total carbon concentrations showed a decrease with depth as can be seen in Fig. 4. Total nitrogen values ranged from 0.12 to 0.27% with an average of 0.18% and showed a decrease in concentration along with the depth. As can be seen from Fig. 4, the concentrations of total carbon and nitrogen follow a similar trend and show a high positive correlation of 0.76.

Concentrations of total carbon and nitrogen are relatively high as compared with other published results in mangrove sediments. For example Alongi et al.

(2000) have reported total carbon in percentage as 6.9±1.4 and 5.2±0.7 in two mangroves sites dominated by Rhizophora species. Similarly total nitrogen concentrations in % are given as 0.23±0.03 and 0.13±0.02 in the same sites. An increase in C/N ratios with depth in the core was also observed and can be interpreted as indicative of preferential loss of nitrogen (N) with depth (Lee, 1990).

These findings are consistent with other parameters reported in the present study. Mangrove ecosystems play a key role in supporting coastal food webs and nutrient cycles in the coastal zone. Their strategic position between the land and the sea make them important sites for land-ocean interactions (Jannerjahn and Ittekkot, 2002). These forests are capable of reaching high aboveground primary production (Twilley et al., 1992) and high instantaneous rates of net photosynthesis in the canopy of up to 25 μmol CO_2 m^{-2} $leaf^{-1}$ (Clough, 1992; Clough et al., 1997). This high aboveground primary production supports high rates of detrital decomposition (Robertson et al., 1992; Alongi et al., 1999) and mangrove forests are usually characterized by sediment accretion as can be seen in the high concentrations of carbon and nitrogen in the sediment profile. Further study on the sediment accumulation rates using ^{210}Pb profiles can be used to establish the chronology for these cores and sedimentation rates for the system.

4.4 Primary Productivity

In coastal areas, especially the oxygen concentration is always moderate and allows a complete aerobic mineralization of the allochthonous organic matter. Another oxygen consuming process can occur in the coastal ecosystems, receiving high loads of ammonium. Nitrification, the conversion of ammonium to nitrate, is the process by which chemoautotrophic bacteria obtain their energy to fix carbon. However, this process demands significant amounts of oxygen and may lead to sub-oxic conditions, especially in eutrophic systems. In order to understand the terrestrial input of nutrients and organic matter to the mangroves and coral reef systems we have analyzed the surface water primary productivity separately, one at Wright Myo Mangrove ecosystem and other at Red Skin coral reef ecosystem.

Both the ecosystems showed distinct characteristics in terms of primary production which may be due to varied anthropogenic loads and also to their local response to the varied physico-chemical conditions. In the Table 5 the measured Gross Primary Production (GPP) from the Wright Myo mangrove creek is –1.12 mgC m^{-2} d^{-1}. The negative GPP indicates the aerobic mineralisation of organic matter in the moderate oxygen conditions. The measured community respiration (CR) is 0.38 mgC m^{-2} d^{-1}. The importance of organic matter metabolism is governed by the negative NPP (Net Primary Production) and the measured NPP of –1.50 mgC m^{-2} d^{-1} suggests that the Wright Myo mangrove system is heterotrophic at present and supported by inputs of allochthonous organic matter. Therefore, the GPP, which is 73% of NEP, indicates the dominance of community respiration over gross production and signifies that the organic matter that remineralised in the water column is 73% of the total terrestrial load.

Unlike mangrove ecosystem the production and respiration scenario in the coral reef ecosystem is quite distinct. From Table 5 the Gross Primary Production (GPP) in the Red Skin coral reef ecosystem was measured as 10.5 mgC m^{-2} d^{-1} and NPP is 7.50 mgC m^{-2} d^{-1}. The measured community respiration (CR) was 3.50 mgC m^{-2} d^{-1}. The positive NPP (GPP>R) indicates that the system produces more organic matter and thereby acts as a sink of inorganic nutrients and CO_2 from the surrounding water. The Red Skin Coral reef system is influenced as the highly productive community, as it receives important input of nutrients from the land and both GPP and NPP elevated relative to the Wright Myo mangrove system and considered as most autotrophic system rather heterotrophic unlike Wright Myo mangrove ecosystem. The NPP that is 71% of the GPP is thus significant to the autotrophic nature of the coral reef system.

Table 3: Different components of Primary Productivity in the mangroves and coral reef ecosystem

Sampling locations	*GPP* *(mgC m^{-2} d^{-1})*	*NPP* *(mgC m^{-2} d^{-1})*	*CR* *(mgC m^{-2} d^{-1})*
Wright Myo Mangrove Creek	-1.12	-1.50	0.38
Red Skin Coral reefs	10.5	7.50	3.50

4.5 Leaf Litter C:N Analysis

Green leaves and litter of two most abundant species *Avicennia officinalis* and *Rhizophora apiculata* were analyzed along with leaf litter of terrestrial origin to find out the variation in the C:N composition and immobilization of either carbon and nitrogen during the course of decomposition. There was observed a decrease in C:N atomic ratio in the litter relative to the senescent leaf. From Table 4 the Avicennia leaf analyzed from the Wright Myo mangrove creek showed that the C:N atomic ratio change from initial (senescent leaf) 78 to 48 in the litter. In Rhizophora the C:N atomic ratio changed from initial (senescent leaf) 85 to 42 in the litter. From Table 5 in the Red skin mangroves the C:N atomic ratio of Avicennia changes from initial (senescent leaf) 76 to 51 in the litter. In Rhizophora the C:N atomic ratio changes from initial (senescent leaf) 89 to 41 in the litter. Since the relative carbon content of the litter is not changing over time, the decrease in C:N ratio is due to a relative enrichment of nitrogen compared to carbon during the leaf decomposition. Part of this nitrogen enrichment can be explained by fixation of atmospheric nitrogen during the decomposition. Thus the main reason for the relative enrichment of leaf litter with nitrogen may be due to the fact that the leaf detritus becomes colonized

Table 4: C:N composition of leaf and litter of different species of mangroves from the Wright Myo

Species	*Leaf type*	*C:N ratio*
Avicennia officinalis	Senescent leaf	78
	Litter	48
Rhizophora	Senescent leaf	85
apiculata	Litter	42

Table 5: C:N composition of leaf and litter of different species of mangroves from Red Skin

Species	*Leaf type*	*C:N ratio*
Avicennia officinalis	Senescent leaf	76
	Litter	51
Rhizophora	Senescent leaf	89
apiculata	Litter	41

by bacteria and fungi immobilizing nitrogen from the aquatic system on decaying leaves as reported by Woitchik et al. (1997) from the Brazilian mangrove system.

5. SUMMARY AND CONCLUSIONS

Andaman tidal ranges were relatively large (3-4 m) and inundation of high water over extensive mangrove forested areas replenished mangrove porewaters and prevented further outflow. As the tidal height of the waters dropped mangrove porewaters flowed into surrounding creeks; this effect is maximum at tidal minima. The Pristine mangroves receive an abundant supply of organic input mainly leaf litter from the terrestrial ecosystem which is the tropical evergreen rainforest and subsequently transferred into the reef system. This organic input far exceeds the nutrient loading and it is evident that from the low P:R (Production: Respiration) ratio the ecosystem is more heterotropic than autotropic and the primary productivity is low in this system. Owing to the high organic input, increased respiration and hydrodynamics the fluxes of trace gases like methane, carbon dioxide and nitrous oxide are high and may be attributed to high nutrient loading as the result of human intervention like agriculture, aquaculture and others.

Although shelf and estuaries represent only about 15.2 and 0.4% of the global ocean area respectively, they account for about 68 and 7.4% of the oceanic CH_4 emission, in that order. However, these estimated values for coastal environment are rather uncertain due to high spatial and temporal variability and the limited data available. The coastal zone with mangroves and estuaries is a region with highly variable physical processes. Lack of long term data and the ability to distinguish natural from anthropogenic change are the major problems in predicting mangrove responses to human impacts. There is a lack of knowledge in terms of mangrove and estuarine status at many levels from gross production and net export of C, whole system balances, role of species diversity and effects of changes in sea level. This study was an attempt to understand a small part of these ongoing processes in mangrove and coral reef ecosystems. Additional data and future studies that focus on whole system C balances from varied coastal ecosystems are however required to better understand the consequence of anthropogenic activities. Such data are imperative to understand the net carbon balance in coastal ecosystems especially because mangroves and coral reefs play a critical role in determining the direction and magnitude of transfer of C for the global shallow water coastal oceans.

REFERENCES

Alongi, D.M., Boto, K.G. and Robertson, A.I. (1992). Nitrogen and phosphorus cycles. *In:* Tropical Mangrove Ecosystems. American Geophysical Union, Washington, D.C., 251-292.

Alongi, D.M. (2002). Present state and future of the world's mangrove forests. *Environmental Conservation*, **29(3)**, 331-349.

Anderson, W.S. and Christensen, B. (1978). Seasonal growth of mangrove trees in Southern Thailand: Phenology of Bruguiera cylindrica, Ceriops tagal, Lumnitzera littorea and Avicennia marina. *Aquatic Botany*, **5**, 383-390.

Balls, P.W. (1994). Nutrient inputs to estuaries from nine Scottish east coast rivers; influence of estuarine processes on inputs to the North Sea. *Estuar. Coast. Shelf Sci*, **39**, 329-352.

Barnes, J., Ramesh, R., Purvaja, R., Nirmal Rajkumar, A., Senthil Kumar, B., Krithika, K., Ravichandran, K., Uher, G. and Upstill-Goddard, R. (2006). Tidal dynamics and rainfall control N_2O and CH_4 emissions from a pristine mangrove creek. *Geophys. Res. Lett.*, **33**, L15405, doi:10.1029/2006GL026829.

Boto, K.G. and Wellington, J.T. (1988). Seasonal variations in concentrations and fluxes of dissolved organic and inorganic materials in a tropical, tidally-dominated, mangrove waterway. *Mar Ecol Prog Ser,* **50**, 151-160.

Dittmar, T. and Lara, R. (2001). Do mangroves rather than rivers provide nutrients to coastal environments south of the Amazon River? Evidence from long-term flux measurements. *Mar Ecol Prog Ser*, **213**, 67-77.

Dittmer, T. (1999). Nutrient dynamics in a mangrove creek (North Brazil) during the dry season. *Mangroves and salt marshes*, **3**, 185-195.

Gattuso, J.P., Frankignoulle, M. and Wollast, R. (1998). Carbon and carbonate metabolism in coastal aquatic ecosystems. *Annual Review in Ecology and Systematics,* **29**, 405-433.

Kawabata, Z., Magendran, A., Palanichamy, S., Venugopalan, V.K., Tatsukawa, R. and Ketchum, B.H. (1967). Phytoplankton nutrients in estuaries. *In:* Lauff, G.H. (ed.) Estuaries American Association for the Advancement of Science, Washington, D.C., 112-119.

Law, C.S., Rees, A.P. and Owens, N.J.P. (1992). Nitrous Oxide: Estuarine Sources and Atmospheric Flux. *Estuarine Coastal and Shelf Science*, **35**, 301-314.

Lee, S.Y. (1995). Mangrove outwelling: a review. *Hydrobiologia,* **295**, 203-212.

Machiwa, J.F., (1999). Lateral fluxes of organic carbon in a mangrove forest partly contaminated with sewage wastes. *Mangroves and Salt Marshes*, **3**, 95-104.

Odum, W.E. and Heald, E.J. (1975). The detritus-based foodweb of an estuarine mangrove community. *In:* Cronin, L.E. (ed.), Estuarine Research, Vol. 1. Academic Press, New York, 329-335.

Ovalle, A.R.C., Rezende, C.E., Lacerda, L.D. and Silva, C.A.R. (1990). Factors affecting the hydrochemistry of a mangrove creek, Sepetiba Bay, Brazil. *Estuar. Coast. Shelf Sci,* **31**, 639-650.

Purvaja, R. and Ramesh, R. (2000). Natural and anthropogenic effects on phytoplankton primary productivity in mangroves. *Chemistry and Ecology*, **17**, 41-58.

Robertson, A.I., Alongi, D.M. and Boto, K.G. (1992). Food chains and carbon fluxes. *In:* Tropical mangrove ecosystems. American Geophysical Union, Washington, D.C, 239-360.

Romigh, M.M. (2005). Organic carbon flux at the mangrove soil-water column interface in the florida coastal everglades. Ph.D. Thesis, Texas A&M University, USA.

Sasaki, K. and Sawada, Y. (1980). Determination of ammonia in estuary. *Bull.Jap.Soc.Sci.Fish,* **46(3)**, 319-321.

Seitzinger, S.P. (1988). Denitrification in freshwater and coastal marine ecosystems: Ecological and geochemical significance. *Limnol. Oceanogr.*, **33**, 702-724.

Trott, L.A. and Alongi, D.M. (2000). The impact of shrimp pond effluent on water quality and phytoplankton biomass in a tropical mangrove estuary. *Mar Poll Bull*, **40**, 947-951.

Trott, L.A. and Alongi, D.M. (1999). Variability in surface water chemistry and phytoplankton biomass in two tropical, tidally-dominated mangrove creeks. *Marine Freshwater Research*, **50**, 451-457.

Twilley, R.R. (1995). The exchange of organic carbon in basin mangrove forests in a southwest Florida estuary. *Estuar. Coast. Shelf. Sci.,* **20**, 543-557.

14

Dissolved Metal Distribution in Indian Mangrove Ecosystem: Case Studies from East Coast of India

Gurmeet Singh, Rajesh Kumar Ranjan, Rita Chauhan and AL. Ramanathan*

School of Environmental Sciences, Jawaharlal Nehru University
New Delhi, India

1. INTRODUCTION

Mangroves represent highly dynamic and fragile ecosystem, which occupy a large fraction of the tropical and subtropical coastline, dominating the intertidal zone of diverse environmental settings. The potential role of mangrove ecosystems as sinks for anthropogenic contaminants in tropical and subtropical areas has been widely recognized. Sediments are important carriers of trace metals in the hydrological cycle and because metals are partitioned with the surrounding waters, they reflect the quality of an aquatic system (Szefer et al., 1995). Heavy metals are natural constituents of the marine environment and some of them, such as Cu, Fe, Zn, and Mn, are biologically essential for normal growth and development (Astorga Espana et al., 2004). Metals such as Cd and Hg have no known use in physiological processes (Darmono and Denton, 1990), but when present in excess have ecological significance due to their toxicity and cumulative behaviour and constitute a potential hazard for environment.

The Indian Sundarban (Latitude 21°32′-22°40′N; Longitude 88°05′-89°E), is located at Ganges-Brahmaputra river systems, in the coastal regions of the Bay of Bengal. It has the potential for being a global biodiversity hotspot and

*Corresponding Author

World Heritage site for its rich and diverse faunal and floral communities (Ramanathan et al., 2008). Hydrology of this estuarine system presents a cyclic pattern, characterized by a large amount of monsoonal precipitation and tidal interplay (Bhattacharya, 1988). During the last few decades this coastal region is under the clutch of environmental degradation due to rapid human settlements, tourist activities, deforestation and increasing agricultural and aquaculture practices. The mangroves ecology of Sundarban are associated with anthropogenic input due to huge discharge of domestic, municipal and industrial wastes, dredged materials, storm water runoff, aerial fall out, oil spills, boating and other sources (Chatterjee et al., 2007) along with the rapid emergence of Haldia Port Complex, a major oil disembarkment terminal in eastern India (Sarkar et al., 2007). In addition, the mangrove receives wastes from urban settlements and a variety of industries including paper and pulp, electrical, pharmaceutical etc. Furthermore, the estuary also receives the raw sewage sludge from 85 km upstream, discharge from highly urbanized metropolitan mega city Kolkata (population 14.5 million) through river Bidhyadhari.

The Pichavaram mangroves are located between the Vellar and the Coleroon estuaries – distributaries of the river Cauvery (Latitude 11°23′-11°30′N; Longitude 79°45′-79°50′E) in the southeast coast of India. The 11 km^2 of Pichavaram, consisting of 51 islets, is 50% covered by forest, 40% by urban waterways (for fishing) and the rest filled by mud and sand flats (Krishnamurthy and Jayaseelan, 1983). The ecosystem faces two monsoon seasons: southwest monsoon from June to July and northeast Monsoon from October to January. The annual temperature variation is 18.2-36°C. The Pichavaram mangrove forest is tide dominated with amplitude ranging from 0.5 to 1 m. Geomorphically, the area is mostly covered by floodplain, sedimentary plain and beach sand. The geology of the area is dominated by Quaternary sediments. Alluvium is dominant in the western part whereas fluvial marine and beach sand covers the eastern part of the mangrove. Thirteen species of true mangroves are reported in the area with *Avicennia* and *Rhizophora* as the most common and dominant (Kathiresan, 2004). The pristine ecosystem has been continuously exposed to the anthropogenic pressure with rapid growth in industrialization as well as degradation of mangrove forest during past few years (Subramanian, 2004; Ramanathan et al., 1999; Ranjan et al., 2008).

Bhitarkanika mangrove is located between Latitude 20°4′-20°8′N and Longitude 86°45′-87°50′E in the Kendrapada district of Orissa, eastern coast of India. It is the second largest mangrove ecosystem of India with immense ecological, geomorphological and biological significance. Bhitarkanika mangrove ecosystem flourishes in the deltaic region, formed by the rich alluvial deposits of Brahmani and Baitarani rivers. It receives inputs of untreated domestic and industrial wastes. The industrial growth in Orissa state results in diversion of fresh river water for industrial growth and future addition of the discharges from such industries deteriorate the water as well as sediment quality (Chauhan et al., 2008).

An attempt has been made to assess the status of heavy metal concentration and elucidate the process governing their relative distribution within this dynamic environment. The present article particularly aims to assess the behaviour of dissolved metals (Fe, Mn, Cr, Cu, Zn and Pb) in the three mangrove ecosystems (namely Sundarban, Bhitarkanika and Pichavaram).

2. MATERIALS AND METHODS

Eighteen surface water samples were collected from Sundarban mangroves in India, representing the entire mangrove ecosystem whereas twelve samples were collected from Pichavaram mangroves (Fig. 1). Fifteen samples were

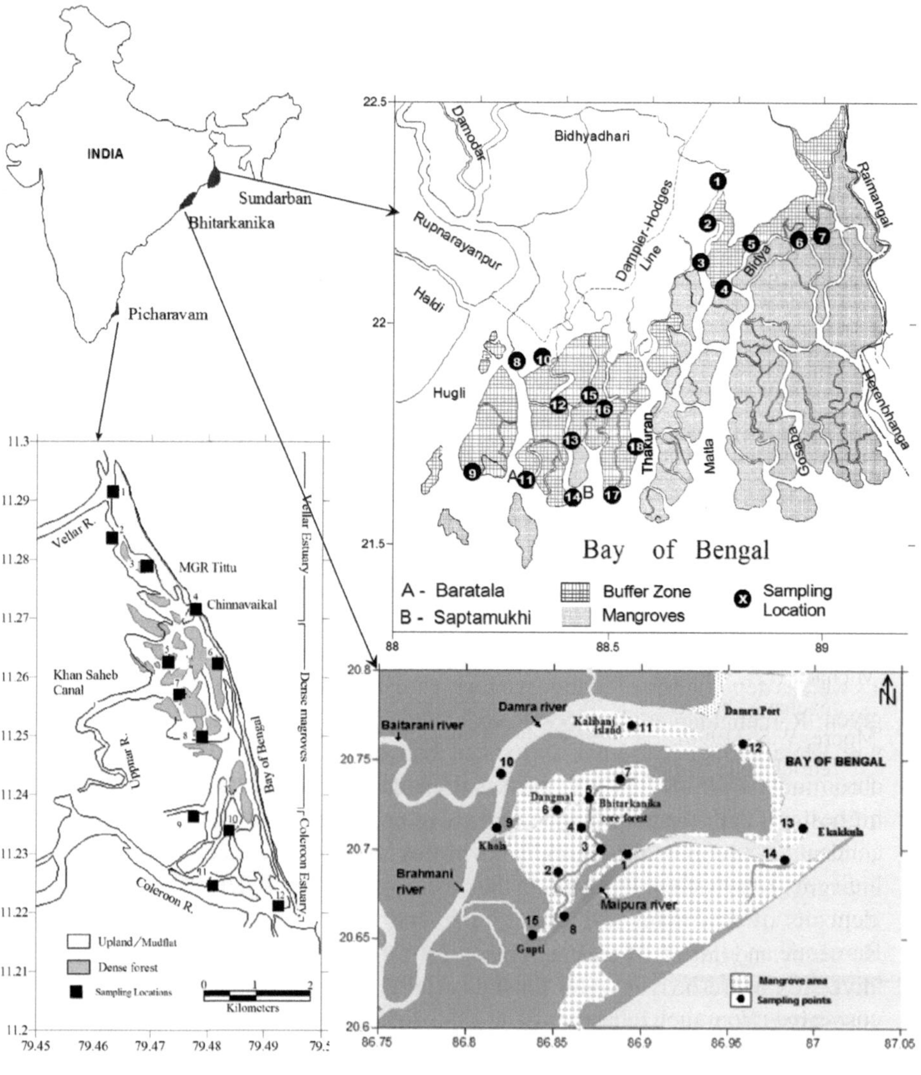

Figure 1: Map showing Sundarban, Bhitarkanika and Pichavaram mangrove ecosystems.

collected from Bhitarkanika mangrove forest. The sampling locations were decided on the basis of land use type, anthropogenic input, vegetation cover and other geomorphologic features (such as sandy beaches, mudflats, etc.).

In Sundarban samples S3, S4, S5, S6, S7, S13 and S14 were collected from pristine area whereas samples S1, S2, S8, S9, S10, S11, S12, S15 and S16 were collected from the area with anthropogenic disturbance. In case of Bhitarkanika mangroves, samples S1-S5 represented estuarine, S6-S10 were from mangrove region and rest of the samples were collected from estuarine region. Samples S1, S2, S3 and S4 were collected from Vellar region whereas S5, S6, S7, S8 and S9 were from dense mangrove forest region. Samples S9, S10 and S12 represented Coleroon estuary region.

The collected samples were filtered with 0.45 μ membrane filter paper and were acidified with HNO_3 to pH <2 at the time of sampling. Collected samples were stored in an ice chest in the field and were transferred to laboratory where they were kept at 4°C until analysis. The dissolved metal concentrations (Fe, Mn, Cr, Cd, Cu, Zn and Pb) in water were determined with Atomic Absorption Spectrophotometer (Shimadzu-AA-6800). The instrument was standardized using multi-elemental standard MESS 2 (Merck). The standards were re-run during the course of analysis and a precision of ≥95% was observed for the measurements.

3. RESULTS AND DISCUSSION

Since rivers are the main source of water requirements of people in India and other parts of the world, water quality needs to be maintained. In recent decades, most rivers are being polluted due to excessive silting by deforestation, urbanization, cultivation, industrialization and various other human activities. The pollution load from the catchment is deposited in the coastal environment making the system vulnerable to the municipal and domestic wastes; industrial effluents and agricultural run-off are the major contributors to coastal pollution. In some areas, the geochemical nature of the land and climatic conditions are also responsible for variation in water quality (Trivedy and Kulkarni, 1988).

The mean concentration of trace metals in surface water of Sundarban are given in Table 1. The concentration of metals exhibited the following order, with little variation: Iron (Fe) manganese (Mn) > lead (Pb) > zinc (Zn) > chromium (Cr) = nickel (Ni) > cadmium (Cd). The metal concentrations were highest in and around the areas with anthropogenic stress (Fig. 2). High concentration of Fe and Mn were observed in mangrove systems and these high concentrations may be due to natural sources such as weathering of rocks from the up streams of rivers and degradation of organic matter (Lacerda, 1997). Ramanathan (1998) has reported 78 mg kg^{-1} of Mn in the leaves of *Avicennia marina* in Sundarban. Fe and Mn leached from mangrove leaves may be quickly converted into colloidal hydroxide at the pH (7.0–8.0) prevailing in the water column. Higher concentration of Zn is due to localised discharges from

Table 1: Heavy metal distribution (in mg L^{-1}) in mangroves of east coast of India

	Sundarban (n=18)		*Bhitarkanika (n=15)*		*Pichavaram (n=12)*	
	Min-Max	*Mean ± S.D.*	*Min-Max*	*Mean ± S.D.*	*Min-Max*	*Mean ± S.D.*
Fe	22.6-90.8	50.5±18.9	20.5-56	29.5±11.1	33.1-90.7	49.34±21.81
Mn	0.2-9.5	2.8±2.4	0.03-0.25	0.16±0.03	0.03-0.45	0.124±0.142
Zn	0.1-2.0	0.5±0.4	0.02-0.17	0.09±0.05	0.03-0.15	0.049±0.034
Cr	0.0-0.9	0.3±0.12	0.35-0.38	0.36±0..3	0.3-1.34	0.53±0.27
Cd	0.0-0.7	0.2±0.08	0.1-0.49	0.29±0.07	0.19-0.80	0.287±0.254
Ni	0.1-0.6	0.3±0.11	0.03-0.04	0.035±0.03	0.01-0.03	0.01±0.008
Pb	0.1-2.2	1.0±0.6	0.37-0.41	0.39±0.12	0.07-0.19	0.138±0.036

S.D. represents standard deviation

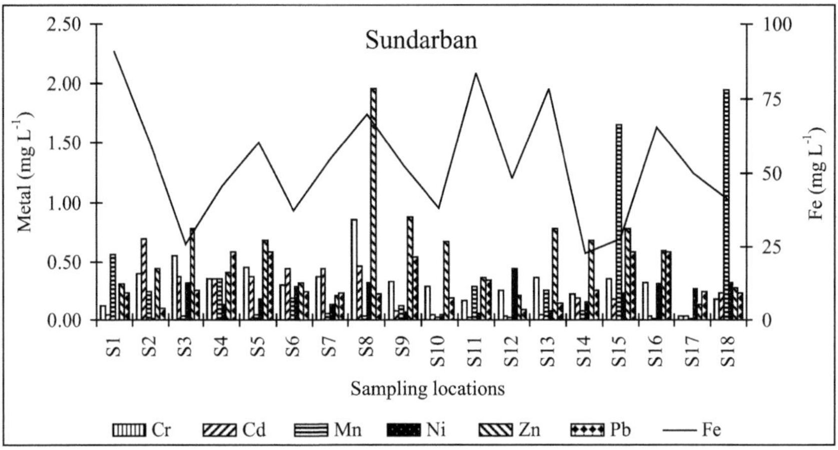

Figure 2: Dissolved metals (in mg L^{-1}) in Sundarban mangrove ecosystem.

industries located in this area (Sarkar et al., 2002), runoff from agricultural lands where it is used extensively in fungicide (Lambert et al., 1981; Murray and Gill, 1978). High concentration of other metals such as Cr, Pb and Zn in the mangrove area respectively may be due to waste discharges from the neighbouring areas. The high concentration of Pb is associated with the use of low quality fuel for the riverine transport (in case of boat etc.). Metals concentration in the coastal environment depends upon a number of factors such as salinity, reducing conditions, pH, organic matter availability etc. Mangrove waters are enriched with dissolved organic matter, low redox potential, and high pH facilitating the precipitation of the heavy metal (Lacerda, 1997, 1998).

In order to understand the importance of these dissolved trace metals, the total data set was subjected to regression analysis and the correlation matrix (at significance level 0.05) is given in Table 2. The significant correlations between Fe and Mn ($r^2 = 0.65$) suggest that these metal have the same origin.

Table 2: Correlation matrix of metals in mangrove water samples from east coast of India

				Sundarban			
	Fe	*Mn*	*Ni*	*Zn*	*Cr*	*Cd*	*Pb*
Mn	0.65						
Ni	0.26	0.03					
Zn	0.53	0.1	0.83				
Cr	0.76	0.22	0.16	0.53			
Cd	0.04	0.06	0.18	0.58	0.01		
Pb	0.37	-0.01	-0.23	-0.1	0	0.06	
				Bhitarkanika			
	Fe	*Mn*	*Ni*	*Zn*	*Cr*	*Cd*	*Pb*
Mn	0.71						
Ni	0.09	0.31					
Zn	0.03	-0.12	0.18				
Cr	0.12	0.07	0.35	-0.17			
Cd	0.05	-0.12	0.15	0.74	-0.05		
Pb	0.07	0	0.37	-0.16	0.91	-0.07	-0.06
				Pichavaram			
	Fe	*Mn*	*Ni*	*Zn*	*Cr*	*Cd*	*Pb*
Mn	0.86						
Ni	0.26	0.69					
Zn	0.06	0.56	0.96				
Cr	0.02	0.36	0.73	0.67			
Cd	-0.21	-0.05	0.26	0.33	-0.32		
Pb	-0.84	-0.56	0.17	0.3	0.48	0.11	

(n=18 for Sundarban, n=15 for Bhitarkanika; n=12 for Pichavaram)
Significance level $p \leq 0.05$

Significant correlation between Fe and Cr (r^2 = 0.76) and Fe-Zn (r^2 = 0.53) indicate that these metals are associated with the oxides and oxyhydroxides of Fe and Mn. No significant correlations between Pb and Cd showed that these are mostly carried by external waste discharges from the neighbouring environment through non-point sources. Further significant correlations between Ni and Zn (r^2 = 0.83), Zn-Cr (r^2 = 0.58) and Zn-Cd (r^2 = 0.58) indicate the possible enrichment of these metals partly through terrigenous origin and partly due to the biological production especially uptake of particulate trace metals by phytoplankton as reported earlier (Knaner and Martin, 1973). The sources and interrelationships between the heavy metals for Sundarban were further evaluated through principal component analysis which segregated 67% of the total data into three factors. Factor 1 is statistically dominant and accounts for 29 percent of variance (Table 3). This factor is characterized by strong loading

Table 3: Principal component analysis of dissolved metals in mangroves of east coast of India

	Factor 1	*Factor 2*	*Factor 3*
	Sundarban		
Fe	-		0.82
Mn	-		0.75
Ni	-	0.79	-
Zn	0.86	-	-
Cr	-	0.57	-
Cd	0.97	-	-
Pb	0.78	-	-
% Variance Explained	29.12	20.67	17.44
Cumulative %	29.12	49.79	67.23
	Bhitarkanika		
Fe	-	-	0.89
Mn	-	-	0.937
Ni	-	0.573	-
Zn	0.975	-	-
Cr	-	0.944	-
Co	0.928	-	-
Cd	0.91	-	-
Pb	-	0.954	-
% Variance Explained	37.34	26.77	22.01
Cumulative %	37.54	65.5	86.15
	Pichavaram		
Fe	0.9	-	-
Mn	0.96	-	-
Ni	-	0.68	-
Zn	-	0.73	-
Cr	-	0.89	-
Cd	-		0.97
Pb	-	0.81	-
% Variance Explained	37.25	36.21	19.6
Cumulative %	37.25	73.47	93.07

Extraction Method: Principal Component Analysis. Rotation Method: Varimax with Kaiser Normalization. Factor greater than eigen value 1 were considered for interpretation.

of Zn, Cd and Pb, because of pollution due to various industrial units located upstream. The pairing of Pb and Cd further suggest that the use of low quality fuel in the local mechanized boats for the transportation adds significantly to the metal pollution (Chatterjee et al., 2007). Factor 2 accounts for 21 percent of variance with positive loadings of Cr and Ni. This metal concentration increase may be due to direct input of nitrate compounds from external sources mainly from the aquaculture effluents, agricultural runoff and domestic sewage (Subramanian, 2004). Factor 3 accounts for 17 percent of variance and depicts the strong biogeochemical affinity of iron and manganese. These two metals play an important role in various metabolic pathways occurring in these coastal ecosystems (Lacerda et al., 1999). Metals are preferentially attached to their oxides and hydroxides in natural system (Jingchun et al., 2006). The stability of these oxides is redox dependent (Clark et al., 1998). Hence this factor may be identified as the inherent biological productivity of the mangroves as well as the redox environment.

A great degree of variability was observed in the concentration of heavy metals in surface water from the Bhitarkanika mangrove ecosystem (Table 1). Average concentration of dissolved metals followed the order Fe>Pb>Cr>Cd>Mn>Zn>Ni. The regions with anthropogenic stress showed the higher concentration for nearly all heavy metals i.e. Fe, Mn, Ni, Zn, Cr, Co, Cd except for Pb (Fig. 3). The higher metal concentration could be due to accumulation along river course by anthropogenic inputs such as industrial pollution, domestic sewage, and agricultural runoff. Earlier studies had reported that these two rivers have extremely variable heavy element concentrations, which is consistently higher than the world river average (Konhauser et al., 1997). The impact of pollution on river chemistry is most evident in the Brahmani River, where high elemental concentrations correspond to the presence of heavily industrialized areas (Chauhan, 2008). It receives effluent discharged directly from NALCO (National Aluminium Company), FCI (Fertilizer Corporation of India), and TTPS (Talcher Thermal Power Station). These industries utilize coal and are active source of fluorides, nitrogen compounds, cyanide, chromium, fly ash, and other suspended solids. The heavy metal concentration was lowest in the dense mangrove region (Fig. 3). Several studies have shown that mangrove sediment can trap metals, thus acting as a filter for the ecosystem (Harbison, 1986; Lacerda, 1998).

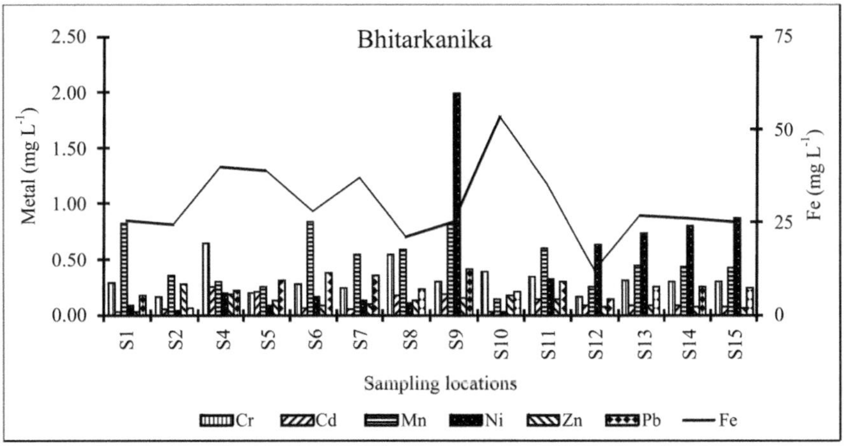

Figure 3: Dissolved metal (in mg/L) in Bhitarkanika mangrove ecosystem.

The statistical analysis of inter-metallic relationship revealed the high degree of correlation and significant relation among the metals, which indicate the identical behaviour of metals during its transport in the mangrove environment. The pairing and affinity between the heavy metals were deduced through a multivariate analysis such as correlation and PCA analysis for all heavy metals. Correlation analysis depicted a positive and significant correlation of Fe and Mn (r^2=0.7) (Table 2). The concentration of Fe and Mn was maximum in mangrove dominated region, followed by bay region and least in estuarine region. The strong affinity of these metals within organic matter depicts the biogeochemical relationship between them. The main anthropogenic source of Ni is the burning of fossil fuel and residual oils while the smaller quantities of Ni can be released through combustion of coal, and the use of fertilizers (Preda and Cox, 2005). The Zn shared a positive and strong correlation with Co (r^2=0.94) and Cd (r^2=0.94). The Zn showed higher concentration in Zone I area than other regions. The high concentration of Co and Cd were mainly due to effect of anthropogenic activities and fertilizers (Tam and Wong, 2000). The Pb and Cr also revealed positive and significant correlation which shows the strong affinity for each other (Table 2). The anthropogenic source of Cr is mainly due to effluents from iron and steel industry and leather tanning industries. The incorporation of Cr in the lattice of sediment as in case of alumino-silicate minerals often leads to lower availability to the system (Shriadah, 1999). Consequently the concentration of Cr was highest in the Zone I region, which receives ample industrial effluents through the river discharges.

The sources and interrelationships between the heavy metals revealed through factor analysis showed that Factor 1 is statistically dominant and accounts for 37 percent of variance (Table 3). This factor is characterized by strong loading of Zn, Co and Cd, due to industrial pollution in the upstream of two rivers. Factor 2 accounts for 26 percent of variance with positive loadings of chromium and lead, which is derived from port activities. The Pb concentration increased may be due to direct input of nitrate compounds from external sources mainly from the aquaculture effluents, agricultural runoff and domestic sewage (Subramanian, 2004). Factor 3 accounts for 22 percent of variance and depicts the strong biogeochemical affinity of iron and manganese. In the deep reducing part of the sediment, iron and manganese hydrous oxides are dissolved and migrate upward through the sediment column and are subsequently precipitated in the oxidizing sediment surface or lost to the overlying water and thus added to water column (Lacerda, 1997).

The physiochemical characteristics of Pichavaram-estuarine mangrove complex water are given in Table 1. The samples from estuarine region represent the highest concentration for nearly all the heavy metals at confluence/entry point of Vellar and Coleroon rivers with mangrove area, which is possibly due to accumulation along river course by anthropogenic inputs like domestic sewage, and agricultural runoff. The heavy metal concentration was lowest in

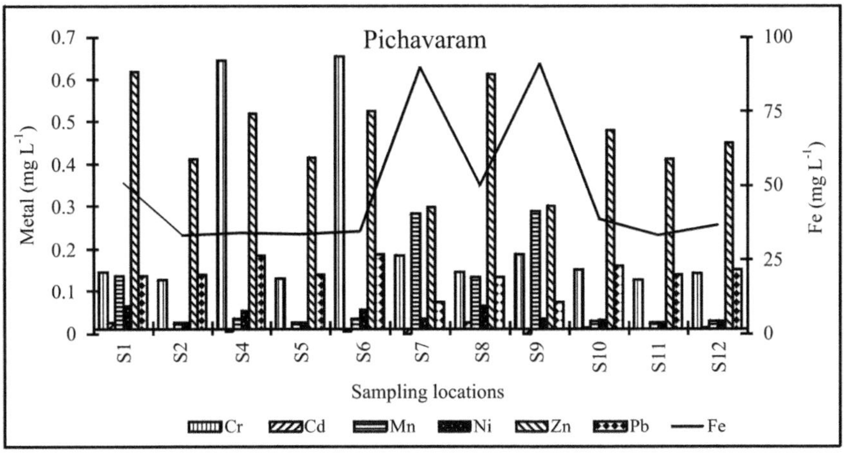

Figure 4: Dissolved metals (in mg L^{-1}) in Pichavaram mangrove ecosystem.

the mangrove region. This is due to the fact that mangrove forest act as filter for the ecosystem, and henceforth sediment act as sink for metal contamination (Machado et al., 2002; Janaki-Raman et al., 2007; Singh et al., 2005). The mean concentration of the dissolved metals followed the order Fe>Cr> Cd>Pb>Mn>Zn>Ni (Fig. 4). The correlation matrix was carried out to delineate the sources and interrelationships between the heavy metals. Positive strong correlation were observed (Table 2) between Fe-Mn, Mn-Ni, Mn-Zn, Zn-Ni, Cr-Ni and Zn-Cr pairs suggesting their identical behaviour in the system. The biogeochemical processes in this ecosystem are governed by the heavy input of sediments and anthropogenic discharges from the Vellar and Coleroon rivers (Ranjan et al., 2008 a,b). Since Coleroon river is a distributary of Cauvery river which is perhaps one of the widely utilized rivers in the world which passes through densely populated industrial cities. Along its course, it carries fertilizer, heavy metals and pesticide from upstream region of the basin (Ramanathan et al., 1999, Selvakumar et al., 2005, Ranjan et al., 2008). Fe, Mn, Zn, Cr and Cd metals are derived from the geochemical weathering of the drainage area as well as from mining activity in the upstream region (Ramanathan et al., 1996). The inverse relation between Pb-Fe, Pb-Mn and Mg-Cr shows that the contribution of Pb is different from the other sources also. This is mainly due to combination of point (industrial) and different non-point sources (agricultural, aquaculture and waste discharge through Khan Sahib canal) of heavy metals contributed in this ecosystem. Thorough principal component analysis was carried out to delineate the sources and interrelationships between the heavy metals. Total variability of 93% was cxplaincd through factor analysis. It was observed that Factor 1 accounts for 37% of variance (Table 3). This factor is characterized by strong loading of Fe and Mn suggesting the dominance of inherent biological productivity as well as redox processes towards controlling the geochemical behaviour of metals in Pichavaram. Factor 2 accounts for 36 percent of variance with positive loadings of Cr, Ni, Zn and Pb. This factor can be attributed to the anthropogenic industrial discharges through Vellar and Coleroon rivers (Ranjan et al., 2008). Factor 3 accounts for 17% of variance and represented Cd.

3. CONCLUSION

Sundarban, Bhitarkanika and Pichavaram mangrove ecosystems situated along deltaic region are under stress of metal contamination due to point as well as non-point sources. The pairing and affinity between the metals are intricately related to their origin and mobility within highly dynamic estuarine environment. The high concentration of Fe and Mn are mainly due to dominance of inherent biological productivity as well as redox processes at sediment-water interface while Cr, Ni, Pb and Zn contaminations are non-point sources due to aquaculture effluents, agricultural runoff, and industrial activities.

ACKNOWLEDGEMENTS

The authors acknowledge Jawaharlal Nehru University for providing necessary facilities to carry out the research.

REFERENCES

Astorga España, M.S., Rodríguez Rodríguez, E.M. and Díaz Romero, C. (2004). Manganese, nickel, selenium and cadmium in mollusc from the Strait of Magellan, Chile. *Food Additives and Contamination*, **21**, 768-773.

Bhattacharya, A. (1988). Intertidal depositional characters in the Hoogly tidal islands, West Bengal. *Indian Journal of Geology*, **60**, 153-164

Chatterjee, M., Silva Filho, E.V., Sarkar, S.K., Sella, S.M., Bhattacharya, A., Satpathy, K.K., Prasad, M.V.R., Chakraborty, S. and Bhattacharya, B.D. (2007). Distribution and possible source of trace elements in the sediment cores of a tropical macrotidal estuary and their ecotoxicological significance. *Environmental International*, **33**, 346-356.

Chauhan, R., Ramanathan, AL. and Kumar, M. (2008). Assessment of spatial variation of heavy metals in an Indian subtropical mangrove dominated estuary. 12th International Conference on Integrated Diffuse Pollution Management (IWA DIPCON 2008). Research Center for Environmental and Hazardous Substance Management (EHSM), Khon Kaen University, Thailand ; 25-29 August 2008. http://www.envicenter.com/iwa2008/web/document/01_048_Rita-Chauhan.pdf.

Clark, M.W., McConchie, D., Lewis, D.W. and Saenger, P. (1998). Redox stratification and heavy metal partitioning in *Aĺicennia*-dominated mangrove sediments: A geochemical model. *Chemical Geology*, **149**, 147-171

Darmono, D. and Denton, G.R.W. (1990). Heavy metal concentrations in the banana prawn, *Penaeus mergyinensis* and leader prawn *P. monodon* in the Townsville region of Australia. *Bulletin of Environmental Contamination and Toxicology*, **44**, 479-486.

Harbison, P. (1986). Mangrove muds: A sink or source for trace metals. *Marine Pollution Bulletin*, **17**, 246-250.

Janaki-Raman, D., Jonathan, M.P., Srinivasalu, S., Armstrong-Altrin, J.S., Mohan, S.P. and Ram-Mohan, V. (2007). Trace metal enrichments in core sediments in Muthupet mangroves, SE coast of India: Application of acid leachable technique. *Environmental Pollution,* **145**, 245-257.

Jingchun, L., Chongling, Y., Macnair, M.R., Jun, H. and Yuhong, L. (2006). Distribution and Speciation of Some Metals in Mangrove Sediments from Jiulong River Estuary, People's Republic of China. *Bulletin of Environmental Contamination and Toxicology*, **76**, 815-822.

Kathiresan, K. (2004). A review of studies on Pichavaram mangrove, southeast India. *Hydrobiologia*, **30**, 185-205.

Knaner, G.A. and Martin, J.H. (1973). Seasonal variations of Cadmium, Copper, Manganese, Lead and Zinc in water and phytoplankton in Monterey Bay, California. *Limnology and Oceanography*, **18**, 597-604.

Konhauser, K.O., Powell, M.A., Fyfe, W.S., Longstaffe, F.J. and Tripathy, S. (1997). Trace element geochemistry of river sediment, Orissa State, India. *Journal of Hydrology*, **193**, 258-269.

Krishnamurthy, K. and Jeyaseelan, M.J.P. (1983). The Pichavaram (India) mangrove ecosystem. *International Journal of Ecology and Environmental Science*, **9**, 79-85.

Lacerda, L.D. (1997). Trace metals in mangrove plants, why such low concentrations? *In:* Kjerfve, B., Lacerda, L.D. and Diop, E.S. (eds), Mangrove Ecological Studies in Latin America and Africa. UNESCO, Paris.

Lacerda, L.D. (1998). Biogeochemistry of trace metals and diffuse pollution in mangrove ecosystems. ISME occasional papers 2. International Society for Mangrove Ecosystems, Okinawa.

Lacerda, L.D., Ribeiro, M.G. and Guiros, B.B. (1999). Manganese dynamics in a mangrove, mud flat tidal creek in SE Brazil. *Mangroves and Salt Marshes*, **3**, 105-115.

Lambert, C.E., Jehanno, C., Silverberg, N., Brun-Cottan, J.C. and Chesselet, R. (1981). Lognormal distributions of suspended particles in the open ocean. *Journal of Marine Research*, **39**, 77-88.

Machado, W., Moscatelli, M., Rezende, L.G. and Lacerda, L.D. (2002). Mercury, zinc, and copper accumulation in mangrove sediments surrounding a large landfill in southeast Brazil. *Environmental Pollution*, **120**, 455-461

Murray, J.W. and Gill, G. (1978). The geochemistry of iron in Puget Sound. *Geochimica Cosmochimica Acta*, **42**, 9-19.

Preda, M. and Cox, M.E. (2002). Trace metal occurrence and distribution in sediments and mangroves, Pumicestone region, southeast Queensland, Australia. *Environment International*, **28**, 433-449.

Ramanathan, AL. (1998) Mangrove forest: A Brief Overview. *In:* Recent trends in Environmental Biogeochemistry (ed.), Subranamian V., ENVIS Publication, New Delhi, India, pp. 195-205.

Ramanathan, AL., Subramanian, V., Ramesh, R., Chidambaram, S. and James, A. (1999). Environmental geochemistry of the Pichavaram mangrove ecosystem (tropical), Southeast coast of India. *Environmental Geology*, **37**, 223-233.

Ramanthan, AL., Singh, G., Majumdar, J., Chauhan, R., Ranjan, R.K., Rajkumar, K. and Santra, S.C. (2008). A study of microbial diversity and its interaction with nutrients in the sediments of Sundarban mangroves. *Indian Journal of Marine Sciences*, **37**, 166-180.

Ranjan, R.K., Ramanathan, AL., Singh, G. and Chidambaram, S. (2008). Assessment of metal enrichments in tsunamigenic sediments of Pichavaram mangroves, southeast coast of India. *Environmental Monitoring and Assessment*, **147**, 389-411.

Sarkar, S.K., Saha, M., Takada, H., Bhattacharya, A., Mishra, P. and Bhattacharya, B. (2007). Water quality management in the lower stretch of the river Ganges, east coast of India: An approach through environmental education. *Journal of Cleaner Production*, **15**, 1559-1567.

Selvakumar, S., Geraldine, P., Shanju, S. and Jayakumar, T. (2005). Stressor-specific induction of heat shock protein 70 in the freshwater prawn Macrobrachium malcolmsonii (H. Milne Edwards) exposed to the pesticides endosulfan and carbaryl. *Pesticide Biochemistry and Physiology*, **82**, 125-132.

Shriadah, M.M.A. (1999). Heavy Metals in mangrove Sediments of the United Arab Emirates Shoreline (Arabain Gulf). *Water Air and Soil Pollution*, **116**, 523-534.

Singh, G., Ramanathan, AL. and Prasad, M.B.K. (2005). Nutrient Cycling in the mangrove environment – A brief review. *International Journal of Ecology and Environmental Sciences*, **31**, 231-244.

Subramanian, A.N. (2004). Status of Indian mangroves: pollution status of the Pichavaram mangrove area, south-east coast of India. *In:* Vannucci, M. (Ed.), Mangrove management & conservation. United Nations Univ Press, Tokyo, pp. 59-75.

Szefer, P., Glasby, G.P., Pempkowiak, J. and Kaliszan, R. (1995). Extraction studies of heavy-metal pollutants in surficial sediments from the southern Baltic Sea off Poland. *Chemical Geology*, **120**, 111-126.

Tam, N.F.Y. and Wong, W.S. (2000). Spatial variation of heavy metals in surface sediments of Hong Kong mangrove swamps. *Environmental Pollution*, **110**, 195-205.

Trivedy, R.K. and Kulkarini, A.Y. (1988). Salient Features of Some Indian Rivers. *In:* Trivedy, R.K. (Ed.), Ecology and Pollution of Indian Rivers. Ashish Publishing House, New Delhi, pp. 395-405.

SECTION 4

Threats to Coastal Aquatic Ecosystems: Developmental and Sustainability Issues

15

Shrimp Culture: Trend, Consequences and Sustainability in the South-western Coastal Region of Bangladesh

Dilip Kumar Datta*, Kushal Roy[1] and Nazia Hassan
Environmental Science Discipline, Khulna University, Khulna 9208 Bangladesh
[1] Ecology, Environment Unit, Unnayan Onneshan, Dhaka 1209, Bangladesh

1. INTRODUCTION

Shrimp culture along the coast of southwest Bangladesh was a traditional practice and usually restricted within land between the levees and the river channels (Deb, 1998; Islam et al., 2005; Nandy et al., 2007). However, since early seventies of the last century farmers and investors are encouraged to involve most in shrimp farming within *polders*—large areas within coastal embankments meant for protecting crop lands from tidal waves, surges and saline intrusion—in response to the demand from international high value customers for shrimp (a luxury food item), low investment and unexpectedly high return where government policies played a favourable role (Alauddin and Tisdel, 1998) and thus ushered the beginning of present shrimp era (BCAS, 2001). Outbreak of disease was a major hindrance for shrimp culture (Alam et al., 2007) and during 1996 Khulna region experienced such impact on approximately 90% of its extensive shrimp farms with a loss of 20% in total shrimp production (Chowdhury and Muniruzzaman, 2003).

*Corresponding Author

Currently shrimp occupies the place after readymade garments as the major export item in the country and contributes ~US$ 300 million as foreign earnings for the country (DoF, 2001) and plays a central role in the fisheries sub-sector of Bangladesh by creating employment opportunities and increasing trade and commerce, processing and marketing, and export as well. Major share of the total shrimp produced in Bangladesh is supplied by the southwestern coastal region of the country (DoF, 2003) which includes Khulna, Bagerhat, Satkhira and southern part of Jessore district. The region is also popularly known as the 'shrimp zone' of the country, a major part of which actually extends into the Sundarbans Impact Zone (SIZ) (Fig. 1).

However, debates, disparities and criticisms followed this sector since its birth. Extensive shrimp culture was not popular among most farmers because it does not secure the food (in terms of grain) availability, and the profit generated from shrimp exports are not shared equitably throughout the chain. The shrimp culture is also characterized by social inequality, disparity, rice versus shrimp conflict, inadequate and ill implementation of policies, conflicts of power and privileges, ecological threats and damages, insufficient structure, lack of clarity in understanding, people's protest and bloodshed, conflicts on saline water

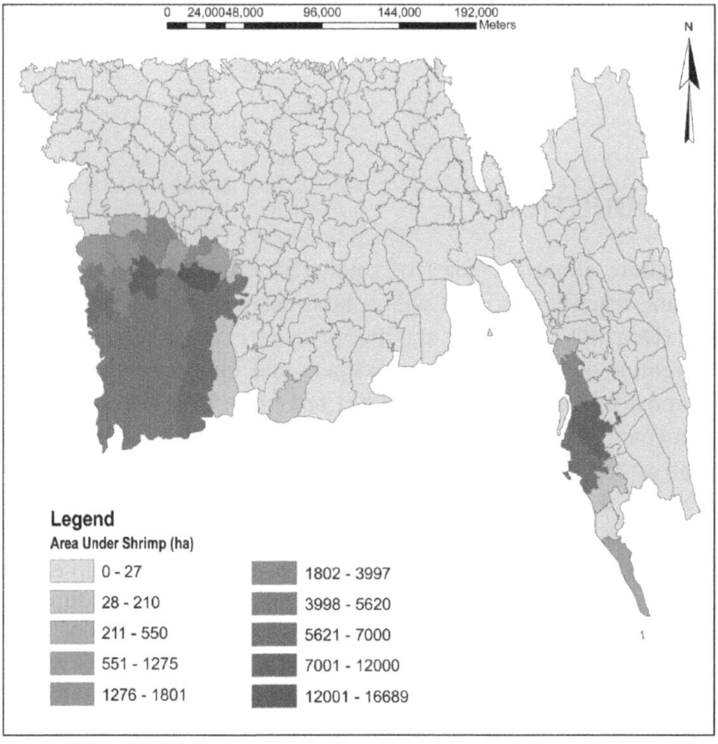

Figure 1: Distribution of area under shrimp culture in the coastal districts of Bangladesh.

breach and so on. Gender disparity, occupational segmentation, wage inequality, and increased job insecurity for women—who are the major work force in this sector—have also been recorded.

With this background the current review investigates the trends and consequences of shrimp culture in the coastal region of Bangladesh with special reference to its southwestern region, and endeavours to decipher the question of its sustainability considering the environmental, societal and economic issues.

2. CURRENT STATE OF SHRIMP CULTURE

The pattern of shrimp culture in the south-western coastal region of Bangladesh is often in rotation with agriculture and reflects the ambient seasonal salinity fluctuations in response to monsoon (Farmer, 1989). The high salinity season (January to July) is characterized by culture of brackish water shrimp and fish, and harvesting of marketable individuals while the low salinity season (August to December) is featured by continued culture of under-sized shrimp and fish subject to the salinity tolerance of the species in low-lying areas, culture of freshwater shrimp and fish, cultivation of slightly salt-tolerant *aman* rice in more elevated areas and harvest of residual brackish water shrimp and freshwater shrimp. *Penaeus monodon* (locally known as *Bagda*)—the brackish water shrimp—comprises 60% of farmed shrimp production, followed by the giant freshwater prawn *Macrobrachium rosenbergii* (locally known as *Golda*) which accounts for 25% (Rosenberry, 1995). More than 70% shrimp culture is favoured in greater Khulna region (Khulna, Satkhira and Bagerhat districts) because both the fresh- and saline-water resources are abundant in almost all seasons and the world's largest single tract mangrove, the Sundarbans, provides food and nursery for the offshore fisheries (Karim and Shah, 2001).

The estimate of total land under shrimp varies considerably among different authorities such as from 143,000 hectares (DoF, 2005) to USAID-Bangladesh (2006) estimates of approximately 203,071 hectares. During 1979-80 the land area under shrimp was slightly more than 20,000 hectares (Ahmed, 1988) which has increased to about 140,000 hectares during 1995 (Metcalfe, 2003). Between 1983-84 and 1996-97 land under shrimp has an average increment of 27% (Barakat and Roy, 2003). The farm size is small in southwestern region compared to that of southeastern region (an average for each *gher* is 5.4 hectares versus 14 hectares). *Bagda* production has increased by 20% per annum in the last fifteen years (Banks, 2002). About 74% of the total area under shrimp was practised by extensive method (traditional) while about 24% was under improved extensive (modified traditional) method (Mazid, 1995). Expansion of *ghers* and production of shrimp over the period of 1983-84 to 1996-97 are summarized in Table 1.

Table 1: Growth of shrimp culture over the period 1983-84 to 1996-97: Gher, land, production and productivity (Barakat and Roy, 2003)

Year	*Number of ghers*	*Land (ha)*	*Production (Mton)*	*Productivity (kg/ha)*
1983-1984	3,171	51,812	4,386	85
1984-1985	3,171	64,246	7,578	118
1985-1986	3,171	87,300	14,658	168
1986-1987	3,778	87,300	14,773	169
1987-1988	3,778	94,010	17,889	190
1988-1989	3,778	108,280	18,625	169
1989-1990	6,581	108,280	18,624	172
1990-1991	6,581	108,280	19,489	180
1991-1992	6,581	108,280	20,335	188
1992-1993	6,581	108,280	26,000	240
1993-1994	6,581	137,996	22,054	160
1994-1995	6,581	137,996	26,277	190
1995-1996	6,581	137,996	68,000	492
1996-1997	6,581	410,000	115,000	821

3. THE TRENDS

3.1 Trends in the Changes of Land-use Pattern

Recent decades are experiencing a rapid conversion of rice fields into shrimp farms and represent the major changes in land use/land cover pattern in coastal Bangladesh (Ali, 2006). Also the duality in farming pattern is prominent along the coast. Small-holders are prompted to intensify rice production by shifting into high yielding varieties (Hossain, 1991; Turner and Ali, 1996) as well increased global market demand for cultured shrimp and its high profit has induced tropical farmers to convert part of their rice fields into shrimp ponds (Fig. 2), and thus transformed them into dual farmers (Pingali et al., 1997).

The lands under crop-agriculture in the southwest Bangladesh were mostly double and triple cropped (Islam, 2007; Karim, 2006). However, this pattern has changed after the introduction of shrimp during 1980s and local varieties of rice have been replaced by high yielding varieties during the same period (Ali, 2006). The *gher* area during 1975 was 20,000 ha which has increased to 141,000 ha during 2000 (BBS, 1975, 2002). The cost of production of shrimp (US$ 4/kg) in Bangladesh is one of the highest in Asia but the net profit from shrimp is 12 times higher than that of high yielding variety (HYV) of rice (Shang et al., 1998). This high profit in shrimp farming is responsible for changes in the land-use pattern in the coastal districts of the country (Hossain et al., 2004). Figure 3 represents a summary of shrimp versus rice yield and income over three decades in a southwestern coastal village. The intent for such changes in the land-use pattern is to increase commodity production to earn quick cash rather than sustainable livelihood development of the coastal communities as a whole (Nandy et al., 2007).

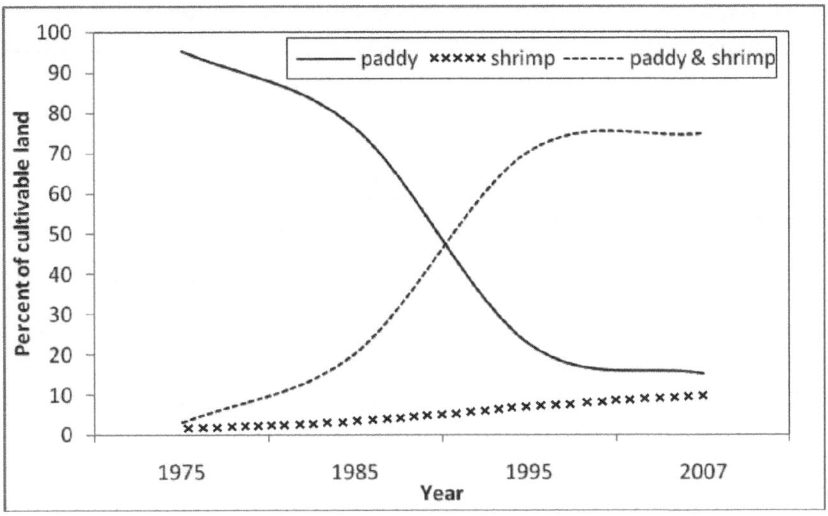

Figure 2: Land-use change in the context of shrimp culture over the period of 1975-2005 in a southwestern coastal village (Islam, 2007).

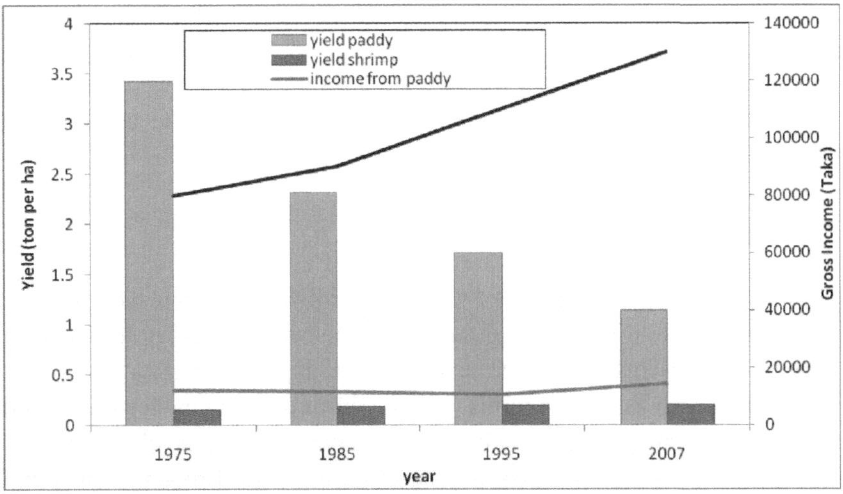

Figure 3: State of paddy versus shrimp yield and income in a southwestern coastal village of Bangladesh (Islam, 2007).

3.2 Trends in Employment Pattern

The total employee in the shrimp sector is estimated to be about 1,593,024 people (DoF, 2003) where 60% are involved in wild post-larvae collection and around 35% are shrimp farmers. The crop-agriculture based employment was dominating (77% to 80%) the total employment regime of the southwest region until the seventies of the last century, and the involvement was drastically reduced to around 25% (BBS, 1999, 2003) during the last decade that is related to a marked change in the land-use pattern. Shrimp sector created a different kind of employment pattern after eighties of the last century, and absorbs a sizeable number of individuals as employees in this sector thus reducing the employment opportunity in the crop agriculture sector.

3.3 Trends in Yield, Export and Earnings

Between 1992-93 and 2001-2002 production of shrimp in Bangladesh showed an increase of 178% along with two-fold increase in export (US $155.48 million to US $297.04 million) and shrimp shares about 90% of the total fish exports (BBS, 2001). Shrimp aquaculture as a whole also contributes 5.2 percent to the National GDP (DoF, 2003b). Bangladesh produced about 3% of world's shrimp and exported worth of nearly 400 million USD in 2007 (BFFEA, 2009), which placed the country as the seventh largest shrimp exporter in the world. After the export oriented Ready Made Garments (RMG) sector shrimp has emerged as the fastest growing export sector in the country.

However, land under shrimp is increasing but not the yield of *bagda* that is declining sharply. This low yield of shrimp in Bangladesh compared to that of Thailand, India and Indonesia may be related to the method of shrimp production which is mostly extensive in nature (Mazid, 1995). The present yield of *bagda* is 150-160 kg per hectare that was 230 kg per hectare in 1995-96 (Aftabuzzaman, 2004).

3.4 Trends in Local Supply Chain

The local supply chain starts at sea with the fry collectors—about half a million in number (DoF, 2005), many of them women—who collect fry for the local middlemen to reach the actual farmers. Haque (1994) identified four sub-sectors in the chain such as shrimp farms (*ghers*), shrimp hatcheries, feed mills and shrimp processing plants while Islam (2008) observed the involvement of at least three major groups such as shrimp farmers, local depot owners and processing factories along with middlemen's presence in every stage. Many small-scale and few medium-scale farmers when facing financial crisis, have to depend on the so-called 'no-interest' loans from local money lenders, who mainly work as middlemen and collect shrimps. These farmers have to agree to sell their produce to particular middleman (locally known as *faria*) with a lower price compared to the actual market price. There are also other groups of people who provide loans (*dadon*) at high interest. They are locally known as '*Dadondar*' (loan providers) and '*Mahajon*' (the great people). Figure 4 elaborates the supply chain and labour force involved in the shrimp sector.

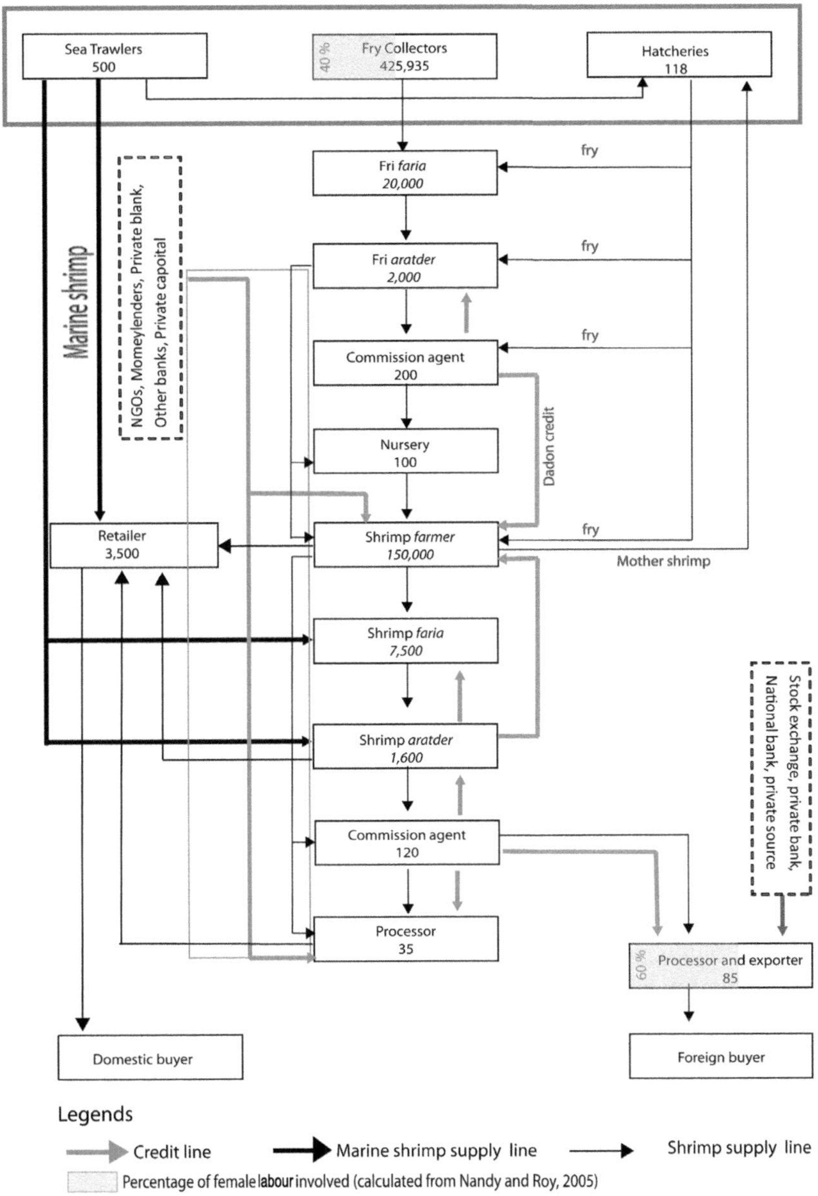

Figure 4: Local supply chain in the shrimp sector (compiled and modified from Rutherford, 1994; Ito, 2004; Nandy and Roy, 2005 and USAID-Bangladesh, 2006).

4. THE CONSEQUENCES

The current trend in shrimp culture has manifold consequences on the physical, biological, societal and economic environment of the region.

4.1 Consequences on the Physical Environment

Widespread waterlogged conditions prevail due to shrimp culture and decline in soil pH has been well reported in shrimp farms (Ali, 2006) and is attributed to the oxidation of exposed iron-sulfide (pyrite) in top soils of shrimp farms. Douglas (1994) attributed this increase in acidity to prolonged waterlog condition, reduced bacterial activities and nitrification of organic C. Depletion of organic C, Ca, Mg and K, and enrichment of N and P are attributed to prolonged waterlogged conditions in shrimp culture areas (Ali, 2006). A significant number of *ghers* exchange water every day to remove waste metabolites the discharge of which is responsible for water and soil pollution. Deb (1998) reported that 10-40% of the *gher* water is exchanged every day containing particulate matters, unused fish feed, various forms of chemical, microorganism, fertilizer, faeces etc. Phillips et al. (1993) reported that the traditional shrimp culture is not expected to represent any significant load of nutrients to the surroundings as such because the cultures rely on natural feed, moderately stimulated by input of manure and fertilizers. However, there are concerns that semi-intensive system is evolving rapidly in recent years in Bangladesh (Islam, 2004) because of the little available room for further horizontal expansion. This involves higher exchange rate and nutrient loading (Hopkins et al., 1993). Farm intensification may lead to adverse consequences such as high mortality from disease outbreak associated with increased ammonia concentrations (Wahab et al., 2003). Semi-intensive shrimp aquaculture may contribute to eutrophication as well with increased nitrogen loading (Jackson et al., 2003).

4.2 Consequences on the Biological Environment

One of the significant reasons for low output from shrimp farms in Bangladesh is high mortality due to disease, often accounting for 70-90% of the stocked shrimp (Wahab et al., 2003) and is attributed to unplanned horizontal expansion of shrimp farming (Hossain and Lin, 2001). During the past few years disease have spread and threatened the sustainability of shrimp aquaculture.

There are also concerns about potential loss of biodiversity in the form of by-catch, considered as wastage, during fry collection from the wild. FAO (1990) estimates that the rate of wastage is significant when compared with the total amount of *bagda* fry collected which is approximately 5000 for every 100 *bagda* fry. Additionally, considerable number of shrimp fry is lost during the sorting processes. The use of drag-nets contribute to over-fishing in the Bay of Bengal and increases pressure on fisheries, which in turn threatens the food security of thousands of coastal people who rely on fish as the main source of protein (Deb, 1998). Some trawlers also harvest brood mothers for sale to hatcheries as a result brood fisheries are declining and marine harvesting techniques are contributing to their loss (DoF, 2003). Deb (1998) suggested that there might be loss of genetic diversity too because of importing shrimp larvae from foreign countries.

Rahman et al. (2002) reported a depletion of livestock resources as a consequence to shrimp culture. With the history of approximately 8750 hectares of mangrove loss due to salt-water intrusion, dike and pond construction and human intervention from the Chokoria Sundarbans in the Cox's Bazar region (Katebi and Habib, 1985; Bhattacharya et al., 2005), present practice of shrimp culture within the Sundarbans Impact Zone (SIZ) is assumed to be a genuine threat to the biodiversity of this mangrove forest. Karim (2006) observed in a shrimp area that about 3.5% of forest and 2.4% of grazing land were converted to shrimp ponds and the remaining vegetation also disappeared fast because of high salinity and inundation due to shrimp farming.

4.3 Societal and Economic Consequences

Shrimp farming in the coastal areas of Bangladesh has been the subject of heated debate and scrutiny due to its negative environmental and societal impacts. The major areas of concern are conflicts between rice and shrimp farming (Islam, 1983), decline in rice yield and increase in soil salinity (Chowdhury, 1988), impact on livelihood support (Alauddin et al., 1995; Alauddin and Tisdel, 1997), social chaos and stress on rural livelihoods (Nijera Kori, 2006) etc. Shrimp farming also is associated with significant changes in land-use and land right patterns that affect traditional agricultural practices (Ahmed, 1996). This change destroys the mangrove forest, increases soil acidity and salinity, and increases water pollution, affects rice cultivation, and increases rural unemployment, inequality, and social violence (de Campos Guimarães, 1989; Rahman, 1994; Deb, 1998; Hossain et al., 2004).

Shrimp culture brought about a substantial economic and social transformation along coastal belt during the 1980s and 1990s (Islam, 1999; Rahman et al., 1994). However, problems of disease and environmental degradation have contributed to a decline in productivity of shrimp which has an uneven impact on shrimp farmers, with many small-scale farmers experiencing a decline in living conditions and economic losses. Many studies highlighted changes in land-use pattern as major conflict in shrimp farming (Karim and Stellwagen, 1998). As agricultural lands were turned into shrimp *ghers*, the share-croppers and landless wage labourers found them losing their livelihoods, and began movements to resist the introduction of shrimp in their areas. This often resulted in violence. During the last two decades, more than 150 people have been killed and thousands injured in shrimp-related violence. To harass the leaders of the anti-shrimp movements rich shrimp farmers also initiate court cases, many of which are still pending. The dominance of non-locals in shrimp farming is a sensitive issue. Coercion is also regular phenomena in shrimp areas. Restriction to movement of local people is generally observed in their village, and the shrimp *gher* owners could blame movement after dusk even in public places as thieves. The *gher* guards watched constantly the local people and sometimes harass them only on suspicion (Sultana, 1994).

The non-locals based in urban centres (towns/cities) entered into shrimp industry because of its profit potential. They have direct link with government bureaucracy and political parties and thus empowered, but are insensitive to local problems. Figure 5 provides a schematic diagram of ownership pattern and control over land in shrimp farms. This has led the local people to be very critical of shrimp culture (Rahman et al., 1995; Nandy et al., 2007). The number of non-locals in shrimp production has declined over time; however the area under their control has created social tension in shrimp areas. Land lease prices are increasing as a result of shift to shrimp cultivation, and landless share-croppers—whose livelihoods depend on the leasing of cultivable land—are deprived of access to paddy fields and fish ponds as well as common property such as grazing lands (Nandy et al., 2007). The customary approach to shrimp farming requires consensus or coordination. If at least 85 percent of the landowners agree to lease out their lands for shrimp farming, the owners of the remaining 15 percent are obliged to lease out their lands for the same purpose or allow the current lessee to inundate the entire area with saline water for farming (Bhattacharya et al., 2005). This practice has led to documented

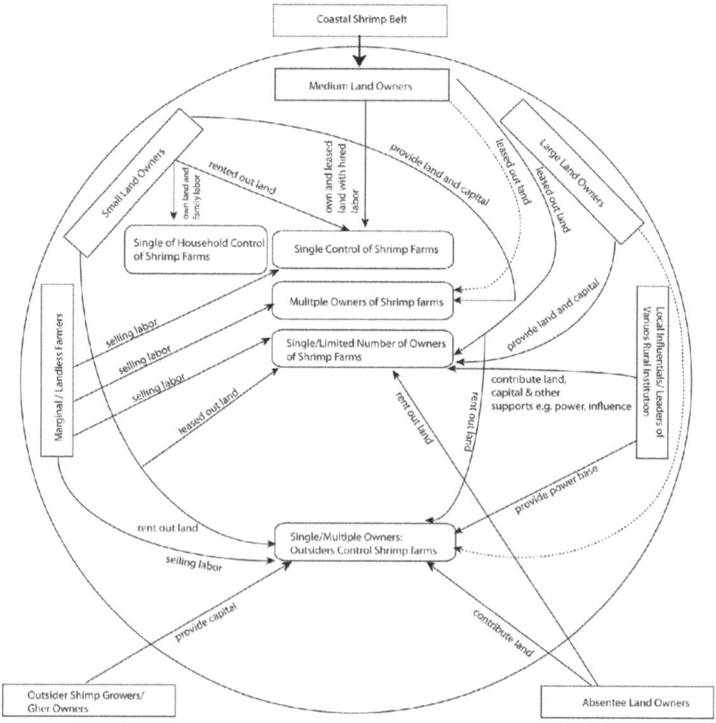

Figure 5: Pattern of land ownership and control over land under shrimp culture areas of Bangladesh (Islam, 2008).

expropriations, evictions, coercion and human rights abuses as small farmers have been displaced and community activists threatened and killed (Nandy et al., 2007). Similar concerns arise where the land in question is *khas* property. *Khas* land is public land that has been set aside for the landless under low cost leasing arrangements. Land is scarce in Bangladesh and inequitably distributed. Many landless households eke out a precarious living from fish and shrimp ponds located on *khas* land and leased from the government. Despite the allocation of *khas* land to the poor and landless, land-grabbing and eviction is widely documented (*Nijera Kori*, 2006; Nandy et al., 2007).

There are about 25,000 workers in shrimp processing factories in Bangladesh and there are records of violation of the ILO rule of working eight hours a day, forced overtime, unpaid overtime, failure to provide adequate health-care, failure to provide child-care and maternal leave, failure to observe the right to organize and collectively bargain, inadequate hygiene and health and safety measures, insufficient bathrooms and toilet facilities etc. (Nandy and Roy, 2005). Concerns have also been expressed about exposure to mercury, cadmium, organo-chlorinated pesticides, dioxins and antibiotics (Barg, 1992). Possibilities of infestations with *salmonella*, or human pathogens such as *Vibrio parahameolyticus* are more likely during peeling, gut removal and cleaning the shrimp before they are frozen (Pullin, 1992).

5. THE *BLUE REVOLUTION*: A REALITY CHECK

The shrimp farming is often regarded as the *blue revolution* and has created huge employment opportunity along the coastal belt of Bangladesh which is one of the strong arguments in favour of shrimp culture. Shrimp culture has also accelerated the trade and commerce in the region. During 1983 the employment on-farm was estimated as 4.1 million and off-farm was 5.1 million person-days which has increased to 250 percent (nearly 22.1 million person-days) within next five years (Ali, 1991). Such employment has been estimated as ~60 million person-days during 2005. While these employment and export gains seem highly impressive, studies (i.e., Majid, 1995; Rahaman et al., 1995) suggest that they have been achieved at a considerable cost such as (a) inequitable share between *gher* owners and land owners, (b) adverse environmental spillovers like loss of vegetation and biodiversity as genetic diversity and abundance, (c) decrease in rice yield, (d) increased social chaos and complexities etc.

The losers due to ecological backlash resulting from shrimp farming are not the shrimp farm owners but the owners of the land under shrimp culture. The shrimp *gher* owners represent the elite and privileged class who has easy access to the urban economic resources, while the land owners represent the rural powerless people who suffer the social and environmental consequences. Therefore at the event of being forced to quit shrimp farming due to some ecological disasters the shrimp *gher* owners can revive easily but not the land owners and the landless people who depend on the employments in the farms (*Nijera Kori*, 2006). Ironically land and water-based activities, such as agriculture, fisheries etc. represent major sources for livelihood support.

6. THE ISSUE OF SUSTAINABILITY: ENVIRONMENTAL CONCERNS

Shrimp aquaculture is now establishing itself as an important aquaculture industry in coastal Bangladesh. However, the benefit of this industry is concentrated to only the interests of a specific group of people who contribute a significant amount of foreign earnings to the nation, while the present development process is posing a challenge to the coastal sustainability, as well as to the industry itself. Shrimp farming is not only the subject of socio-economic impact; it has a great impact on coastal environment while it is actively acting on different coastal processes (e.g., salinity, sedimentation etc.). Chowdhury et al. (2006) observed that the integrated shrimp-rice farming is more sustainable than only shrimp farming. They developed a triangular model of sustainability where a shrimp-rice mixed culture has been observed to be more environmentally sound compared to semi-intensive method of shrimp culture (Fig. 6).

Islam et al. (2003) observed that the sh=rimp-salt farming is more profitable compared to shrimp-rice farming. The decision on increased shrimp production depends on economic considerations (Islam, 2007). However, recently marginal farmers are frequently demonstrating on 'back to rice' (Nandy et al., 2007)

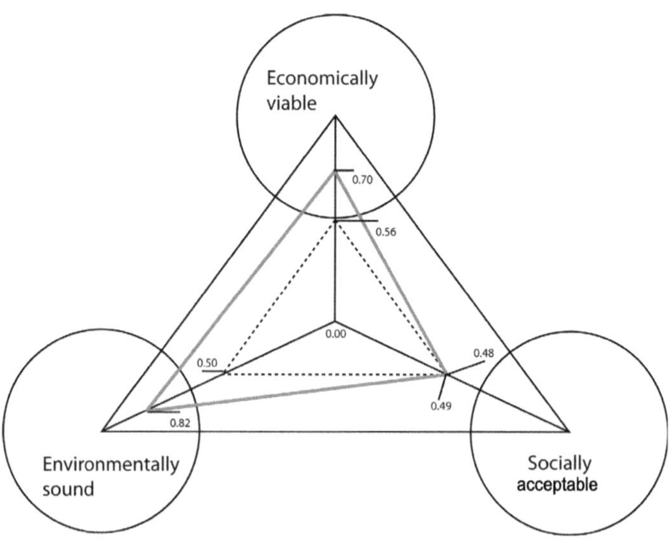

Figure 6: Triangular model of shrimp culture and its sustainability (Chowdhury et al., 2006).

because shrimp production failed to provide guarantee for food security in most regions of southwest Bangladesh.

The relationship between economic return from shrimp culture and international market demand could be established by analyzing the history of the evolution of the shrimp sector industry in Bangladesh. Bangladesh shrimp sector is an export-oriented sector; therefore depends on international market demand for its survival. A backup plan must be needed to survive if the international market fails.

Sustainability of the shrimp sector depends on mainly two functional questions: how much the coastal environment would support shrimp culture? And how long the international market demand would be active? Uptil now we lack any research on these issues.

7. THE MEASURES FOR SUSTAINABILITY: POLICY AND LEGAL ISSUES

A range of options may be available to the shrimp sector to improve environmental outcomes and to address social and human-right concerns. These will require a more favourable policy and legal environment and review of all pertinent laws and regulations.

These laws and regulations include the Fish Act (1950), Shrimp *Mohal* Management Policy (1992), the Shrimp Cultivation Tax Act (1992), and the National Fish Policy. Additional documents such as the Bangladesh Environment Conservation Act (1995), the National Environment Management Action Plan (NEMAP) and the National Conservation Strategy provide guidelines that are relevant for sustainable shrimp culture (WWF, 2006). There is also a need of 'Corporate Social Responsibility (CSR)' in Bangladesh to build a hygienically safe, economically viable, environmentally sustainable and socially responsible shrimp industry. Sustainability of shrimp culture depends on collaboration of all stakeholders throughout the product chain. Corporate Social Responsibility and Environmental Stewardship are contemporary key issues in the international shrimp market. Events like shrimp by-catch, therefore, would lessen market demand and devalue the products, which is not in the best interest of the producers.

Efforts such as the Hazard Analysis at Critical Control Point (HACCP) food safety standards show that governments can play a stronger role in developing industry standards. However, implementation and enforcement of HACCP remains inadequate and incomplete. Effective government-led efforts to improve industry practices in the area of worker rights will require not only more resources but also a willingness to hold companies accountable for their actions. Governments also need to accept responsibility for punishing companies that violate labour laws, as well as those with an active role in abuses like forced labour or human trafficking. Not only should governments on both ends of the supply chain step up inspections and commit to the enforcement of labour laws, they must also use the criminal justice system to adequately compensate workers and punish egregious exploiters. Industry-led codes of conduct have had some success in similar industries.

8. CONCLUSION

Analysis and integration of the current trend in land-use pattern, employment opportunities, yield, export and earning due to shrimp farming shows that the culture has reached its spatial limitations, restriction in scope for employment and shrinkage in yield, export and earning. A disturbance in local supply chain pattern is also noted. It is also evident that such limitations in shrimp sector are the result of the consequences on bio-physical and socio-economic environment. To avoid these limitations and for a sustainable shrimp culture practice along the coastal southwest Bangladesh an expressed shared responsibility is essential to be exercised by all players in this sector.

REFERENCES

Aftabuzzaman, M. (2004). Organic Aquaculture. National Shrimp Farmers' Association, Dhaka, 18 p.

Ahmed, A.T.A. (1996). Policy guidelines for management of coastal shrimp culture development. Paper presented at the World Aquaculture 1996 Conference, Bangkok.

Ahmed, M. (1988). Bangladesh Agriculture: Towards Self Sufficiency. Ministry of Information, Government of Bangladesh, Dhaka, 24 p.

Alam, S.M.N., Pokrant, B., Yakupitiyage, A. and Phillips, M.J. (2007). Economic returns of disease-affected extensive shrimp farming in southwest Bangladesh. *Aquaculture International*, **15,** 363-370.

Alauddin, M., Mujeri, M.K. and Tisdel, C.M. (1995). Technology-employment-environment linkages and rural poor of Bangladesh: Insights from farm level data. *In:* Ahmed, I. and Doelemean, J. A. (eds.) Beyond Rio: the environmental crisis and sustainable livelihoods in the third world. McMillan Publishing, London.

Alauddin, M. and Tisdel, C. (1997). The environment and economic development of South Asia: An overview from Bangladesh. McMillan Publishing, London.

Alauddin, M. and Tisdel, C. (1998). Bangladesh's Shrimp Industry and Sustainable Development: Resource Use Conflict and Development. *Asian Fisheries Science*, **11**, 97-110.

Ali, A.M.S. (2006). Rice to shrimp: Land use/land cover changes and soil degradation in Southwestern Bangladesh. *Land Use Policy*, **23**, 421-435.

Ali, M.Y. (1991). Towards sustainable development: Fisheries resources of Bangladesh. Bangladesh Ministry of Environment and Forest and National Conservation Strategy Secretariat, Dhaka.

Banks, R. (2002). Brackish and Marine Water Aquaculture. Memo. Poseidon Aquatic Resource Management Limited for DFID-Bangladesh, 41 p.

Barakat, A. and Roy, P.K. (2003). Bangladesh: Community based property rights and human rights, and overview of resources, and legal and policy developments. *In:*

Torre, I. and Barnhizer, D. (eds.) The blues of revolution: The damaging impacts of shrimp farming. The Industrial Shrimp Action Network and the Asia Pacific Environmental Exchange, ISA Net, WA, USA, 39 p.

Barg, U. (1992). Guidelines for the promotion of environmental management of coastal aquaculture development. FAO Fisheries Technical Paper 328, Rome, 32 p.

BBS (1975). Bangladesh Statistical Year Book. Bangladesh Bureau of Statistics, Dhaka, Bangladesh.

BBS (1999). Bangladesh Statistical Year Book. Bangladesh Bureau of Statistics, Dhaka, Bangladesh.

BBS (2001). Bangladesh Statistical Year Book. Bangladesh Bureau of Statistics, Dhaka, Bangladesh.

BBS (2002). Bangladesh Statistical Year Book. Bangladesh Bureau of Statistics, Dhaka, Bangladesh.

BBS (2003). Bangladesh Statistical Year Book. Bangladesh Bureau of Statistics, Dhaka, Bangladesh.

BCAS (2001). The Coastal Shrimp Sector in Bangladesh: Review of the Literature with Annotated Bibliography. Bangladesh Centre for Advanced Studies (BCAS), Dhaka, 144 p.

BFFEA (2009). Data obtained from Bangladesh Frozen Food Exporters Association, Khulna, Bangladesh.

Bhattacharya, D., Rahman, M. and Khatun, F. (2005). Environmental Impact of Structural Adjustment Policies: The Case of Export-Oriented Shrimp Culture in Bangladesh. *In:* Bhattacharya, D., Rahman, M., Khatun, F., Raihan, A. and Deb, U.K. (eds.) Bangladesh in the Global Trade Regime, Labour, Environment. A Pro-Poor Analysis of the Shrimp Sector in Bangladesh Agriculture, Export and Trade Negotiations, Dhaka: Pathak Shamabesh, Dhaka, pp. 26-49.

Chowdhury, M.A.M. (1988). Socio-economic consequences of shrimp cultivation in Bangladesh: a case study of Satkhira, Chakaria and Maheshkhali. *Bangladesh Journal of Public Administration*, **2(2)**, 49-76.

Chowdhury, M.B.R. and Muniruzzaman, M. (2003). Shrimp disease and its consequences on the coastal shrimp farming in Bangladesh. *In:* Wahab, M.A. (ed.) Environmental and socio-economic impacts of shrimp farming in Bangladesh. Technical proceeding. BAU-NORAD Workshop, Dhaka, Bangladesh. pp. 39-48.

Chowdhury, M.A., Shivakoti, G.P. and Salequzzaman, M. (2006). A conceptual framework for the sustainability assessment procedures of the shrimp aquaculture industry in coastal Bangladesh. International Journal of Agricultural Resources, Governance and Ecology, **29(3)**, 198-201.

de Campos Guimarães, J.P. (1989). Shrimp culture and market incorporation: A study of shrimp culture in paddy fields in Southwest Bangladesh. *Development and Change*, **20(4)**, 653-682.

Deb, A.K. (1998). Fake Blue Revolution: Environmental and Socio-Economic Impacts of Shrimp Culture in the Coastal Areas of Bangladesh. *Ocean and Coastal Management*, **4**, 63-88.

DoF (2001). Fish Week Compendium. Department of Fisheries, Ministry of Fisheries and Livestock, Dhaka, Bangladesh, 50 p.

DoF (2003). Annual Report. Department of Fisheries, People's Republic of Bangladesh, Dhaka, Bangladesh, 36 p.

DoF (2005). Fishery Statistical Yearbook of Bangladesh 2003–2004. Fisheries Resources Survey System, Department of Fisheries, Ministry of Fisheries and Livestock, Matshya Bhaban, Dhaka, 46 p.

Douglas, I. (1994). Land degradation in the humid tropics. *In:* Roberts, N. (ed.), The Changing Global Environment. Blackwell, Oxford, UK, pp. 332-350.

FAO (1990). Shrimp Seed Collectors of Bangladesh. Food and Agricultural Organization (FAO), United Nations, India.

Farmer, A. (1989). Proposed research programme for the brackishwater station to be established at Paikgacha. Final Report submitted to World Bank second agricultural research project, Fisheries Research Institute, GoB. Available online at: http://www.fao.org/docrep/field/003/AC370E/AC370E21.htm#ch1.2.2, date accessed: 12 July 2009.

Haque, S.M. (1994). Annual report of Bangladesh Frozen Foods Exporters Association (BFFEA). Dhaka, BFFEA Special Bulletin, 61 p.

Hopkins, J.S., Hamilton II, R.D., Sandifer, P.A., Browdy, C.L. and Stokes, A.D. (1993). Effect of water exchange on production, water quality, effluent characteristics and nitrogen budgets of intensive shrimp ponds. *Journal of World Aquaculture Society*, **24**, 304-320.

Hossain, M.S. and Lin, C.K. (2001). Land use zoning for integrated coastal zone management: Remote sensing, GIS and RRA approach in Cox's Bazar Coast, Bangladesh. ITCZM Publication Series No 3. Integrated Tropical Coastal Zone Management, Asian Institute of Technology, Thailand, 25 p.

Hossain, M. (1991). Green Revolution in Bangladesh. University Press Limited, Dhaka, 144 p.

Hossain, S., Alam, S.M.N., Lin, C.K., Demaine, H., Khan, Y.S.A., Das, N.G. and Roup, M.A. (2004). Integrated management approach for shrimp culture development in the coastal environment of Bangladesh. *World Aquaculture*, **35(4)**, 35-44.

Islam, A. (1983). A report on aquatic culture in Bangladesh. *Fisheries Information Bulletin*, **1(2)**.

Islam, M.R. (2004). Managing Diverse Land Uses in Coastal Bangladesh: Institutional Approaches. *In:* Hoanh, C.T., Tuong, T.P., Gowing, J.W. and Hardy, B. (eds.) Environment and Livelihoods in Tropical. CAB International 2006, pp. 237-248.

Islam, M.S. (1999). Social and institutional aspects of shrimp-rice farming in Bangladesh. Paper presented in the workshop on Economic, social and environmental implications of shrimp-rice integrated farming system in Bangladesh. Dhaka, Bangladesh.

Islam, M.S. (2008). From Pond to Plate: Towards a Twin-driven Commodity Chain in Bangladesh Shrimp Aquaculture. *Food Policy*, **33(3)**, 209-223.

Islam, M.S., Milstein, A., Wahab, M.A., Kamal, A.H.M. and Dewan, S. (2005). Production and Economic Return of Shrimp Aquaculture in Coastal Ponds of Different Sizes and with Different Management Regimes. *Aquaculture International*, **13**, 489-500.

Islam, M.S., Wahab, M.A., Miah, A.A. and Kamal, A.H.M.M. (2003). Impacts of shrimp framing on socio-economic and environmental conditions in the coastal region of Bangladesh. *Pakistan Journal of Biological Sciences*, **6(24)**, 2058-2067.

Islam, N. (2007). Exploring the Drivers for Conversion of Agricultural Land to Shrimp Farms in South West Region of Bangladesh. Unpublished MSc Thesis, Environmental Science Discipline, Khulna University, 87 p.

Ito, S. (2004). Globalisation and Agrarian Change: A Case of Freshwater Prawn Farming in Bangladesh. *Journal of International Development*, **16**, 1003-1013.

Jackson, C., Preston, N., Thompson, P.J. and Burford, M. (2003). Nitrogen budget and effluent nitrogen components at an intensive shrimp farm. *Aquaculture*, **218**, 397-411.

Karim, M. and Stellwagen, J. (1998). Shrimp Aquaculture. Final Report, Vol. 6, Fourth Fisheries Project, Department of Fisheries, Dhaka, Bangladesh, 101 p.

Karim, M.R. and Shah, M.S. (2001). Brackish Water Shrimp Cultivation Restricts Coastal Agriculture. *Khulna University Studies*, Bangladesh, **2(1)**, 123-134.

Karim, R. (2006). 5 Brackish-water Shrimp Cultivation Threatens Permanent Damage to Coastal Agriculture in Bangladesh. *In:* C.T. Hoanh, T.P. Tuong, J.W. Gowing and B. Hardy (eds.) Environment and Livelihoods in Tropical Coastal Zones. CAB International 2006, pp. 61-70.

Katebi, M.N.A. and Habib, G. (1985). Sundarbans and Forestry. *In:* Proceedings of the national workshop on Coastal area resource development and management. CARDMA, Dhaka, Bangladesh, pp. 79-100.

Mazid, M.A. (1995). Bangladesh: Country Paper. *In:* FAO/NACA Regional study and Workshop on the environmental assessment and management of Aquaculture development (TCP/RAS/2253). NACA Environment and Aquaculture Development Series, No. 1.

Metcalfe, I. (2003). Environmental Concerns for Bangladesh, South Asia. *Journal of South Asian Studies*, **26(3)**, 423-438.

Nandy, G., Ali, S. and Farid, T. (2007). Shrimp and Popular Economics: Who gains and who pays (*in Bengali*). ActionAid, Bangladesh, 144 p.

Nandy, G. and Roy, K. (2005). Status of women labors in shrimp processing industries in Khulna, Social Activities For Environment (SAFE), Khulna, Bangladesh, 24 p.

Nijera Kori (2006). Annual Report, Nijera Kori, Dhaka, Bangladesh, 120 p.

Phillips, M., Lin, C.K. and Beveridge, M.C.M. (1993). Shrimp culture and the environment: lessons from the world's most rapidly expanding warm water aquaculture sector. *In:* Pullin, R.S.V., Rosenthal, H. and Ma-en, S.N. (eds.), Environment and Aquaculture in Developing Countries. ICLARM Conference Proceedings, vol. 312, pp. 171-197.

Pingali, P., Hossain, M. and Gerpacio, R. (1997). Asian Rice Bowls: the Returning Crisis. CAB International, Wallingford, United Kingdom, 341 p.

Pullin, R.S.V. (1992). Aquaculture and Biodiversity. Paper presented at the Centenary Symposium of the Port Erin Marine Biological Station, University of Liverpool (Isle of Man, 17-18 September 1992), International Center for Living Resources Management (ICLARM), Metro Manila, 1992.

Rahaman, A., Islam, A, Roy, I, Azad, L. and Islam, K.M. (1995). Shrimp culture and environment in the coastal region. Research Report. Bangladesh Institute of Development Studies, Dhaka.

Rahman, A., Islam, M.A., Roy, I. and Azad, L. (1994). Shrimp culture and the environment in the coastal region. Research Report, Bangladesh Institute of Development Studies, Dhaka, Bangladesh, 124 p.

Rahman, A. (1994). The impact of shrimp culture on the coastal environment. *In:* Rahman, A.A., Haider, R., Huq, S. and Jansen, E.G. (eds.), Environment and Development in Bangladesh. University Press Ltd., Dhaka, pp. 490-524.

Rahman, M.Z., Khan, M.J., Hossain, M.J., Sarkar, M.S.K. and Hasanuzzaman, M. (2002). Effect of Shrimp Culture on Livestock Production. *Online Journal of Biological Sciences*, **2(10)**, 703-705.

Rosenberry, B. (ed.) (1995). World Shrimp Farming 1995: Annual Report. San Diego, Shrimp News International, 68 p.

Rutherford, S. (1994). CARE and Gher: Financing the Small Fry. Report prepared for CARE International in Bangladesh with support from Bangladesh Aquaculture and Fisheries Resource Unit (BAFRU).

Shang, Y.C., Leung, P. and Ling, B.H. (1998). Comparative economics of shrimp farming in Asia. *Aquaculture*, **164(1-4)**, 183-200.

Sultana, M. (1994). Gender, Poverty, Natural Resource Decline and Sustainable Development in Bangladesh. Mimeo, UN Research Institute for Social Development/WWF, Switzerland, April 1994.

Turner II, B.L. and Ali, A.M.S. (1996). Induced Intensification: Agricultural change in Bangladesh with implications for Malthus and Boserup. Proceedings of the National Academy of Science, USA, **93**, 14984-14991.

USAID Bangladesh (2006). A Pro-poor Analysis of the Shrimp Sector in Bangladesh. Arlington, USA: Greater Access to Trade Expansion, Development & Training Services, U.S. Agency for International Development (USAID), 93 p.

Wahab, M.A., Berghaim, A. and Braaten, B. (2003). Water quality and partial mass budget in extensive shrimp ponds in Bangladesh. *Aquaculture*, **218**, 413-423.

16

Role of Sand Dunes and Mangroves in the Mitigation of Coastal Hazards with Reference to 2004 Tsunami

K. Rajani Priya*, K. Santhi Kiran[1] and Ugyen Pema[1]

Department of Geology, Osmania University, Hyderabad

[1]International Internship on Disaster Management, Centre for Environment Education (CEE), and South Asia Youth Environment Network (SAYEN), Ahmedabad, India

1. INTRODUCTION

"*Sand dunes and mangroves seem to have sheltered (their) neighbours from the force of the disaster. Only nature seemed to have been able to stand up to nature.*" The above statement given by Channa Bambaradeniya, Programme Co-ordinator, IUCN Sri Lanka, signifies the role of coastal natural resources as barriers during natural hazards. The importance of coastal dunes, coral and granitic reefs and mangroves came into limelight during the aftermath of the severe tsunami on December 26, 2004. Concomitantly, emphasizing the importance of these natural resources, a word "bio-shields" is coined recently to coastal natural resources that depict how naturally occurring plants, trees, dunes in coastal areas shield a vulnerable region from the ferocity of storms and tsunami (NIO, 2006).

The impact of tsunami was so severe that it ravaged the coastal regions of South East Asian nations (Indonesia, Sri Lanka, India, Thailand, Malaysia, Maldives, Seychelles) surrounding the Indian Ocean (UNEP, 2006). The transformation of a natural event into a human and economic disaster depends on the fundamental problems of development that a region faces and these

*Corresponding Author

problems contribute to the region's vulnerability to the catastrophic effects of natural hazards. On one hand the increasing socio-economic development and technological advancement of the world to counter any disaster is at our disposal and on the other hand the compassion towards nature and the measures taken to conserve it are deteriorating, which result in magnifying the severity and magnitude of the natural event into a disaster (UNEP, 2005).

Healthier ecosystem is the less affected region during any kind of natural disasters. Natural barriers have their own role to play in reducing the impact of the hazard or can virtually nullify the effect. The preliminary environmental survey of United Nations Environment Programme after the Asian tsunami revealed that the extensive damage had occurred to natural barriers such as mangroves, coral reefs, and sand dunes of the areas that acted as the first line of defense from the tsunami. Remote sensing images taken before and after the tsunami event provided evidence to support the claims that coral reefs, mangroves, sand dunes and other coastal vegetation, as well as peat swamps, protected life from the tsunami (UNEP, 2006). Generally, coastal dunes favour the growth of salt tolerant plants, whose thick root systems help to bind the sand and prevent erosion. As a resilient natural barrier to destructive forces of winds and waves, sand dune systems are now accepted globally as vital in the defense of coastal areas against storm-surges, floods, wave erosion, and tsunami and also in the control of the global warming (IGNZ, 2005). Similarly, mangroves play an important role in coastal land protection as they reduce the coastal erosion. Mangroves buffer coastal waters from undesirable land-based influences, such as sediment, contaminant or nutrient runoff (NTHMP, 2001).

On December 26th, after the mega thrust earthquake triggered the tsunami in the Indian Ocean, nature's ferocity had no bounds for its destruction activity and the Indian Oceanic rim countries bore the brunt. However, the effect of tsunami was insignificant in certain places, where sand dunes and mangroves are undestructed. The presence of sand dunes dissipated the force of the energy of the enormous waves and lessened the destruction at the reaches of habitable places. Field investigations and remote sensing imagery clearly indicated that in places where natural barriers were present, the degradation and destruction was minimal, whereas the places that lost these barriers were seriously affected. Places in Nagapattinam district in Tamilnadu (India) were the victims of tsunami, where sand dunes and mangroves were absolutely absent and no other sea belt protection was present on the entire coastal belt. To multiply the chaos, the houses and structures were constructed in the vulnerable regions. Conversely, undestructed Pichavaram mangrove forest protected hamlets that include T.S. Pettai, Vadakku Pichavaram, Therkku Pichavaram, Meenavar Colony, MGR Nagar and Kalaingar Nagar against the fury of tsunami (MSSRF, 2006). Sri Lanka offers similar evidences where intact coastal ecosystems helped to buffer against furious waves. Many areas of Yala and Bundala National Parks were protected as they were buffered by vegetated coastal sand dunes (UNEP, 2009).

Severe damage occurred in some areas of Sri Lanka where coral reefs had been damaged and mined drastically. Similar observations were made in the provinces of Phang Nga in Thailand that were protected by mangrove forests and sea grass beds and were less affected (UNEP, 2005). The life saving role of mangroves was conclusively evident from earlier cases of natural hazards, such as the Orissa Super cyclone of 1999 in India.

Thus, the significance of these natural barriers is highlighted during the mitigation of natural hazards and it invokes the need for their conservation. This article intends to study the role of coastal dunes and mangrove ecosystems in mitigating the coastal disasters and emphasizes on their preservation by adopting proper conserving measures, enacting policies and generating public awareness and community-based approach and thus sustaining the livelihood.

2. SAND DUNES

Formation of sand dunes is a natural phenomenon, which is not manmade and henceforth prime importance lies on their conservation. Coastal barrier dunes are formed by wave and wind action that blows dry sand particles landward from the beach (Dutta, 1981). With steady winds, the sand drifts act as a barrier to moving sand and the drift gradually grows into a sizeable mound. Until a dune is completely veiled and stabilized by plant cover, sand may be blown away by the winds.

Sand dunes protect low lying backshore areas and preserve the integrity of low barrier islands, by holding sand that replaces erosion of beaches from storms and high tidal waves. Moreover, it is such an ecosystem that integrates and supports a wide range of flora and fauna. Sand dunes are the only form of defense against sea flooding along the stretch of the coast. Sand dunes increase the height of the coastal land and stand as strong defense against the tsunami or high tidal waves. The effects of present climatic change include sea level rise which aggravates the problem of saline water intrusion and submergence of inland areas in the absence of any barrier mechanism (MOE, 2005). The self-repairing capacity of natural dune systems is an important characteristic in the mitigation of coastal erosion (GLA, 2005).

The growing population, construction activities, and excess recreational use of the barrier islands in the coastal areas affect the stability of the dune environment. Human activities that damage the dune system make it more vulnerable to winds and sand erosion. Flooding, sand winning, beach cleansing and other activities are not only the threats to the sand dunes but also cause damage to their existence and lead to ecological imbalance and diminishing aesthetics of the coastal system. Thus, sand dunes need to be protected from development activities like mineral extraction, and also excess recreational use that causes inherent damage, and fragmentation.

2.1 Sand Dunes Saved the Villages

The village Palayazhar, situated between Buckingham canal and Kollidam River (channel of Cauvery River) in Tamilnadu state in India is vulnerable to sea hazards. It is the place where the river Kollidam enters into the sea and is susceptible to damage. Similarly, village Kodiampalem itself is an island that can be easily attacked by the waves. In Palayazhar and Kodiampalem our observations revealed that the houses were built very near to the coast and some of the houses were built on the sand dunes by levelling them down. This action destroyed the natural barriers i.e. sand dunes and thereby exposed the village to sea hazards. During the tsunami of 2004, entire houses in Kodiampalem were flooded and those very near to the sea were completely swept off. This is the example of unconstructive development against the course of nature. Traces of sand dunes can be seen there with casurinas trees protecting the sand dunes but people have cut down those trees. Small pockets of low lying islands called Pillumedu and Chinnavaikkal near Palazhyar situated at low elevations were completely innundated by waters.

On the contrary, in the same villages the places where houses were constructed behind the dunes had also been affected but the destruction was much lesser. The casualties and loss of properties could not have been prevented by man, but appropriate measures taken could have lessened the toll of lives, property and livelihood (Figs 1 and 2).

Figure 1: Casurina and other trees on the sand dunes near the village Palayzhar and Kodiambalem.

Figure 2: Destruction of houses built on sand dunes in coastal villages.

2.2 Saving the Sand Dunes

Natural sand dunes are the cheapest and strongest barriers against floods and tsunami. Artificial sand dunes are expensive and can be easily swept away by the tidal waves and the winds. The might of sand dunes can be strengthened by increasing the height and stability of existing sand dunes and building new dunes can be done through restoration of damaged dunes using vegetation and beach nourishment. Mangroves, casurinas, coconut trees, marram and creeping grasses, thorny trees, etc., can be planted in accordance with the nature of the soil and considering the favourable factors. Government must take initiatives for the protection of the dunes; however the participation and efforts of the people who inhabit the coastal areas are more vital. Further, through adaptive management of dunes such as grass planting, dune thatching, dune fencing, beach recycling and reprofiling, constructions like sandbag structures, gabion revetments, artificial headlands, artificial reefs, groynes, beach drainage, timber revetments, sea walls and using novel protection methods, conservation is possible. Although they act as bio-shields against natural disasters, dunes only serve as temporary barriers during storm tides of short duration and cannot act effectively against persistent beach recession, which is caused by rising sea level, migrating inlets, or changing shoreline dynamics (NCCES, 2006). Methods that encourage the growth and sustenance of dunes in many coastal areas are detailed hereunder:

2.2.1 Dune Grass Planting

Dune grass planting, an ideal method of managing coastal dune maintains dune growth by trapping and stabilising blown sand. Planting mangroves and even coconuts, casurinas, palm trees, marram grass (sand-binding species), creeping grass, etc., stabilize dunes (Arun et al., 1999). These plants support the growth of new fore dunes along the toe of existing dunes. Native sand binding vegetation such as spinifex and pingao grown on the seaward face of the frontal dune (SCHB, 2001) also gives good results. Dune grass planting is important in the maintenance of natural dune building and repairing processes and also in the prevention of wind erosion damage to the dune. However, the plantation will not be successful in the active fore dune environment.

Planting mangroves will not prevent erosion, but will accelerate natural recovery after storm damage by creating a reservoir of sand within the fore dunes that enables the dunes to withstand the next period of erosion. Thorny trees can also be planted along the coast so as to deter the pedestrians walk on those dunes. Marram grass is particularly effective as it positively thrives on growing dunes, and is perhaps the easiest to transplant. It is tolerant of salt spray but not immersion, and should be planted above the expected run-up limit of the storm waves.

New vegetation cover on an eroding dune face will encourage recovery without damaging the ecological and geomorphologic interest or the aesthetic appeal of the shoreline. The approach is potentially self-sustaining, if the grasses become well established.

2.2.2 Dune Thatching

Dune thatching is another method of saving dunes. Thatching of exposed dune faces or blowouts using waste cuttings from forests, or other low cost materials, is a traditional way of stabilizing sand that reduce trampling and protect vegetation. Thatching will encourage dune recovery and resist erosion (Fig. 3). It reduces surface speed of winds and encourages sand deposition. Thatching is a low cost method that has degradable materials and the failure of the process also does not jeopardize long-term environmental interests.

Figure 3: Sand dunes with coconut trees.

2.2.3 Sand Bag Structures

Sand bag structures are feasible in many areas. Establishing these structures does not require costly equipment or skilled labour. These can be used to form any shape of shoreline structure. However, they will have a short life expectancy due to lack of resistance to physical damage. They are potentially most useful as buried revetments under the dune face, where they form a final line of protection after the overlying sand had been eroded by storm waves (SNHP, 2000).

Buried sand bags can form a useful and low cost line of defense in areas that are subjected to mild, seasonal erosion. They are also useful for temporary headland protection, but are easily removed with no significant long-term impacts on the physical or natural environment.

2.2.4 Public Awareness and Community-based Participation

By generating awareness and education about natural hazards and their mitigation measures, people living near the shores and coastal areas can do a lot for their own protection from natural disasters like sea flooding and tsunami. They should be encouraged to take initiatives to protect themselves by protecting sand dunes and planting of mangroves. A system could be formed, whereby each individual is responsible for the maintenance of these barriers in their surroundings. Their responsiveness and contribution would be the greatest of all forms of efforts made by any other organization or government bodies. Community-based environmental management programmes are considered effective in raising community awareness and changing attitudes and behaviour of people towards the protection of nature.

3. MANGROVES

Mangrove forests, naturally resilient, have tremendous power to withstand nature's fury and also provide many benefits in the form of subsistence and economy. A mangrove is a tropical maritime tree or shrub, which lives between the sea and the land in areas inundated by tides. Mangrove is considered as a species as well as a community of plants with approximately 70 species belonging to 20 families of angiosperms, worldwide. Mangrove forests are coastal plant communities that are part of a larger coastal ecosystem that typically includes mudflats, sea grass meadows, tidal marshes, salt barrens and even coastal upland forests and fresh water streams and rivers. Mangroves form important ecological and geographical links between land and sea and help protect coastlines from erosion, storm damage and wave action (Tomlinson, 1986). Mangroves exhibit distinctive genetic diversity, high intrinsic natural productivity and unique habitat value. In more tropical climates, coral reefs may also be a part of this ecosystem.

Mangroves play a significant role in protecting the hinterland against cyclones, typhoons, and tsunamis and also from the ingress of seawater during tidal surge. They protect and stabilize landmass from coastal erosion by dissipating the wave energy (Warnitchai, 2005). Mangroves are frequently inundated by the tides and thus have special physiological adaptations to deal with salt in their tissues. They also have adaptations within their root systems to support themselves in soft mud sediments and can serve as coastal protection from floods and tsunami, and riverine mangroves help remove pollutants before they enter adjacent coastal waters. The stability of mangroves is of immense importance in preventing shoreline erosion. Mangroves are important carbon sinks that may help to mitigate the global impacts of climate change with their high primary production, carbon accumulation and exchange with adjacent ecosystems. Moreover, they are repositories of diverse biological diversity and are nursery and breeding grounds for vast marine life, and thus serve as excellent breeding grounds for the subsistence of coastal dwellers.

As they are now being exploited by modern encroachments, mangrove forests are currently among the most threatened habitats in the world and are disappearing at an accelerated rate. Approximately 15.9 million hectares (over 60,000 sq. miles) of mangrove forests are present in the warm waters of tropical oceans all over the world. The destruction rate of the mangrove habitat is now at about 1% every year with 50% lost over the last 50 years (Ong, 1994). According to the latest assessment on Indian forest cover, mangroves occupy

0.49 m.ha, i.e., 0.15% of the entire forest cover (SFR, 1999). The extent of Indian mangroves covered is about 7% of the world mangroves and 8% of the Indian coastline. However, Indian Remote Sensing data showed that the total area of the mangroves decreased to 4474 square km (Nayak, 1993).

3.1 Causes of Depletion

Mangroves are under pressure from agriculture, industry, fuel wood extraction, aquaculture, and diversion of water for irrigation, which leads to reduction in fresh water inflow and increase in soil salinity. The intense chemical and biological activities in mangroves cause them to act as sinks, which concentrate pollutants. Over a period of time, environmental stress can kill large number of mangrove trees that reduce species diversity. In addition, the charcoal and timber industries, as well as the developing tourism industry and other coastal development activities also have severe impact on mangrove forests.

Prawn aquaculture is often carried out at the expense of mangroves (Franklin and Palanivelu, 1989). The conversion of former mangrove areas to aquaculture is a widespread activity, particularly in South East Asia, East Africa and Latin America that destroyed kilometres of mangroves (Unesco, 1990). It has been estimated that thousands of hectares of mangroves have been cleared to make room for artificial shrimp ponds. The shrimp aquaculture industry has grown exponentially over the last decades, leaving impact on wetlands around the world. Their effects range from complete destruction to disruption of ecosystem functioning, leading to chronic poisoning of mangroves and associated fauna and to the loss of biodiversity and productivity (PDO-ICZMP, 2005). Overuse of the resources is the most widespread problem although with varied intensity. Human population pressure is the major driving cause for overuse and thus degrading the ecosystem and their environment conditions to the extent that subsequent natural re-establishment is prevented or becomes very difficult and expensive.

Hydrological modification also causes damage to mangrove existence. The prime causes of hydrological modifications are damming, water extraction, land-use changes and deforestation in the mangrove upstream that led to changes such as build-up of salinity in water and soils, changes in sedimentation patterns, which in turn affect coastal configuration and the structure of navigational channels (ISME, 2004). Changes in inter tidal areas can reduce the productivity of the mangrove ecosystems. Coastal land use changes include the direct irreversible physical loss of mangrove-dominated areas as a result of the development of infrastructure and conversion of land for agriculture, cattle breeding and the building of saltpans. These major impediments not only lead to direct losses, but also to the rapid and often uncontrolled degradation and elimination of adjacent mangrove systems. Mangrove roots are highly susceptible to clogging by crude oil and other pollutants, and prolonged flooding from artificial dikes and sea walls. Ship breaking is a dangerous activity that threats the mangrove sustenance.

3.2 Mangroves – Life Saving Role

Pichavaram mangrove wetland, the area that had survived from the fury of tsunami, occupying an area of about 1400 ha, is located about 280 km down south of Chennai, in the Cuddalore district. Pichavaram mangrove is one of the rare mangrove forests in India and it represents 14 exclusive mangrove species (ISME, 2004; Kannupandi and Kannan, 1998). 17 hamlets belonging to this wetland utilize the resources of the belt. According to various sources of information, out of these 17 hamlets, six hamlets are under physical coverage of the mangrove wetlands, five hamlets are located on or near the open beach and not protected by mangroves and remaining six hamlets are far away from mangrove forest. Visualized damage had taken place to hamlets that were not covered by mangroves and hamlets that have been located near to the beach were totally devastated, but no damage had occurred to six hamlets that are physically protected by the mangroves. The hamlets that were under water were damaged by salinization of land and water resources, including drinking water (Fig. 4).

M.S. Swaminathan Research Foundation Centre for Research on Sustainable Agriculture and Rural Development identified hamlets namely, T.S. Pettai, Vadakku Pichavaram, Therkku Pichavaram, Meenavar Colony, MGR Nagar and Kalaingar Nagar protected by Pichavaram mangrove forests against the fury of tsunami. In addition to it, various reports revealed that people had used the public information dissemination system to evacuate villages before the arrival of tsunami. These hamlets are located between 100 m and 1000 m from mangroves. In these hamlets seawater had not entered into the houses and there was no loss of property (Fig. 5).

Figure 4: Mangrove forests along the side of Buckingham Canal.

Figure 5: Mangrove trees protected the village Pichavaram.

Various reports conclusively confirmed that mangrove trees located in rows close to the sea got uprooted due to the impact of the tsunami and beyond that there was no damage. It was evidenced that mangrove forests reduced the impact of the tsunami in two ways: (a) velocity of the tsunami water greatly reduced after it entered into the mangroves due to friction created by thick mangrove forest and (b) volume of water reaching a point that was greatly reduced after entering into the mangroves, and was distributed to all the canals and creeks that were present all over the mangroves.

3.3 Protecting Mangroves

The objectives in saving mangroves include improving the areal extent of mangroves, maintaining diverse species and protecting them from human interventions. In view of it, various activities have to be planned and depending on each area the key issues have to be considered, by identifying the problems and finally formulating the solutions.

Area under mangrove cover, closed or open canopy, has to be reviewed with the support of survey and monitoring based on remote sensing. Fieldwork in national as well as international level such as Global Forest Resources Assessment Scheme by Food and Agricultural Organization (FAO) is to be carried out to generate data and find the status and trends in mangrove area extent by each region throughout the world. In India, a multi-departmental National Natural Resources Management System (NNRMS) has been set up, and one of its objectives is to conserve special ecosystems like mangroves and wetlands.

Role and performance of mangrove ecosystems has to be examined by monitoring indicators of their growth like assessing health, evolution changes, progressive and regressive, etc. Landward buffer areas have to be established to allow retreat of mangrove ecosystems. Whether dunes, mangroves, or hardened barriers are used to reduce the impact of tsunami and storm waves, the expected reduction in wave forces should be modelled so that the anticipated protective effects are not overestimated in a manner that places coastal residents in danger.

Value of mangroves both as ecosystems and from a socio-economical point of view should be given importance. Public awareness has to be raised through establishment of pilot management areas, demonstration sites and encouraging education on laws and regulations and approaches for recommended practices. A 'code of conduct' has to be considered for the management and sustainable use of mangrove ecosystems (Dennis Hwang et al., 2005).

Mangroves can be utilized in a sustainable manner by controlling use through zoning with a range of reserved areas. Ecological and cultural sites have to be covered under a network of strict no-use zones such as Southeast Asia Biosphere Reserve Network. Encouraging community involvement in protecting the ecosystem with the view to enhance local benefits such as fisheries, coastal protection and biological filtering is a sustainable method. Development schemes that support replacement of mangrove fuel wood with alternate fuel sources can be adopted to protect mangrove plantation. Enforcing the principle of 'no-use without replacement' gives good results.

Aquaculture can be promoted by changeover to mixed practices and promote alternatives to the use of antibiotics and growth hormones in it. Pollution has to be identified as a prime culprit of mangrove degradation and strict enforcement of existing legislation has to be encouraged. Recognized methods have to be used for ensuring purification. Encouraging the maintenance of environmental flows upstream of mangrove areas, notably through the control of water diversion, impoundment and release is a method that ensures purification.

4. MODEL PLAN

A model plan has been envisaged to protect seashore habitats from coastal hazards. This model is designed by considering human habitat, their livelihood and business transactions that take place in coastal areas in a sustainable way. This will help as a proper mitigation strategy in reducing the vulnerability of coastal sensitive areas to coastal hazards (Figs 6 and 7).

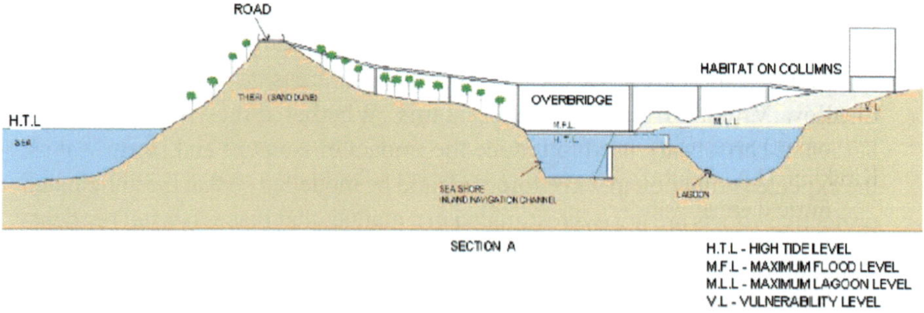

Figure 6: Schematic diagram of a typical seashore protection arrangement including fishing and inland navigation development and rehabilitation as seashore people affected by tsunami of 26-12-2004.

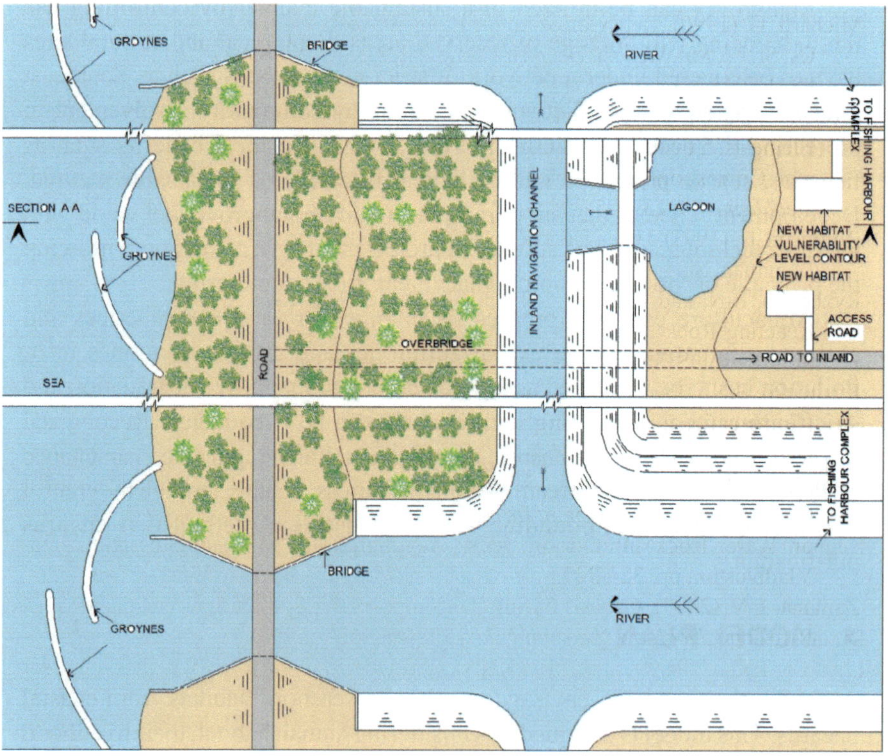

Figure 7: Model plan of the mitigation strategies to be adopted in island villages.

5. CONCLUSION

Improper land use practices due to lack of appropriate legislation and enforcement of regulations and policies related to their management are the major drawbacks in the conservation of coastal areas. In addition to it, lack of exchange of knowledge, education, expertise among professionals, and non-participation of local public are the major inadequacies for the conservation of these coastal resources. Inadequate communication, public unawareness and non-participation in their development are aggravating the quandary. In addition, a shortage of professionals to work in the field of disaster management in several scientific and management institutions, law enforcement agencies and local communities is summing up the problem.

Coastal zones and environments are conserved by adopting 3-C policy. It includes coastal habitat management, coastal erosion management and coastal pollution control (Hwang et al., 2005). These three interrelated processes should go hand-in-hand with active community participation, and a strong commitment in every citizen to preserve nature. From the government side, restrictions have to be strictly imposed on setting up any activity in Coastal Regulation Zone (CRZ). Collaboration with universities, research institutions, forest departments, fisheries, and technicians has to be maintained. Training and publications at international and national levels has to be encouraged and they should be adapted to local needs. With the lessons learnt from the present catastrophe, disaster prevention, mitigation and preparedness are proved to be better than disaster response in achieving the goals and objectivity of vulnerability reduction. Thus, Yokohama message of May, 1994 emphasizing on the need for an emphatic shift in the strategy for disaster mitigation will be fulfilled.

REFERENCES

Arun, A.B., Beena, K.R., Raviraj, N.S and Sridhar, K.R. (1999). Coastal Sand dunes – A Neglected Ecosystem. *Current Science*, **77(1)**, 19-21.

Dennis Hwang, Reinwald O'Connor and Playdon (2005). Mitigating the Risk from Coastal Hazards: Strategies & Concepts for Recovery from the December 26, 2004 Tsunami.

Dutta, A.K. (1981). Physical Geology. Modern Book Agencies Pvt. Ltd, Calcutta.

Franklin, T. and Palanivelu, R. (1989). Indian prawn culture in tropical waters of India, Vol 2. *In:* Coastal Zone '89. Proc. 6th Symp. Coastal Zone Management, Charleston, South Carolina. American Society of Civil Engineers, New York. pp. 1089-1094.

GLA (2005). Community-based Dune Management for the Mitigation of Coastal Hazards and Climate Change Effects: A Guide for Local Authorities, New Zealand.

IGNZ (2005). Community-based Dune Management for the Mitigation of Coastal Hazards and Climate Change Effects: A Guide for Local Authorities.

ISME (2004). Mangrove Action Plan for the sustainable management of mangroves 2004-2009. ITTO Project PPD17/01 Rev.1 (F).

Kannupandi, T. and Kannan, R. (1998). Hundred years of Pichavaram mangrove forest. *In:* An Anthology of Indian mangroves. ENVIS Special Publication, Annamalai University, India. p. 66.

MOE (2005). Planning for Climate Change Effects on Coastal Margins. Report of Ministry of Environment, New Zealand.

MSSRF (2006). M.S. Swaminathan Research Foundation Centre for Research on Sustainable Agriculture and Rural Development.

Nayak, S.T. (1993). Remote sensing applications in the management of wetland ecosystem with special emphasis on management of mangrove ecosystem and coastal protection, Visakhapatnam, India. p. 21.

NCCES (2001). Restoration and Management of Coastal Dune Vegetation. Public Document printed by North Carolina Cooperative Extension Service.

NIO (2006). National Workshop on Formulation of a Science Plan for Coastal Hazard Preparedness. NIO, Goa.

NTHM (2001). Designing for Tsunamis – Seven Principles for Planning and Designing for Tsunami Hazards. National Tsunami Hazard Mitigation Program, 2001. p. 60.

Ong, J.E. (1994). The Status of Mangroves in ASEAN. *In:* Living Coastal Resources of Southeast Asia: Status and Management. Australian Institute of Marine Science, Townsville. pp. 52-55.

PDO-ICZMP (2005). Pollution and its Management Approach. Working paper, WP043.

SCHB (2000). Sand Dune Erosion, Soil Conservation – Technical Handbook. Soil Conservation Hand Book. (ISBN 0478240333)

SNHP (2000). A Guide to Managing Coastal Erosion in Beach/Dune Systems. Scottish Natural Heritage Publication.

State of Forest Report (SFR) (1999). Forest Survey of India, Ministry of Environment and Forests.

Tomlinson, P.B. (1986). The Botany of Mangroves. Cambridge University Press, New York.

UNEP (2005). National Rapid Environmental Assessment – Sri Lanka. UNEP Report.

UNEP (2005). National Rapid Environmental Assessment – Thailand. UNEP Report.

UNEP (2006). After the Tsunami Rapid Environmental Assessment. UNEP Report.

UNESCO (1990). Mangrove Programme, 1990.

Warnitchai, P. (2005). Lessons learned from the 26 December 2004 tsunami disaster in Thailand. Proc. of Scientific Forum on Tsunami, its Impact and Recovery, 6 to 7 June 2005, AIT Conference Center, Bangkok, Thailand.

17

Assessment of Potential Health Risk through Arsenic Flow in Food Chain—A Study in Gangetic Delta of West Bengal

A.C. Samal, S. Kar, Piyal Bhattacharya and S.C. Santra

Department of Environmental Science, University of Kalyani
Kalyani, Nadia, West Bengal-741235 (India)

1. INTRODUCTION

West Bengal is one of the severely arsenic affected states in India. Over 500 million people are at risk through arsenic poisoning in the Ganga-Meghna-Brahmaputra plain (Chakraborti et al., 2004; Pal et al., 2007). Nine out of eighteen districts of the state of West Bengal are reported to have groundwater arsenic contamination (Chakraborty et al., 2002). Among the affected districts, North 24-Parganas deserves special mention. More than 95% of the people here depend on the groundwater for drinking, cooking and other domestic uses and major amount of groundwater is also used for irrigation of crops specially during non-monsoon period. Thus there is a chance of land contamination and accumulation of arsenic through agricultural products grown in contaminated soil and water. The exposure to arsenic may involve a number of pathways through ecosystem. This indicates that water-soil-crop-food transfer as well as cooking and direct intake of drinking water may be the major pathway of arsenic entry in man and higher animal system. A number of people are thus suffering from arsenic induced skin lesions and other symptoms.

Several crops were reported to accumulate arsenic in substantial quantities (ICAR, 2001; Panda and Das, 2001; Das et al., 2004; Samal, 2005). Arsenic levels in rice are also elevated in crop grown on arsenic contaminated soils

(Meharg and Rahman, 2003). Abedin et al. (2002) have reported the accumulation of arsenic and transformation of arsenic species in rice plant. Several other factors also influence the toxicity of arsenic viz. quantity consumed per day through food, concentration of arsenic in dietary and cooking water, poor socio-economic condition and subsequent malnutrition, illiteracy and food habits. All these factors are important with respect to the accelerated risk of arsenic toxicity. Alam et al. (2003) studied on the implications of contamination of agricultural soils from long-term irrigation with arsenic-contaminated groundwater for phyto-accumulation in food crops, and hence dietary exposure to arsenic and other metals.

The mechanism of arsenic toxicity in mammals is very complex. The fate of arsenic compounds in the human body varies with the type of compound. It depends in vivo on oxidation and reduction reactions between As(V) and As(III) and consequent methylation reaction in liver. Arsenate is generally reduced to arsenite, which is afterwards partly methylated. The epidemiological studies show that the people who are chronically exposed to arsenic compounds by orally or respiratory routes show both basal cell and squamous cell carcinomas of the skin as well as lung cancer (Nagvi et al., 1994). Arsenic seems to be a cancer promoter rather than a cancer initiator (Lee-Feldstein, 1986). Due to increased consumption of arsenic-contaminated drinking water by the people, skin lesions would appear and there will be direct toxic effect as well as chromosomal changes, increased rates of skin cancer, risks of lung and bladder, disturbance of the cardiovascular and nervous system functions and eventually leading to death (WHO, 1993).

Soil contamination with arsenic input from various sources may prove to be detrimental to plant through its accumulation but the entry of arsenic in food chain induce the possibility of biomagnifications. Bhumbla and Keefer (1994) observed in plants that arsenic is not generally translocated readily to shoots, but most of the arsenic, taken up by crops, tends to remain in the roots. There are wide variations in the tendency of different crop plants to accumulate different amounts of arsenic when subjected to the same soil level of arsenic.

Higher levels of arsenic were also observed in cooked food ingredients compared to raw materials. Daily dietary intakes of arsenic from such foodstuffs for adults were estimated to be 171.20 μg and 189.13 μg and for children are 91.89 μg and 101.63 μg in the Jalangi and Domkal blocks of West Bengal respectively (Roychowdhury et al., 2002, 2003). Identically in another investigation, Mandal et al. (1998) showed that despite having safe water for drinking and cooking, the study group could not avoid an intake of arsenic through edible herbs grown in contaminated water, food materials contaminated through washing, and the occasional drinking of contaminated water. The concentration of arsenic in cooked rice was higher than that in raw rice and absorbed water combined, suggesting a chelating effect by rice grains, or concentration of arsenic because of water evaporation during cooking, or both. The method of cooking and water used can affect the amount of arsenic in cooked rice, which will have implications for the assessment of the health risks of arsenic (Bae et al., 2002; Pal et al., 2009).

In the present investigation a micro level study was carried out to evaluate the arsenic level in food composites and dietary intake in some arsenic contaminated villages of North 24 Parganas, West Bengal. A considerable amount of arsenic bioaccumulation is found in the crops and vegetables through arsenic contaminated groundwater irrigation. The potential health risk is thus compounded with dietary intakes through cooked foods and arsenic-contaminated drinking water.

2. MATERIALS AND METHODS

In the present investigation, selected villages of Habra-II block of North 24 Parganas (West Bengal) were taken up where the crops and vegetables were cultivated by arsenic-contaminated groundwater. Water samples were collected from tube wells in pre-washed (with 1:1 HNO_3) polyethylene bottles after pumping up at least 3-5 minutes. After collection, concentrated HNO_3 (1.0 ml L^{-1}) was added as preservative. Vegetable and crop samples were collected from arsenic-contaminated fields and stored in polyethylene bags, and kept cool until processed in laboratory. Samples were washed thoroughly with tap water to remove soil and other particles and finally washed with de-ionized water with continuous shaking for several times until the washed water was free from arsenic. The samples were dried in hot air oven at 60°C for about 72 hours for yielding of dry matter. Dry samples were made fine powder by pestle and mortar, sieved and stored in polyethylene bags with proper labelling at room temperature.

2.1 Teflon Bomb Digestion of Vegetable and Crop Samples

The vegetable and crop samples are digested in Teflon bomb for arsenic analysis in FI-HG-AAS (Das et al., 1995; Mandal et al., 1997). About 0.2-0.5 gm of dry powder of vegetable or crop samples was placed in a Teflon bomb and added 2 ml of concentrated HNO_3 (E. Merk, India) and 1 ml of H_2O_2 (30% V/V, E. Merk, India) drop by drop and mixed thoroughly and allowed to stand for a few minutes at room temperature and agitated to avoid bubbling. The Teflon bomb was kept overnight. Next day it was placed in a hot air oven at a temperature of about 120°C for eight hours. After cooling the Teflon bomb was opened and the cover washed with deionised water very carefully for transferring the washed water inside the Teflon. Then the sample volume was reduced to about 0.5-1 ml on a hot plate at a temperature 60-80°C. After cooling the volume was made to 2 ml with repeated washing with deionised water. Then the liquid was filtered through a millipore membrane (0.45 mm pore size) filtering apparatus and finally kept in refrigerator at 4°C for arsenic analysis.

2.2 Sample Analysis

The water and crop samples were analyzed by Flow Injection Hydride Generation Atomic Absorption Spectrometry (FI-HG-AAS) for determination of total arsenic. The detail technique of FI-HG-AAS followed is described by Chatterjee et al. (1995) and Samanta et al. (1999). Preserved water samples were analyzed against arsenite and arsenate mixture (1:1) as the standard. For crop and vegetable, digested samples were analyzed against arsenate as standard. To check the accuracy of the techniques various types of Standard Reference Material (SRM) samples were analyzed in the same procedure (Table 1). The results show good agreement with the certified values.

Table 1: Particulars of Standard Reference Material (SRM) by FI-HG-AAS

Reference sample	*Certified value* ($\mu g\ g^{-1}$)	*Found value* ($\mu g\ g^{-1}$)
NIES, CRM-10 (Rice flour unpolarised low level Cd)	0.17	0.172±0.009
NIES, CRM-10 (Rice flour unpolarised high level Cd)	0.15	0.158±0.018
NIST, SRM 1570a (Spinach leaves)	0.068±0.012	0.062±0.003

3. RESULTS

In the study area, different crops and vegetables are cultivated in different seasons. Mainly three seasonal crops are Aus/Pre-khari or summer crop, Aman/Khari or monsoon crops and Ravi/Boro or winter crops. In pre-monsoon period (March-May) pre-kharif cultivation were done through arsenic-contaminated groundwater irrigation. Aus paddy and different types of summer vegetables were cultivated during this period. The monsoon cultivation is mainly rain-fed. The maximum groundwater irrigated cultivation is done in post-monsoon period (November-February), i.e. the Rabi/Boro crops of winter season. The HYV-Boro paddy is cultivated massively in this season. The other crops and vegetables cultivated in this season are mainly wheat, different pulses, different types of oil seeds, spices and different types of vegetables, mainly cabbage, cauliflower, brinjal, parbal, basella (Pui) potato etc. Some area was used for different types of winter flowers cultivation in this season.

3.1 Bioaccumulation of Arsenic in Cereals and Pulses

As the rice is the main crop of the study area, the accumulation of arsenic in irrigated and rain-fed rice crops of the study area were analyzed separately and data were presented in Table 2. The data thus reveals the fact the average accumulation of arsenic in Boro rice grain found is 492.97±178.47 $\mu g\ kg^{-1}$ and that of Aman rice is 161.19±23.91 $\mu g\ kg^{-1}$ dry weight basis. The other cereals like wheat grain accumulate low arsenic, i.e. 94.52±22.79 $\mu g\ kg^{-1}$. The pulses

Table 2: Bioaccumulation of arsenic in cereals, pulses and oilseeds

Cereals	*Total number of sample analyzed (n)*	*Arsenic concentration ($\mu g\ kg^{-1}$ dry weight basis±SD)*
Rice (Boro)	12	492.97±178.47
Rice (Amon)	6	161.19±23.91
Wheat	3	94.52±22.79
Pea	3	101.47±41.93
Lentil	4	21.86±9.66
Green gram	3	31.52±4.35
Mustard seed	3	88.94±10.42

also showed very low accumulation of arsenic. i.e. green gram (31.52±4.35 $\mu g\ kg^{-1}$) and lentil (21.86±9.66 $\mu g\ kg^{-1}$). The oilseed mustard seed, analyzed and found that it accumulates 88.94±10.42 $\mu g\ kg^{-1}$ of arsenic. Thus significant bioaccumulation of arsenic was found in rice.

3.2 Bioaccumulation of Arsenic in Vegetables

Varieties of vegetables are cultivated in this area in different seasons with varied degree of irrigations. The bioaccumulation of arsenic in the cultivated vegetables are shown in Table 3. It is found that Arum (*Colocasia* sp.)

Table 3: Bioaccumulation of arsenic in vegetables

Vegetables	*Total number of sample analyzed (n)*	*Arsenic concentration ($\mu g\ kg^{-1}$ dry weight basis±SD)*
Arum	4	908.48±164.41
Radish	6	784.40±407.43
Potato	6	606.93±97.73
Pich of Plantain tree	2	521.4±101.62
Spinach	3	378.0±199.07
Basella (Pui)	3	370.67±216.0
Amaranth	4	366.24±315.17
Bitter gourd	4	283.77±211.40
Cabbage	10	248.96±79.109
Papaya	2	221.671±98.42
Parbal	5	221.40±119.24
Brinjal	10	220.16±70.95
Ladies finger	6	205.10±104.37
Turnip	3	190.58±27.58
Pumpkin	3	149.84±31.14
Onion	3	113.46±22.18
Bean	3	96.95±11.11
Cauliflower	4	90.66+36.0
Kidney bean	4	79.13±39.84
Tomato	6	75.08±42.57
Green chili	3	32.85±7.41

(908.48±1164.41 μg kg^{-1}) showed highest accumulation and green chili (32.85±7.41 μg kg^{-1}) showed lowest accumulation. It is found that the tuberous vegetables found higher accumulation than the leafy vegetables and also the leafy vegetables showed higher accumulation than the fruity vegetables.

3.3 Food Habits and Arsenic Intake from Diverse Sources

Rice and vegetable are the main food among the people of the study area. To study the arsenic risk exposure from food, we have studied the arsenic in rice, dal (pulses) and vegetables. Due to recent arsenic toxicity awareness, the people try to drink safer water from far distant places which is available in their locality (mainly from deep tube wells of about 400 to 500 feet) but they were not aware of risk from food-chain arsenic toxicity.

Normally the people of our study area eat rice, vegetable and dal three times per day. As such an adult (male and female) normally eat 150 gm of rice with water in morning and 250 gm of rice in lunch and dinner at each meal, i.e. 650 gm of rice per day. The children (< 15 years of age) eat about 400 gm of rice per day. Adults eat about 150 gm of vegetable and 25 gm of dal at each meal and children around 100 gm of vegetable and 20 gm of dal. The people ate a little quantity of fish, egg and meat (monthly 3 to 4 times), so these are not taken for consideration in dietary intake computation.

3.4 Daily Arsenic Intake through Food Stuff

Rice is the main caloricious food among the surveyed population. The average concentration of arsenic in rice is 492.9 μg kg^{-1} The total arsenic intake through rice among different groups of people per day basis are: adult (304.36 μg) and child (187.30 μg) (Table 4). The villagers of our study area mostly take their meal as rice with mixed vegetable curry and dal (pulses). The total arsenic intake through vegetable and pulses in different groups in the study area are shown in Table 4. The average vegetables taken by adults and children are 450

Table 4: Daily average arsenic intake through food chain (adult and children)

Source	*Group*	*Wet weight consumed day^{-1} (in gram)*	*Dry weight consumed day^{-1} (in gram)*	*Arsenic concentration (μg kg^{-1})*	*Total arsenic (μg)*
Rice	Adult	650	617.5	492.9	304.36
	Child	400	380	492.9	187.30
Vegetables	Adult	450	90	276.35	24.87
	Child	300	60	276.35	16.58
Pulses	Adult	75	71.25	26.69	1.90
	Child	60	57.0	26.69	1.52

and 300 gm respectively i.e. 90 gm and 60 gm in dry weight (about 80% moisture content in vegetables). The average arsenic content in vegetables is 276.35 μg kg^{-1} on dry weight basis. So the total arsenic consumed from vegetables in the study area among different groups per day basis are 24.87 μg (adult) and 16.58 μg (child). Likewise we have studied the arsenic intake through pulses. Low levels of arsenic content in pulses are found. The average arsenic content in pulses is 26.69 μg kg^{-1}. The total arsenic intake through pulses is 1.90 μg (adult) and 1.52 μg (child).

In our study we have examined the daily basis consumption of arsenic from different sources. The average arsenic intake from drinking water is 23.7 μg L^{-1} (the deep tube wells). Estimated total arsenic intake from drinking water in adults and children in the studied area is shown in Table 5. It is calculated that adult consumed 94 μg (@ 4 lit. day^{-1}) and child consumed 58.75 μg (@ 2.5 lit. day^{-1}) arsenic from drinking water alone per day.

In addition a large volume of water is used for cooking purpose. The villagers used their nearby or own shallow arsenic-contaminated tube well water for cooking purpose as the deep tube wells are not sufficient. In cooking procedure for rice, curry and dal etc., they often use contaminated water. The adult consumes about two litre of water per day and the children consume about 1.5 litre of water per day through cooking alone. The total amount of arsenic from cooking water is also shown in Table 5. The average arsenic content in cooking water (except deep tube well) of the studied areas is 120 μg L^{-1} and the total arsenic consumption through cooking water is 260 μg (adult) and 180 μg (child). Thus the total daily intake of arsenic among the people of the study area is also estimated. It is found that the all total average arsenic intake from food in different groups of people per day basis is 331.13 μg (adult) and 205.05 μg (child). The total arsenic body burden (μg day^{-1} $person^{-1}$) of the different group of people in the study area from different sources like drinking water, cooking water, rice, vegetable and pulses are shown in Table 6. Thus the estimated total arsenic intake in adult was about 685.13 μg day^{-1} and 444.15 μg day^{-1} in child.

Table 5: Average daily intake of arsenic from drinking and cooking water in the study area

Group	*Drinking water*			*Cooking water*		
	Water per day (in litre)	*Average arsenic concentration (μg L^{-1})*	*Intake of total arsenic day^{-1} (μg)*	*Water per day (in litre)*	*Average arsenic concentration (μg L^{-1})*	*Intake of total arsenic day^{-1} (μg)*
Adult	4	23.5	94	2	120	260
Child	2.5	23.5	58.75	1.5	120	180

Table 6: Daily total arsenic intake through drinking, cooking water and foodchain in the study area (μg day^{-1} $person^{-1}$)

Group	*Total arsenic from drinking water (μg)*	*Total arsenic from cooking water (μg)*	*Total arsenic from rice*	*Total arsenic from vegetables*	*Total arsenic from pulses*	*All total arsenic intake from food*
Adult	94	260	304.36	24.87	1.90	685.13
Child	58.75	180	187.30	16.58	1.52	444.15

4. DISCUSSION

Arsenic contaminations of soil from groundwater irrigation sources may prove detrimental to plant through its uptake to the toxic limit and more specifically detrimental to society as it enters in food chain. The possibility of biomagnifications cannot be avoided in many cases and their biotransformation in the food chain. The bioaccumulation of arsenic is found in mainly crops and vegetables cultivated in and around our study locations. As rice is the dominant crop in our study area, we analyzed the accumulation of arsenic in different rice varieties (Table 2). The arsenic accumulation in Boro paddy is higher than Aman crop as Boro rice is highly irrigated crop. There were similar reports of accumulation of arsenic in rice plant noted in the past (Abedin et al., 2002).

Analyzing the accumulation of arsenic in other cereal crops, wheat and pulses (Table 2), it is found that in comparison with rice the arsenic accumulation level is very low. This is perhaps due to the fact that these crops require less amounts of irrigation water and also their physiological nature relating to arsenic uptake by root system.

We have also analyzed the accumulation of arsenic in the consumable parts of different vegetable cultivated in the study area (Table 3). The data reveal that a significant amount of arsenic accumulate in different vegetables. It also observed that the underground tuber and leafy vegetables accumulate more arsenic than the fruity fleshy vegetables. This relationship was previously studied by some workers (Roychoudhry et al., 2002; Alam et al., 2003) in different arsenic-contaminated places. We found that the vegetable arum, radish, basella, amaranth and cabbage accumulate more arsenic and other vegetables like kidney bean and tomato accumulate less arsenic. This may be due to the fact that higher accumulated vegetables require more water for irrigation and the lower accumulated vegetables (kidney bean and tomato) require less water for irrigation. In contrast, higher accumulation in potato is also noted, as it requires less water for irrigation; this may due to the fact that potato is an underground vegetable cultivated in soil and the tuber can also absorb arsenic from soil directly.

Considering the total dietary intake except water, the exposed population in our study area is consuming high amount of arsenic from their foodstuffs, and this demonstrates the high risk of suffering from arsenical skin lesions and cancer.

5. CONCLUSION

The cultivated crops and vegetables in the study area accumulate a significant amount of arsenic. The higher arsenic accumulating vegetables are arum, radish, potato, pich of plantain tree, basella, cabbage, amaranth and spinach (more than 300 μg kg^{-1}) and the less arsenic accumulating vegetables are tomato, kidney bean and cauliflower (less than 100 μg kg^{-1}). The pulses, lentil and green gram accumulate less arsenic (less than 50 μg kg^{-1}).

It has been observed from our survey that the main source of irrigation in the area, during winter month, is the groundwater. There are three or four crops harvested in each year at present with the extensive irrigation. High quantity of groundwater was withdrawn every year for irrigation. In this process of intensive irrigation a huge amount of arsenic is deposited on the surface soil in this area over the year. The people of the affected area also consume a considerable amount of arsenic from their diets as they eat the crops and vegetables which are cultivated in the arsenic-contaminated soils with arsenic-contaminated groundwater irrigation.

The people of these areas are suffering from arsenicosis and other types of allied diseases due to arsenic toxicity. The signs of chronic arsenic toxicity manifested as melanosis, oedema, skin lesions, keratosis and the other diseases, viz. gastro-intestinal problem, asthma, bronchial troubles, cancer of liver, skin and limbs are common. It is also reported that the economically weaker people are living in this area. They are too weak and malnourished to fight the arsenic scourge and die at young age. The people of young age are reporting the deterioration of health and inability to perform hard works. There are many social problems arising due to arsenicosis. The affected families in the arsenic-contaminated area are not acceptable by the other community. People reported that the agricultural products from their land are difficult to sell in market with the proper price. A form of social dejection has developed towards the people of arsenic affected areas.

Now groundwater is the major source for irrigation in Gangetic delta region of West Bengal. As the population is increased, higher food production is required. Thus the people are cultivating more and more cereal crops using intensive irrigation. In this process a higher exploitation of groundwater irrigation is done and the potential of arsenic contamination is increased day to day in this area. Proper watershed management should be done in the arsenic-contaminated areas and immediate action is needed to minimize the groundwater withdrawal. In these regions a substantial amount of surface water is also available in the river Ganga and beels. These surface water sources should be properly managed and can be used for irrigation. In addition the current agricultural practices should be changed in arsenic-prone areas or the crops requiring maximum irrigated water should be replaced with crops requiring low irrigation. There is need for avoidance of more arsenic accumulating crops and vegetables in agricultural practices. Thus an integrated approach of groundwater use in domestic and agricultural practice is highly essential.

ACKNOWLEDGEMENTS

The authors would like to thank the School of Environmental Studies, Jadavpur University, India, for providing FI-HG-AAS facility and University of Kalyani, West Bengal, India, for providing infrastructural facility and financial support.

REFERENCES

Abedin, M.J., Cresser, M.S., Meharg, A.A., Feldmann, J. and Cotter-Howells, J. (2002). Arsenic accumulation and metabolism in rice (*Oryza sativa* L.). *Environmental Science & Technology*, **36(5)**, 962-968.

Alam, M.G., Snow, E.T. and Tanaka, A. (2003). Arsenic and heavy metal contamination of vegetables grown in Samta village, Bangladesh. *Science of the Total Environment*, **308(1-3)**, 83-96.

Bae, M., Watanabe, C., Inaoka, T., Sekiyama, M., Sudo, N., Bokul, M.H. and Ohtsuka, R. (2002). Arsenic in cooked rice in Bangladesh. *Lancet*, **360(9348)**, 1839-1840.

Bhumla, D.K. and Keefer, R.F. (1994). Arsenic mobilization and Bioavailability in Soils. *In:* Nriagu, J.O. (Ed.), Arsenic in the environment. Part 1: Cycling and Characterization. John Wiley and Sons, Inc., New York, USA, pp. 51-81.

Chakraborty, D., Rahaman, M.M., Paul, K., Chowdhury, U.K., Sengupta, M.K., Lodh, D., Chanda, C.R., Saha, K.C. and Mukherjee, S.C. (2002). Arsenic calamity in the Indian Subcontinent: What lessons have been learned? *Talanta*, **58**, 3-22.

Chakraborti, D., Sengupta, M.K., Rahman, M.M., Ahamed, S., Choudhury, U.K., Hossain, M.A., Mukherjee, S.C., Pati, S., Saha, K.C., Dutta, R.N. and Zaman, Q.Q. (2004). Groundwater arsenic contamination and its health effects in the Ganga-Meghna-Brahmaputra plain. *Journal of Environmental Monitoring*, **6**, 74N-83N.

Chatterjee, A., Das, D., Mandal, B.K., Roy Chowdhury, T., Samanta, G. and Chakraborti, D. (1995). Arsenic in Ground Water in Six Districts of West Bengal, India: The Biggest Arsenic Calamity in the World. Part 1: Arsenic Species in drinking water and urine of the affected people. *Analyst*, **120**, 643-650.

Das, D., Chatterjee, C., Mandal, B.K., Samanta, G. and Chakraborty, D. (1995). Arsenic in groundwater in six districts of West Bengal, India: The Biggest Arsenic Calamity in the World, Part 2: Arsenic concentration in drinking water, hair, nails, urine, skin scale and liver tissue (biopsy) of the affected people. *The Analyst*, **120**, 917-924.

Das, H.K., Mitra, A.K., Sengupta, P.K., Hossain, A., Islam, F. and Rabbani, G.H. (2004). Arsenic concentrations in rice, vegetables, and fish in Bangladesh: A preliminary study. *Environment International*, **30(3)**, 383-387.

ICAR (2001). Final Report: Status causes and impacts of arsenic contaminant in groundwater in parts of West Bengal vis-à-vis management of agricultural systems. [Adhoc scheme executed (1998-2001) by BCKV (Nodal Center); NBSS & LUP, Regional Centre, Kolkata; NDRI, Kalyani; GSI, Government of India, Kolkata; CSSRI, Regional Research Station, Canning Town; SWID, Government of West Bengal, Kolkata; Principal Investigator, Sanyal, S.K.].

Lee-Feldstein, A. (1986). Cumulative exposure to arsenic and its relationship to respiratory cancer among copper smelter employees. *Journal of Occupational Medicine*, **28**, 296-302.

Meharg, A.A. and Rahman, M.M. (2003). Arsenic contamination of Bangladesh paddy field soils: Implications for rice contribution to arsenic consumption. *Environmental Science & Technology*, **37(2)**, 229-234.

Mandal, B.K., Chowdhury, T.R., Samanta, G., Basu, G.K., Chowdhury, P.P., Chanda, C.R., Lodh, D., Karan, N.K., Dhar, R.K., Tamili, D.T., Das, D., Saha, K.C. and Chakraborti, D. (1997). Chronic arsenic toxicity in West Bengal, *Current Science*, **72(2)**, 114-117.

Mandal, B.K., Chowdhury, T.R., Samanta, G., Mukherjee, D.P., Chanda, C.R., Saha, K.C. and Chakraborti, D. (1998). Impact of safe water for drinking and cooking on five arsenic-affected families for 2 years in West Bengal, India. *Science of the Total Environment*, **218(2-3)**, 185-201.

Nagvi, S.M., Vaishnavi, C. and Singh, H. (1994). Toxicity and metabolism of arsenic in Vertebrates. *In:* Nriagu, J.O. (ed.), Arsenic in the environment. Part II: Human Health and Ecosystem Effects. John Wiley and Sons, Inc., New York, USA, pp. 55-91.

Pal, A., Nayak, B., Das, B., Hossain, M.A., Ahamed, S. and Chakraborti, D. (2007). Additional danger of arsenic exposure through inhalation from burning of cow dung cake laced with arsenic as a fuel in arsenic affected villages in Ganga-Meghna-Brahmaputra plain. *Journal of Environmental Monitoring*, **9**, 1067-1070.

Pal, A., Chowdhury, U.K., Mondal, D., Das, B., Nayak, B., Ghosh, A., Maity, S. and Chakraborti, D. (2009). Arsenic burden from cooked rice in the populations of arsenic affected and nonaffected areas and Kolkata city in West Bengal, India. *Environmental Science & Technology*, **43**, 3349-3355.

Panda, S. and Das, N.C. (2001). Impact of cultivation of pointed gourd (*Trichosanthes dioica*) on mobilization of arsenic in surface soil and water – A report. Proceedings of International Conference on Changing Environmental Scenario of India and Adjacent Countries since Independence. Calcutta, India, p. 107.

Roychowdhury, T., Uchino, T., Tokunaga, H. and Ando, M. (2002). Survey of arsenic in food composites from an arsenic-affected area of West Bengal, India. *Food and Chemical Toxicology*, **40(11)**, 1611-1621.

Roychowdhury, T., Tokunaga, H. and Ando, M. (2003). Survey of arsenic and other heavy metals in food composites and drinking water and estimation of dietary intake by the villagers from an arsenic-affected area of West Bengal, India. *Science of the Total Environment*, **308(1-3)**, 15-35.

Samal, A.C. (2005). An Investigation on Accumulation of Arsenic in Ecosystem of Gangetic West Bengal and Assessment of Potential Health Risk. Ph.D. Thesis, University of Kalyani, India, p. 234.

Samanta, G., Choudhury, T.K., Mondal, B.K., Biswas, B.K., Choudhury, U.K, Basu, G.K., Chanda, C.R., Lodh, D. and Chakrabarti, D. (1999). Flow injection hybrid generation atomic absorption spectrophotometry for determination of arsenic in water and biological samples from arsenic affected districts of West Bengal, India and Bangaladesh. *Microchemical Journal*, **62**, 174-191.

World Health Organization (1993). Guidelines for Drinking Water Quality, 2nd Edition. Vol. 1. Recommendations. WHO, Geneva, pp. 41-42.

Author Index

Subject Index

GPSR Compliance

The European Union's (EU) General Product Safety Regulation (GPSR) is a set of rules that requires consumer products to be safe and our obligations to ensure this.

If you have any concerns about our products, you can contact us on ProductSafety@springernature.com

In case Publisher is established outside the EU, the EU authorized representative is:

Springer Nature Customer Service Center GmbH
Europaplatz 3
69115 Heidelberg, Germany

Batch number: 10815920

Printed by Printforce, the Netherlands